AF358719

Fermented Beverages Revisited: From Terroir to Customized Functional Products

Fermented Beverages Revisited: From Terroir to Customized Functional Products

Editors

Maria Dimopoulou
Spiros Paramithiotis
Yorgos Kotseridis
Jayanta Kumar Patra

Basel • Beijing • Wuhan • Barcelona • Belgrade • Novi Sad • Cluj • Manchester

Editors

Maria Dimopoulou
Department of Vine, Wine
& Beverages Sciences
West Attica University
Athens
Greece

Spiros Paramithiotis
Department of Biological
Applications and Technology
University of Ioannina
Ioannina
Greece

Yorgos Kotseridis
Food Science and
Human Nutrition
Agricultural University
of Athens
Athens
Greece

Jayanta Kumar Patra
Research Institute of
Integrative Life Sciences
Dongguk University-Seoul
Goyangsi
Korea, South

Editorial Office
MDPI
St. Alban-Anlage 66
4052 Basel, Switzerland

This is a reprint of articles from the Special Issue published online in the open access journal *Fermentation* (ISSN 2311-5637) (available at: www.mdpi.com/journal/fermentation/special_issues/ ZB1H4H6X29).

For citation purposes, cite each article independently as indicated on the article page online and as indicated below:

Lastname, A.A.; Lastname, B.B. Article Title. *Journal Name* **Year**, *Volume Number*, Page Range.

ISBN 978-3-7258-0312-5 (Hbk)
ISBN 978-3-7258-0311-8 (PDF)
doi.org/10.3390/books978-3-7258-0311-8

Contents

About the Editors

Maria Dimopoulou

Dr. Dimopoulou Maria is a graduate of the Department of Chemistry at the Aristotle University of Thessaloniki. She has obtained a postgraduate specialization in Food Science and a PhD in Oenology from the University of Bordeaux in France. She has worked as a postdoctoral researcher at the R&D Department of BioLaffort Company in France as well as a scientific associate at the Aristotle University and the Agricultural University of Athens. She has been working since 2021 as an Assistant Professor at the University of Western Attica in the Department of Vine, Wine, and Beverages, teaching the courses of General Microbiology and Wine Microbiology. She has published over 40 papers in international journals and conferences and has participated in European and international research projects.

Her research interests deal with: The evolution of microflora in all stages of winemaking; the interaction of microorganisms and their effect on the qualitative characteristics of wine; the metabolic and biochemical pathways of wine microorganisms; wine spoilage microorganisms and treatment methods and techniques.

Spiros Paramithiotis

Spiros Paramithiotis graduated from the Department of Food Science and Human Nutrition of the Agricultural University of Athens, Greece, in 1996. In 1998, he received a scholarship for postgraduate studies from the State Scholarship Foundation of Greece, and in 2002, he received a PhD degree in Food Microbiology from the Agricultural University of Athens. In 2003, he joined the Agricultural University of Athens, Department of Food Science and Human Nutrition, as a member of the scientific staff. In 2023, he was appointed as an Assistant Professor of Microbiology by the University of Ioannina, Department of Biological Applications and Technology.

His research interests lie mainly in the fields of food fermentations and foodborne diseases, with particular emphasis on microbial taxonomy, metabolism, physiology, symbiotic patterns, and the underlying molecular mechanisms. He has participated in several research projects funded by the EU, the Greek Secretariat of Research and Technology, and the food industry. He has authored and co-authored more than 150 publications in peer-reviewed journals and book chapters. He has edited and co-edited six books on aspects related to food microbiology and received more than 4000 citations.

Yorgos Kotseridis

Georgios Kotseridis is an Enology Professor. He has received his Diploma in Food Science at the School of Agriculture from the Aristotle University of Thessaloniki. He then moved to France for postgraduate studies in Enology. He received his Diplôme Nationale d'Oenologue, his master D.E.A., and finally his PhD degree in Enology and Viticulture from the University of Bordeaux, Faculty of Enology, in 1994, 1995, and 1999, respectively. He joined the Food Science Department of the Agricultural University of Athens in 2003, and he founded the first research university Laboratory of Enology and Alcoholic Drinks (L.E.A.D.) in Greece. He also initiated the MSc in Enology for graduate students, educating some of the most successful young winemakers in the Greek wine industry. Besides, from 2011 to 2013, he worked at the Cool Climate Enology and Viticulture Institute (Ontario, Canada), assisting the wineries of Ontario with winemaking problem solving.

His research interests include dealing with the impact of viticultural and enological practices on wine flavor. He has co-authored more than 90 publications and presented research results at more than 50 wine conferences all over the world.

Dr. Kotseridis has solid experience in winemaking, as he started working in commercial winemaking as an assistant winemaker for J.P. Moueix Ets. in Chateaux Magdelaine, La Fleur Petrus, later in Calon Segur Medoc, and then returned to Greece in Domaine Hatzimichalis, Estate Porto Carras, Estate Avantis, Estate Theopetra Tsililis, Anhydrous Estate Santorini, and other numerous wineries producing high-quality wines.

Jayanta Kumar Patra

Dr. Jayanta Kumar Patra, M.Sc., Ph.D., PDF, is currently working as an Associate Professor at Dongguk University, the Republic of Korea. His current research is focused on nanoparticle synthesis by green technology methods and their potential applications in biomedical and agricultural fields. He has about 14 years of teaching and research experience in the fields of pharmacology and nanobiotechnology. Dr. Patra has published more than 180 papers in various national and international peer-reviewed journals, along with around 40 book chapters in different edited books. Dr. Patra has also authored 20 books on the topic related to medicinal plants, nanomaterials, functional foods, ethnopharmacology, biotechnology, etc., in various publications.

Preface

The production of fermented beverages dates back to ancient times. Humans utilized the available raw materials and practical wisdom to create nourishing products. This practical knowledge passed through generations and reached our modern times. Scientific and technological progress allowed the thorough assessment and characterization of the production procedure, the composition of the fermenting microcommunity, the fate of bioactive compounds during fermentation, as well as the effect of their consumption on human health. Nowadays, a wealth of knowledge has accumulated and is constantly supplemented by new data that improves our understanding of the biotic and abiotic factors that affect the production and quality of fermented beverages. Standardization was the first step that was achieved for products that have met public acceptance. The customization of functional products is the next challenge that creates exciting new prospects.

The Guest Editors are grateful to the researchers who published their work in this Special Issue and the MDPI editorial team for their support.

Maria Dimopoulou, Spiros Paramithiotis, Yorgos Kotseridis, and Jayanta Kumar Patra
Editors

fermentation

Editorial

Fermented Beverages Revisited: From Terroir to Customized Functional Products

Spiros Paramithiotis [1,*], Jayanta Kumar Patra [2], Yorgos Kotseridis [3] and Maria Dimopoulou [4]

[1] Department of Biological Applications and Technology, University of Ioannina, 45110 Ioannina, Greece

[2] Research Institute of Integrative Life Sciences, Dongguk University-Seoul, Goyangsi 10326, Republic of Korea; jkpatra@dongguk.edu

[3] Department of Food Science & Human Nutrition, Agricultural University of Athens, 11855 Athens, Greece; ykotseridis@aua.gr

[4] Department of Wine, Vine and Beverage Sciences, University of West Attica, 12243 Athens, Greece; mdimopoulou@uniwa.gr

* Correspondence: paramithiotis@uoi.gr

Citation: Paramithiotis, S.; Patra, J.K.; Kotseridis, Y.; Dimopoulou, M. Fermented Beverages Revisited: From Terroir to Customized Functional Products. *Fermentation* **2024**, *10*, 74. https://doi.org/10.3390/fermentation10020074

Received: 14 January 2024
Accepted: 20 January 2024
Published: 24 January 2024

The production of fermented beverages dates back to antiquity. The substrates most frequently employed are milk, fruits and cereals, resulting in a wide range of products, some of which are currently recognized as characteristic of certain geographical areas. The early studies on fermented beverages aimed to document the production procedures and characterize the driving microbiota. As data accumulated and the need for standardization became more evident, assessment of the metabolic activities associated with the quality characteristics of each product towards the selection of the most suitable starter cultures dominated research. At the same time, the beneficial attributes for human health were also recorded, highlighting the importance of the thorough characterization of the factors that determine these activities. Nowadays, a wealth of knowledge has accumulated, which, combined with the conceptual, methodological and technological advances of recent years, enable us to study the factors that affect the development of these microbial communities, elucidate the mechanisms by which their metabolic activities direct the fate of bioactive compounds and unravel how the latter modulate human health.

In recent years, research in the field of fermented beverages has been concentrated mainly in the following four areas: 1. physicochemical and microbiological characterization, 2. the effect of climate change and mitigation strategies, 3. quality improvement and 4. the development of new products.

Our knowledge regarding the microbial communities that drive spontaneous fermentations, their dynamics during fermentation and the storage of the final products, as well as their metabolic activities and how their metabolites shape the physicochemical properties and the nutritional value of fermented beverages, has been significantly improved over the recent years. This improvement has been facilitated by the technological advances that enabled the rapid characterization of microbial communities to the species level, along with their functional metabolic pathways. Thus, a series of studies have improved our understanding of already established fermented beverages, such as kombucha [1,2], and they have introduced traditional products, such as the Peruvian "Chicha de siete semillas" [3], and ethnic practices, such as the epop of India [4].

The effects of climate change on fermented beverages, along with possible mitigation strategies, have drawn scientific attention, particularly in viticulture. Global warming accelerates the ripening of grapes but alters their composition; they contain more sugars and fewer organic acids. From a cultivation point of view, several agricultural practices have been proposed to address this issue [5]. From a winemaking perspective, the greater carbohydrate content could lead to higher ethanol production, which is not always perceived positively by consumers, and the lower acidity would be detrimental to the stability of the product. Collectively, these changes would result in a significant deviation from the typical characteristics of wine. Among the possible mitigation strategies, biological acidification by suitable non-*Saccharomyces* yeast strains that would utilize carbohydrates to produce lactic acid instead of ethanol seems to be a realistic solution. Their application in a wide range of grape varieties, along with the effect on the physicochemical characteristics and the sensorial profile of wine, has been extensively assessed, highlighting the feasibility of this approach.

The term "quality" may refer to a wide range of properties that may be affected by a series of biotic and abiotic factors, such as the type and characteristics of the raw materials employed, the metabolic capacity of the microorganisms driving the fermentation, the conditions of fermentation etc. Not all fermented beverages have the factors affecting their quality features fully elucidated. Therefore, the recent advances in the field of metagenomic analyses, coupled with analytical tools that enable the integration of metagenomic with metabolomic approaches, have improved our understanding of the factors that affect the fermentation process; the role of the different members of the microcommunity and how they direct the quality of the final product, such as in the case of green kombucha [6] and Huangjiu fermentation [7]. The fate of bioactive compounds has also been the epicenter of intensive study. Thus, the factors that affect the concentration of compounds, such as phenolics, amino acids, peptides and fatty acids, during fermentation, as well as their biological activities, have been extensively assessed. Especially regarding the latter, a wealth of data has been generated through in vitro experiments, indicating significant biological activity that leads to specific health claims [8]; in many cases, these are accompanied by adequately described modes of action. As a result, there is a constant effort to enhance the nutritional value and the functional activities of fermented beverages, mostly through the use of carefully selected starter cultures [9] or the direct addition of specific nutrients [10].

Finally, the development of new products with customized functional potential has also been a field of intensive effort. Research has mostly focused on the customization of kefir and kombucha beverages, either by using alternative raw materials or by using microbial strains with specific metabolic activities. These have led to the development of a wide range of kefir and kombucha analogs [8], many of which hold very promising capacities. In addition, many fermented beverages with interesting properties have also been developed by using non-conventional materials or underexploited strategies [11,12].

The aim of this Special Issue was to contribute to the aforementioned research areas with studies of high quality. This has been achieved, as very interesting manuscripts, addressing research gaps and providing new insights, have been published. These manuscripts improve our understanding of fermented beverages and form a basis for future research.

Author Contributions: Conceptualization, S.P. and M.D.; writing—original draft preparation, S.P.; writing—review and editing, S.P., J.K.P., Y.K. and M.D. All authors have read and agreed to the published version of the manuscript.

Funding: This research received no external funding.

Acknowledgments: The Guest Editors would like to express their gratitude to all authors and reviewers for their contributions.

List of Contributions:

1. Bouloumpasi, E.; Skendi, A.; Soufleros, E.H. Survey on Yeast Assimilable Nitrogen Status of Musts from Native and International Grape Varieties: Effect of Variety and Climate. *Fermentation* **2023**, *9*, 773.

2. Guo, M.; Deng, Y.; Huang, J.; Zeng, C.; Wu, H.; Qin, H.; Zhang, S. Integrated Metagenomics and Network Analysis of Metabolic Functional Genes in the Microbial Community of Chinese Fermentation Pits. *Fermentation* **2023**, *9*, 772.

3. Pinu, F.R.; Stuart, L.; Topal, T.; Albright, A.; Martin, D.; Grose, C. The Effect of Yeast Inoculation Methods on the Metabolite Composition of Sauvignon Blanc Wines. *Fermentation* **2023**, *9*, 759.

4. Du, P.; Li, Y.; Zhen, C.; Song, J.; Hou, J.; Gou, J.; Li, X.; Xie, S.; Zhou, J.; Yan, Y.; et al. Effect of Microbial Reinforcement on Polyphenols in the Acetic Acid Fermentation of Shanxi-Aged Vinegar. *Fermentation* **2023**, *9*, 756.
5. Frolova, Y.; Vorobyeva, V.; Vorobyeva, I.; Sarkisyan, V.; Malinkin, A.; Isakov, V.; Kochetkova, A. Development of Fermented Kombucha Tea Beverage Enriched with Inulin and B Vitamins. *Fermentation* **2023**, *9*, 552.
6. Antal, E.; Kállay, M.; Varga, Z.; Nyitrai-Sárdy, D. Effect of *Botrytis cinerea* Activity on Glycol Composition and Concentration in Wines. *Fermentation* **2023**, *9*, 493.
7. Sherman, E.; Yvon, M.; Grab, F.; Zarate, E.; Green, S.; Bang, K.W.; Pinu, F.R. Total Lipids and Fatty Acids in Major New Zealand Grape Varieties during Ripening, Prolonged Pomace Contacts and Ethanolic Extractions Mimicking Fermentation. *Fermentation* **2023**, *9*, 357.
8. Guido, L.F.; Ferreira, I.M. The Role of Malt on Beer Flavour Stability. *Fermentation* **2023**, *9*, 464.
9. Brugnoli, M.; Cantadori, E.; Arena, M.P.; De Vero, L.; Colonello, A.; Gullo, M. Zero- and Low-Alcohol Fermented Beverages: A Perspective for Non-Conventional Healthy and Sustainable Production from Red Fruits. *Fermentation* **2023**, *9*, 457.
10. Sánchez-Suárez, F.; Peinado, R.A. Use of Non-*Saccharomyces* Yeast to Enhance the Acidity of Wines Produced in a Warm Climate Region: Effect on Wine Composition. *Fermentation* **2024**, *10*, 17.
11. Paramithiotis, S.; Patra, J.K.; Kotseridis, Y.; Dimopoulou, M. Fermented Beverages Revisited: From Terroir to Customized Functional Products. *Fermentation* **2024**, *10*, 57.

References

1. Barbosa, C.D.; Uetanabaro, A.P.T.; Santos, W.C.R.; Caetano, R.G.; Albano, H.; Kato, R.; Cosenza, G.P.; Azeredo, A.; Goes-Neto, A.; Rosa, C.A.; et al. Microbial–physicochemical integrated analysis of kombucha fermentation. *LWT-Food Sci. Tehnol.* **2021**, *148*, 111788. [CrossRef]
2. Li, S.; Wang, S.; Wang, L.; Liu, X.; Wang, X.; Cai, R.; Yuan, Y.; Yue, T.; Wang, Z. Unraveling symbiotic microbial communities, metabolomics and volatilomics profiles of kombucha from diverse regions in China. *Food Res. Int.* **2023**, *174*, 113652. [CrossRef] [PubMed]
3. Rebaza-Cardenas, T.; Montes-Villanueva, N.D.; Fernandez, M.; Delgado, S.; Ruas-Madiedo, P. Microbiological and physical-chemical characteristics of the Peruvian fermented beverage 'Chicha de siete semillas': Towards the selection of strains with acidifying properties. *Int. J. Food Microbiol.* **2023**, *406*, 110353. [CrossRef] [PubMed]
4. Mili, R.; Sundriyal, R.C. EPOP (traditional starter culture): A complex composition of plant resources prepared by the Misings of Assam, Northeast India. *J. Ethn. Foods* **2023**, *10*, 47. [CrossRef]
5. Santos, J.A.; Fraga, H.; Malheiro, A.C.; Moutinho-Pereira, J.; Dinis, L.-T.; Correia, C.; Moriondo, M.; Leolini, L.; Dibari, C.; Costafreda-Aumedes, S.; et al. A Review of the potential climate change impacts and adaptation options for European viticulture. *Appl. Sci.* **2020**, *10*, 3092. [CrossRef]
6. Dartora, B.; Hickert, L.R.; Fabricio, M.F.; Ayub, M.A.Z.; Furlan, J.M.; Wagner, R.; Perez, K.J.; Sant'Anna, V. Understanding the effect of fermentation time on physicochemical characteristics, sensory attributes, and volatile compounds in green tea kombucha. *Food Res. Int.* **2023**, *174*, 113569. [CrossRef] [PubMed]
7. Liu, S.; Zhang, Z.F.; Mao, J.; Zhou, Z.; Zhang, J.; Shen, C.; Wang, S.; Marco, M.L.; Mao, J. Integrated meta-omics approaches reveal *Saccharopolyspora* as the core functional genus in huangjiu fermentations. *NPJ Biofilms Microbiomes* **2023**, *9*, 65. [CrossRef] [PubMed]
8. Chong, A.Q.; Lau, S.W.; Chin, N.L.; Talib, R.A.; Basha, R.K. Fermented beverage benefits: A comprehensive review and comparison of kombucha and kefir microbiome. *Microorganisms* **2023**, *11*, 1344. [CrossRef] [PubMed]
9. Kim, H.; Hur, S.; Lim, J.; Jin, K.; Yang, T.-H.; Keehm, I.-S.; Kim, S.W.; Kim, T.; Kim, D. Enhancement of the phenolic compounds and antioxidant activities of Kombucha prepared using specific bacterial and yeast. *Food Biosci.* **2023**, *56*, 103431. [CrossRef]
10. Isakov, V.A.; Pilipenko, V.I.; Vlasova, A.V.; Kochetkova, A.A. Evaluation of the efficacy of Kombucha-based drink enriched with inulin and vitamins for the management of constipation-predominant irritable bowel syndrome in females: A randomized pilot study. *Curr. Dev. Nutr.* **2023**, *7*, 102037. [CrossRef] [PubMed]
11. Cardinali, F.; Osimani, A.; Milanovic, V.; Garofalo, C.; Aquilanti, L. Innovative fermented beverages made with red rice, barley, and buckwheat. *Foods* **2021**, *10*, 613. [CrossRef] [PubMed]
12. Eidt, G.; Koehler, A.; Cortivo, P.R.D.; Ayub, M.A.Z.; Flôres, S.H.; Arthur, R.A. Development and consumer acceptance testing of a honey-based beverage fermented by a multi-species starter culture. *Food Biosci.* **2023**, *56*, 103182. [CrossRef]

fermentation

Article

Use of Non-*Saccharomyces* Yeast to Enhance the Acidity of Wines Produced in a Warm Climate Region: Effect on Wine Composition

Fernando Sánchez-Suárez and Rafael A. Peinado *

Department of Agricultural Chemistry, Edaphology and Microbiology, Building Marie Curie 3rd Floor, University of Córdoba, 14014 Córdoba, Spain; g62sasuf@uco.es
* Correspondence: qe1peamr@uco.es

Abstract: One of the most notable effects of climate change, especially in warm regions, is the decrease in acidity (i.e., increase in pH) of wines and a reduction in their aromatic profile. To address this issue, must from a white grape variety with low acidity were inoculated with two non-*Saccharomyces* yeasts (*Lachancea thermotolerans* and *Torulaspora delbrueckii*) to enhance the acidity of the resulting wines. Basic oenological variables and major volatile compounds and polyols of the wines were analyzed, and the results were compared with those obtained through a *Saccharomyces cerevisiae* strain. Through multiple regression analysis, we found relations between the production of lactic acid to compounds involved in yeast metabolism and redox balance, including glycerol, acetic acid, isobutanol, isoamyl alcohols, and 2-phenylethanol. By means of principal component analysis, we obtained three components that explain more than 89% of the observed variability. The first component differentiates wines produced by *L. thermotolerans*; the second differentiates wines obtained by *S. cerevisiae* from those obtained by *T. delbrueckii*; and the third component is related to the temperature of fermentation. Organoleptic wines produced with *S. cerevisiae* were the best valuated, but taste was a highlight of the wines produced with *L. thermotolerans* due to possessing the best acidity.

Keywords: climate change; *Lachancea thermotolerans*; lactic acid; redox balance; wine

check for **updates**

Citation: Sánchez-Suárez, F.; Peinado, R.A. Use of Non-*Saccharomyces* Yeast to Enhance the Acidity of Wines Produced in a Warm Climate Region: Effect on Wine Composition. *Fermentation* **2024**, *10*, 17. https://doi.org/10.3390/fermentation10010017

Academic Editors: Maria Dimopoulou, Spiros Paramithiotis, Yorgos Kotseridis and Jayanta Kumar Patra

Received: 7 November 2023
Revised: 22 December 2023
Accepted: 23 December 2023
Published: 26 December 2023

1. Introduction

Climate change affects all types of crops and environmental processes, resulting in higher temperatures and reduced annual rainfall in Mediterranean climate regions [1]. Furthermore, the precipitation is often intense, causing water to run off instead of being absorbed into the soil. As temperatures continue to rise, crop evapotranspiration (ETo) increases significantly. This will increase the water demand in the plant and subsequently in the soil, which is exacerbated by reduced precipitation and a dry atmosphere [2,3].

The effect of climate change on wine production can be grouped into two areas, i.e., viticulture and oenology. In the domain of viticulture, an advancement in the date of ripening, harvesting, and overall phenology stands out [4,5], resulting in ripening occurring during a warmer time of the year that will negatively affect grape quality and reduce yields, both in terms of quality and quantity. In addition, the increase in temperatures and reduction in precipitation adds to heat waves during the ripening period, which can have very harmful effects on the plant, causing ripening blockages and berry dehydration and shriveling, collectively resulting in a decrease in both quality and quantity of the product [6].

Regarding the oenological aspect, it is worth noting that musts and, consequently, wines show reduced quality, mainly based on an increase in pH and a reduction in titratable acidity because of the malic acid degradation by the high temperatures and harvest delay [4]. As a result, a mismatch between industrial maturity and phenolic maturity is observed. To minimize this gap, grapes are often allowed to overripen, resulting in wines with higher alcohol content. Additionally, lower quantities of phenolic compounds, responsible for color and astringency in red wines, and odorant compounds in red and white wines are synthesized. An increase in temperatures and lower thermal oscillations between day and night also contribute to these properties [7,8].

To mitigate such effects, different strategies can be carried out. Among the viticultural ones, these include a reduction of bunch exposure to sunlight by means of sprawl trellis systems [9]; late pruning to delay the phenological cycle [10,11]; a reduction of the leaf/fruit ratio to delay the harvest date [12,13]; grapevine irrigation [14]; and the application of sun protectants [15–17]. In newly established vineyards, it is recommended that one should use grapevine varieties and clones that stand out for their resistance to high temperatures, drought, and delayed vegetative cycle [4,18].

Regarding oenological strategies, the purposes are (i) to increase the acidity of the resulting wines, (ii) to reduce ethanol concentration, and (iii) to enhance the aroma of the wines. These objectives can be achieved with new technologies, such as electrodialysis for the reduction of K^+, nanofiltration for the elimination of sugars of the must, dealcoholization of wines, and through the use of non-*Saccharomyces* yeasts [19–23].

The use of non-*Saccharomyces* yeasts has multiple effects, one of the most notable being the increase in aroma compounds produced by certain yeasts, such as *Torulaspora delbrueckii* or *Metschnikowia pulcherrima* [24–30]. Specifically, in some studies on the latter yeast, co-inoculations have been found to produce lower amounts of higher alcohols, esters, and fatty acids, but much higher amounts of polyfunctional thiols, such as 3-sulfanylhexan-1-ol, 3-sulfanylhexyl acetate, and 4-methyl-4-sulfanylpentan-2-one. These wines had more positive scores in tastings than those inoculated solely with *S. cerevisiae* [30].

Another notable effect is the increase in acidity and reduction of pH produced by *L. thermotolerans* [31], which can produce significant amounts of lactic acid from sugars [32]. This acid is mainly present in wines due to the malolactic fermentation. The increase in lactic acid by this yeast will reduce the addition of tartaric acid to correct must and wine acidity, a common practice in winegrowing regions of warm climates. Several authors have highlighted the improvement in sensory properties related to acidity in wines produced by *L. thermotolerans* [31–38].

The aim of this paper is to analyze the effect of two non-*Saccharomyces* yeast strains on the acidity and the production of volatile aroma compounds related to yeast metabolism. To this end, a white grape variety that is scarcely aromatic and low in acidity was used.

2. Materials and Methods

2.1. Experimental Design

White grapes (*Vitis vinifera* L. cv. Cayetana Blanca) cultivated in Extremadura, Spain (region V, Amerine-Winkler), were used.

Grapes were manually harvested, destemmed, and pressed. The must obtained was divided in eighteen batches of 12 L and was introduced in fermentation tanks of 15 L. Previously, starter cultures of *L. thermotolerans* (Lt, Level2 Laktia™, Lallemand, Montreal, QC, Canada), *T. delbrueckii* (Td, Viniferm NSTD™, Agrovin, Alcázar de San Juan, Spain), and *S. cerevisiae* (Sc, Viniferm Pasión™, Agrovin, Alcázar de San Juan, Spain) were prepared according to the instructions of the provider. The batches were inoculated at a rate of 30 g of yeast per hL. Fermentations were carried out in triplicate and two different temperatures (14 and 21 °C) were assayed for each yeast strain. To control the temperature, the fermentation tanks were provided with a temperature sensor and a double shell through which cold water circulates.

2.2. Determination of Enological Parameters in Must and Wines and Fermentation Kinetics

The pH, volatile and titratable acidity, ethanol, and reducing sugars were measured using official methods [39]. Lactic acid was determined by reflectometry by using the Reflectoquant® test (Merck, Darmstadt, Germany).

The initial must had a sugar concentration of 200 g/L, a pH value of 3.4, and the titratable acidity was 5.3 g/L (expressed as tartaric acid). However, the volatile acidity and lactic acid were under the detection limit.

The fermentation kinetics were monitored by daily measurement of the density using two hydrometers, Pobel, Madrid, Spain with scales of 0.900–1.000 g/cm^3 and 1.000–1.100 g/cm^3.

2.3. Volatile Compounds Determination

A gas chromatograph and a flame ionization detector (GC-FID) were used for the analysis of major volatile compounds according to the conditions described by Peinado et al. [30]. Based on this protocol, 0.5 μL aliquots of wine samples (10 mL), previously supplied with 1 mL of 4-methyl-2-pentanol as an internal standard (1 g/L), were directly injected in an HP-6890 gas chromatograph, Agilent Technologies, Santa Clara CA, USA, with a capillary column CP-WAX 57 CB (50 m in length, 0.25 mm in internal diameter, and 0.4 μm in coating thickness). Tartaric acid was previously removed from the wine sample by precipitation with 0.2 g of calcium carbonate and subsequent centrifugation of the sample.

To identify the analyzed compounds, standards were injected under the same conditions as the samples. Additional information about LRI to identify volatile compounds is detailed in Table S1. Quantification was carried out by means of the corresponding calibration curve of the standards.

The identification and quantification of aroma compounds were carried out with standard solutions of pure compounds of analytical grade, purchased from Sigma-Aldrich and Merck. Pure water was obtained from a Milli-Q purification system (Millipore, Bedford, MA, USA).

2.4. Organoleptic Analysis

To evaluate the effect of fermentation on the sensory characteristics of the musts, a sensory assessment was carried out. The guidelines outlined in the standard UNE 87-020-93 [40] were adhered to for the preparation of the samples. The test consists of evaluating the wines (color, aroma, and taste) using three quality levels: undesirable, acceptable, and desirable. Furthermore, as recommended by the UNE 87-020-93 standard, the quality levels were assessed based on an increasing scale of scoring for each level [40]. The results were analyzed in accordance with the indications provided by the UNE 87-020-93 standard [40].

The tasting panel consisted of 8 women and 7 men from the research group of the authors who have wide experience in the topic of enology and in the organoleptic analysis of the wines.

2.5. Statistical Analysis

To determine if there were statistically significant differences between the variables analyzed, a multifactorial analysis of variance was performed using the studied variables (yeast strain and the temperature of fermentation). In addition, a multiple regression analysis was carried out to relate the production of lactic acid with some of the determined parameters. Lastly, to identify similarities and differences between the wines obtained in the assayed conditions, cluster and principal component analyses were also performed.

All these statistical analyses were carried out using the Statgraphics Centurion XVII v.2 software, developed by STSC, Inc. (Rockville, MD, USA).

3. Results and Discussion

3.1. Fermentation Kinetics

Figure 1 shows the fermentation kinetics of the assays. Fermentations conducted at 14 °C were slower than those conducted at 21 °C. Particularly, treatments involving must inoculated with *S. cerevisiae* at 14 °C end 4 days after those conducted at 21 °C. The differences observed in the fermentation duration at both temperatures are more noticeable in the case of non-*Saccharomyces* yeasts. Some studies have also provided evidence of a similar behavior due to the temperature of fermentation [41].

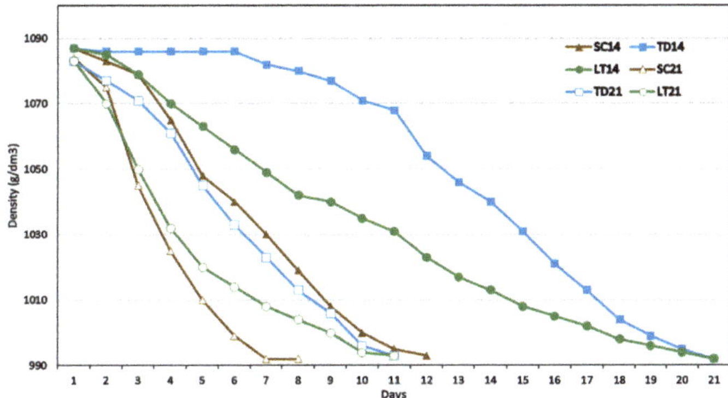

Figure 1. Fermentation kinetics of *Saccharomyces cerevisiae* (Sc), *Torulaspora delbrueckii* (Td), and *Lachancea thermotolerans* (Lt) at two different temperatures, 14 and 21 °C.

In relation to non-*Saccharomyces* yeasts, the fermentation kinetics put in evidence that *T. delbrueckii* requires a longer adaptation time to the medium. However, once fermentation is initiated, its kinetics are faster than those observed in *L. thermotolerans*.

3.2. Oenological Parameters

Table 1 shows the oenological parameters of the obtained wines and the results of the multivariate analysis of variance. The factors analyzed were temperature and the yeast strain.

Table 1. Oenological parameters determined in wines produced by *Saccharomyces cerevisiae* (Sc), *Torulaspora delbrueckii* (Td), and *Lachancea thermotolerans* (Lt) at 14 °C and 21 °C. Multivariate analysis of variance (MANOVA) involved the factors of temperature (T) and the yeast used (Y).

Parameter	14 °C			21 °C			MANOVA	
	Sc	Td	Lt	Sc	Td	Lt	T	Y
pH	3.20 ± 0.02	3.23 ± 0.02	3.25 ± 0.01	3.31 ± 0.02	3.34 ± 0.02	3.22 ± 0.02	***	**
Buffer capacity (meq/L)	31 ± 1	34 ± 2	47 ± 2	30.0 ± 0.9	28.5 ± 0.7	49 ± 1	ns	***
Titratable acidity Tartaric acid (g/L)	5.95 ± 0.08	5.7 ± 0.1	7.9 ± 0.1	4.9 ± 0.1	4.85 ± 0.09	7.65 ± 0.06	***	***
Lactic acid (g/L)	0.50 ± 0.02	0.48 ± 0.01	3.7 ± 0.1	0.15 ± 0.07	0.33 ± 0.04	4.8 ± 0.2	**	***
Volatile acidity Acetic acid (g/L)	0.5 ± 0.1	0.31 ± 0.02	0.74 ± 0.05	0.22 ± 0.05	0.52 ± 0.03	0.63 ± 0.01	***	***
Ethanol (% v/v)	11.5 ± 0.1	11.8 ± 0.1	11.7 ± 0.2	11.6 ± 0.1	11.2 ± 0.2	11.2 ± 0.1	**	ns
Reducing sugars (g/L)	1.8 ± 0.2	1.7 ± 0.3	1.9 ± 0.2	2.5 ± 0.4	2.3 ± 0.3	1.8 ± 0.2	ns	ns

** denote significant differences at 99% confidence level; *** denote significant differences at 99.9% confidence level; ns = no significant differences.

Most enological parameters are dependent on both factors, except for residual sugars, which did not exhibit significant differences due to either yeast strain or temperature. Parameters related to wine acidity (pH, titratable acidity, and volatile acidity) showed significant differences for both factors. This was also the case for lactic acid production. As was reported by several authors [34,38,42–44], wines fermented with *L. thermotolerans* showed higher values of titratable acidity and buffering capacity compared to other wines. This observation is mainly due to the ability of this yeast to produce lactic acid from pyruvate through the action of lactate dehydrogenase enzyme, bypassing conversion into ethanol [32]. This yeast also produced a higher amount of acetic acid in wines, increasing their volatile acidity. The concentration of acetic acid is 0.3 g/L below the established legal limit. Although its concentration is close to the perception threshold, as described later, the tasters did not detect it. Therefore, we can conclude that this increase in volatile acidity compared to the other wines does not pose a problem for organoleptic quality. In this sense, some authors reported, as in our study, an increase in this parameter, whereas others observed that the volatile acidity decreased [24–28,45].

Regarding the ethanol concentration, it was lower in wines produced at 21 °C. Considering that residual sugars did not show significant differences due to the fermentation temperatures, the minor concentration of ethanol could be attributed to its higher volatility at higher temperatures [36].

3.3. Major Volatile Aroma Compounds and Polyols

Major volatile aroma compounds were grouped into four families: higher alcohols and methanol, esters, polyols, and carbonyl compounds (Table 2).

Table 2. Major volatile compounds and polyols (mg/L, except where indicated) were determined in wines fermented with *Saccharomyces cerevisiae* (Sc), *Torulaspora delbrueckii* (Td), and *Lachancea thermotolerans* (Lt) at two different temperatures. Multivariate analysis of variance (MANOVA) involved taking variations in factors of temperature (T) and the yeast used (Y).

Compound	14 °C			21 °C			MANOVA	
	Sc	Td	Lt	Sc	Td	Lt	T	Y
Methanol	31 ± 3	32 ± 5	26.5 ± 0.4	31 ± 2	25 ± 3	31 ± 1	*	ns
Propanol	33 ± 1	55.3 ± 0.7	55 ± 4	28 ± 1	33 ± 2	58.1 ± 0.1	***	***
Isobutanol	28.6 ± 0.8	68 ± 1	58.8 ± 0.1	38 ± 2	70 ± 2	106 ± 1	***	***
Isoamyl alcohols	195 ± 4	270 ± 6	245 ± 2	261 ± 8	258 ± 3	254 ± 3	***	***
2-Phenylethanol	24.6 ± 0.4	44.7 ± 0.2	27 ± 2	45 ± 1	58 ± 2	29 ± 5	***	***
Ethyl acetate	32 ± 1	34 ± 2	158 ± 0	38 ± 1	53.1 ± 0.2	152 ± 1	***	***
Ethyl lactate	nd	nd	73 ± 3	nd	nd	91 ± 3	***	***
Diethyl succinate	8.5 ± 0.2	15 ± 1	11.6 ± 0.8	9.7 ± 0.4	12.8 ± 0.3	13.5 ± 0.2	ns	***
2,3-Butanediol (*levo*)	224 ± 2	256 ± 31	213.2 ± 0.1	249 ± 11	231 ± 19	210 ± 10	ns	ns
2,3-Butanediol (*meso*)	71 ± 5	209 ± 23	68 ± 1	78 ± 5	179 ± 10	46 ± 4	ns	***
Glycerol (g/L)	7.6 ± 0.1	11 ± 1	7.1 ± 0.2	6.8 ± 0.3	6.4 ± 0.5	9.0 ± 0.4	*	*
Acetaldehyde	59 ± 2	177 ± 2	165 ± 2	36 ± 4	72 ± 2	103 ± 13	***	***
Acetoin	4.4 ± 0.3	13.2 ± 0.9	11.4 ± 0.1	3.3 ± 0.3	6.3 ± 0.5	18.3 ± 0.6	ns	***

* denote significant differences at 95% confidence level; *** denote significant differences at 99.9% confidence level; ns = no significant differences; nd = not detected.

All compounds, except for methanol and 2,3-butanediol (*levo*), were present in different amounts depending on the yeast strain. Additionally, the temperature significantly influenced the concentration of the analyzed compounds.

Higher alcohols are produced during fermentation due to the metabolism of sugars and amino acids. As reported by Mauricio et al. [37], the yeasts may be able to use amino acids not only as nitrogen sources but also as redox agents to balance the oxidation−reduction potential under conditions of restricted oxygen. In this sense, the production of higher alcohols could depend on the fermentation conditions and on the metabolic pathways of each specific yeast. In line with Azzolini et al. [27], the production of 2-phenylethanol stands out in *T. delbrueckii*, whose main descriptor is the floral (rose) aroma [46]. The rest of the higher alcohols show similar concentrations between both non-*Saccharomyces* yeasts.

The second group consists of esters. They can be formed either chemically or through yeast metabolism. Specifically, they are produced as the final step in fatty acid synthesis, which involves the release of Coenzyme A (if it occurs via alcoholysis instead of hydrolysis). This group includes ethyl acetate, ethyl succinate, and ethyl lactate, which are characterized by fruity aromas, except for ethyl acetate, which at high concentrations has an unpleasant aroma of glue and chemicals [46].

L. thermotolerans produced the highest amount of ethyl acetate and ethyl lactate. This can be explained by the highest concentration of both acetic and lactic acids. Probably, lactate formation via NAD-dependent lactate dehydrogenase (LDH) serves to replenish oxidized NAD+ depleted through glycolysis. This fact could alter the redox balance, resulting in the production of acetic acid.

The third group included 2,3-butanediol (*levo* and *meso* forms) and glycerol, which is the third most abundant compound in wines after water and ethanol. Glycerol, as described for amino acids, is related to redox balance. It serves as a redox valve to eliminate excess cytosolic NADH under anaerobic conditions and is coupled with acetic acid production [23]. Moreover, *T. delbrueckii* stood out in the production of polyols, especially 2,3-butanediol (*meso*), which was not affected by the fermentation temperature. On the other hand, the production of glycerol did not follow a clear trend, as it was produced in higher quantities by *T. delbrueckii* at low temperatures, and the same was true for *L. thermotolerans* at 21 °C. Authors, such as Belda et al. [19] and Puertas et al. [35], described a higher production of glycerol by *T. delbrueckii* than by *S. cerevisiae*, while, in the review of Benito et al. [24], disparate results were found in the case of *L. thermotolerans*.

Lastly, the concentration of carbonyl compounds was higher in wine produced with non-*Saccharomyces* yeasts. Regarding the effect of temperature, it has been observed that there was a significant reduction in these compounds as temperature increased. This may have been due to metabolic deviations at low temperatures, favoring the synthesis of secondary compounds associated with glycerol–pyruvic fermentation.

3.4. Multiple Regression Analysis for Lactic Acid

As described above, the production of lactic acid impacts the redox balance. Additionally, the production of compounds, such as higher alcohols, glycerol, or acetic acid, is also related to the redox balance. To establish a link between the production of lactic acid and some of the compounds involved in the redox balance, we conducted a multiple regression analysis. The results showed that lactic acid production is related to volatile acidity (acetic acid), glycerol, and the higher alcohols isobutanol, isoamyl alcohols, and 2-phenylethanol (see Table 3). Propanol has been excluded because its synthesis is related to α-ketobutyric acid and not to any amino acid.

Table 3. Coefficients, R-squared, *p*-value, and Durbin–Watson statistic of the multiple regression analysis carried out to relate lactic acid concentration with some compounds produced during alcoholic fermentation.

	Coefficient	*p*-Value
Constant	1.564	>0.05
Volatile Acidity	1.379	<0.05
Glycerol	−0.371	<0.05
Isobutanol	0.0436	<0.05
Isoamyl alcohol	0.0323	<0.05
2-Phenyletanol	−0.136	<0.05
R^2	99.10%	
Durbin–Watson	3.033	>0.05
Model		<0.05

Based on the results provided by the model, the *p*-value is less than 0.05, which indicates a statistically significant relationship between the variables with a confidence level of 95.0%.

The R^2 statistic indicates that the adjusted model explains 99.10% of the variability in lactic acid values. The Durbin–Watson statistic examines the residuals to determine if there is any significant correlation based on their order in the data file. Since the *p*-value is greater than 0.05, there is no indication of serial autocorrelation in the residuals with a confidence level of 95.0%. Furthermore, considering the sign of the coefficients for each independent variable, we can conclude that as lactic acid production increased, so did the contents of acetic acid, isobutanol, and isoamyl alcohols, while glycerol and 2-phenylethanol decreased.

3.5. Cluster and Principal Component Analysis

3.5.1. Cluster Analysis

Cluster analysis is an exploratory technique used to classify objects or cases into groups (clusters) based on their similarity. To this end, a group of variables is chosen as classifying factors. The smaller the distance between two clusters, the greater the similarity that exists between the samples that comprise both clusters [47].

Here, cluster analysis, according to Ward's method, was carried out, with the classifying variables being the major volatile compounds (except methanol because it is not produced by yeasts), polyols, lactic acid, and acetic acid (Figure 2).

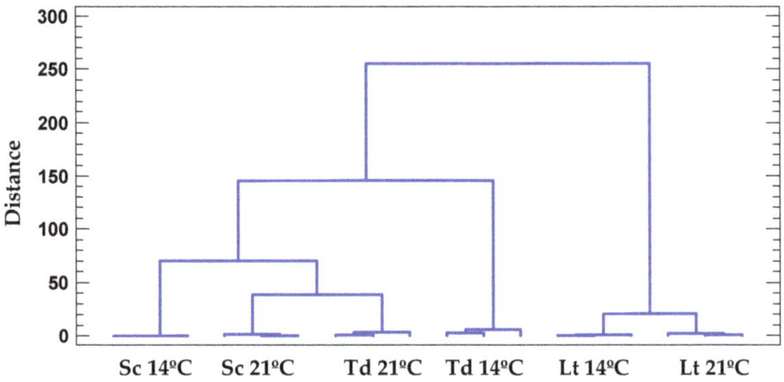

Figure 2. Cluster analysis according to Ward's method. *Saccharomyces cerevisiae* (Sc), *Torulaspora delbrueckii* (Td), and *Lachancea thermotolerans* (Lt) are depicted at two different temperatures (14 and 21 °C).

It can be observed that wines produced by *L. thermotolerans* were quite different than those produced with the two other yeasts. According to the distance of separation between 14 and 21 °C for *L. thermotolerans*, we can assume that there were scarcely any differences between both wines. The other cluster groups the rest of the wines and, in this case, there were notable differences among wines produced at 14 and 21 °C independently of the yeast used. In this sense, at high fermentation temperatures, wines produced by *S. cerevisiae* and *T. delbrueckii* are more similar than those obtained at low temperatures.

3.5.2. Principal Component Analysis

Principal component analysis (PCA) is a multivariate technique of analysis used to reduce the dimensionality of a dataset. To this end, the original variables are transformed into new uncorrelated variables (components). A reduced number of components should explain the greatest variability. Subsequently, the analyst must relate the selected components to one of the sources of variation. In our case, these are the yeast species and the temperature. Several authors have recently used this statistical procedure to relate wine composition with the variables involved in wine production [48].

Here, a PCA was performed, using as classifying variables the major volatile compounds (except methanol, because it is not produced by yeast), polyols, lactic acid, and acetic acid (Figure 2). The first three components explained 89.26% of the observed variability.

Specifically, the first principal component explained 49.29% of the variability and distinguished between wines produced by *L. thermotolerans* and the rest of the wines (Figure 3a). The variables with the highest weight in such differentiation were lactic acid, propanol, ethyl acetate, ethyl lactate, and acetoin (Table 4), all of which were positively correlated with *L. thermotolerans*.

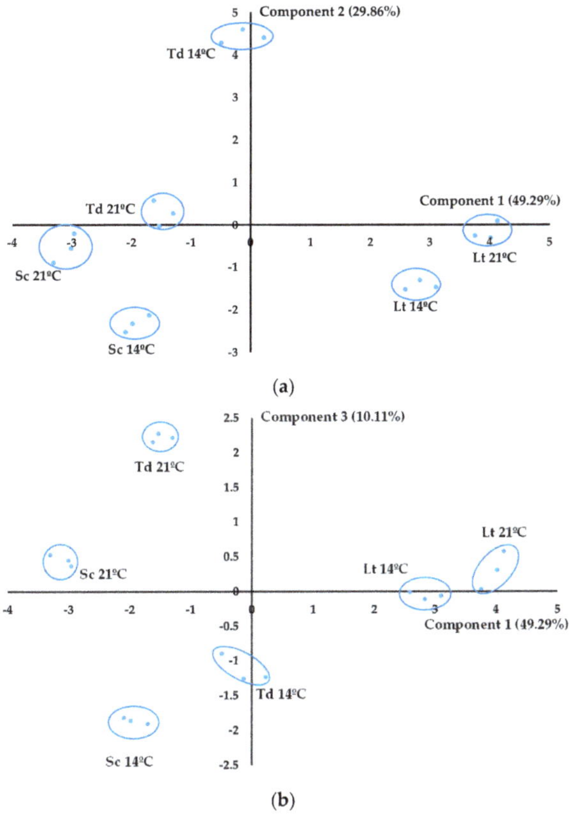

Figure 3. (**a**) Two-dimensional representation of the components 1 and 2 obtained in the principal component analysis. (**b**) Two-dimensional representation of the components 1 and 3 obtained in the principal component analysis.

Table 4. Weight of the variables in each one of the selected principal components (Pc) obtained in the principal component analysis.

Classifying Variables	Pc1 (49.29%)	Pc2 (29.86%)	Pc3 (10.11%)
Lactic acid	**0.3613**	−0.1223	0.0818
Acetic acid	0.2829	−0.2076	0.1287
Propanol	**0.3405**	0.1663	−0.2235
Isobutanol	0.2922	0.1927	0.3037
Isoamyl alcohols	0.0635	**0.3739**	0.3934
2-Phenylethanol	−0.1877	0.3039	**0.5004**
Ethyl acetate	**0.3473**	−0.1333	0.1873
Ethyl lactate	**0.3579**	−0.1308	0.1069
Diethyl succinate	0.1890	**0.4120**	0.0905
2,3-Butanediol (*levo*)	−0.2341	0.2925	−0.1563
2,3-Butanediol (*meso*)	−0.1276	**0.4110**	0.0149
Glycerol (g/L)	0.1376	0.3108	**−0.5006**
Acetaldehyde	0.2350	0.2446	−0.3080
Acetoin	**0.3453**	0.1716	−0.0679

The second component (29.86% of the variability) differentiates between wines produced with *T. delbrueckii* (positive values of the second component) and those obtained with *S. cerevisiae*. Isoamyl alcohols, diethyl succinate, and 2,3-butanediol (*meso*) were the compounds involved in such differentiation (Table 4).

The third component (10.11% of the variability) discriminates between wines produced with *T. delbrueckii* and *S. cerevisiae* at different temperatures. As can be seen in Figure 3b, positive values of the third component contain wines obtained at 21 °C and negative values contain the wines obtained at 14 °C. For these wines, the compounds related to the fermentation temperature were 2-phenylethanol and glycerol (Table 4).

3.6. Organoleptic Characterization

Figures 4 and 5 show the distribution of frequencies of the evaluations given by the tasting panel at 14 and 21 °C, respectively. Additionally, the medians (dark bars) and the trend lines are provided. At both temperatures, wines obtained with the *S. cerevisiae* yeast strain received higher ratings. Taking into account the number of responses categorized as acceptable and desirable, it appears that wines produced at 21 °C were rated higher than those produced at 14 °C.

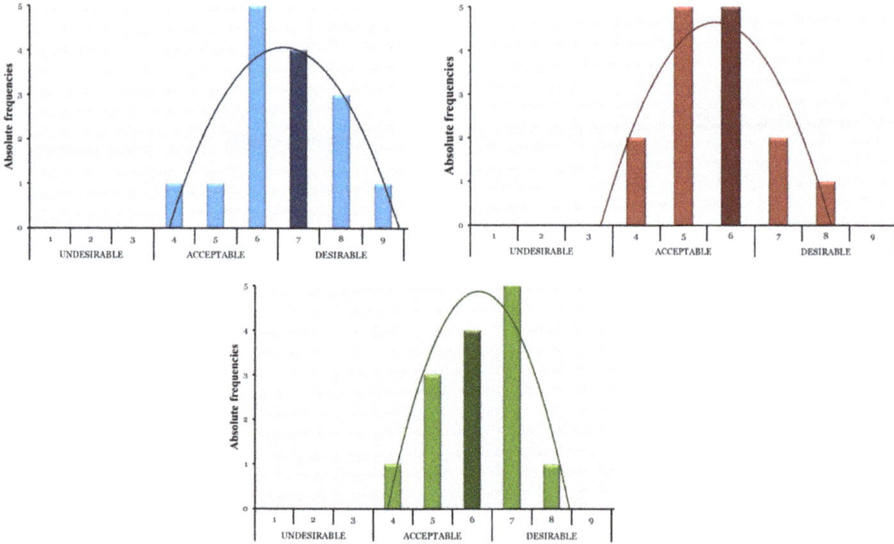

Figure 4. Distribution of absolute frequencies and trend lines obtained in the evaluation of wines obtained at 14 °C by *Saccharomyces cerevisiae* (blue bars); *Torulaspora delbrueckii* (red bars); and *Lachancea thermotolerans* (green bars). Dark bars indicate the median value.

Visually, tasters prefer wines produced with *L. thermotolerans* at 14 °C as they were characterized by greater clarity and brightness. All wines showed color nuances from green to pale yellow, except for those fermented with *T. delbrueckii* at 21 °C, which exhibited a slight turbidity problem and a more brownish hue.

Regarding the aroma notes, the tasters were asked to give the aromatic descriptor that was detected (Supplementary Material Figures S1 and S2). In the aromatic phase, control wines stood out positively compared to those fermented by non-*Saccharomyces* yeasts. *S. cerevisiae* showed positive floral and fruity notes at both temperatures, and *T. delbrueckii* showed fruity, herbaceous, and vegetal notes at 21 °C. On the other hand, *L. thermotolerans* stood out for aromas, such as vegetal and nuts at 14 °C and toasted notes at 21 °C, while negative aromas, such as chemical and pungent aromas, were detected at both temperatures.

In the taste phase, wines fermented with *L. thermotolerans* stood out in terms of acidity, freshness, and persistence at both temperatures.

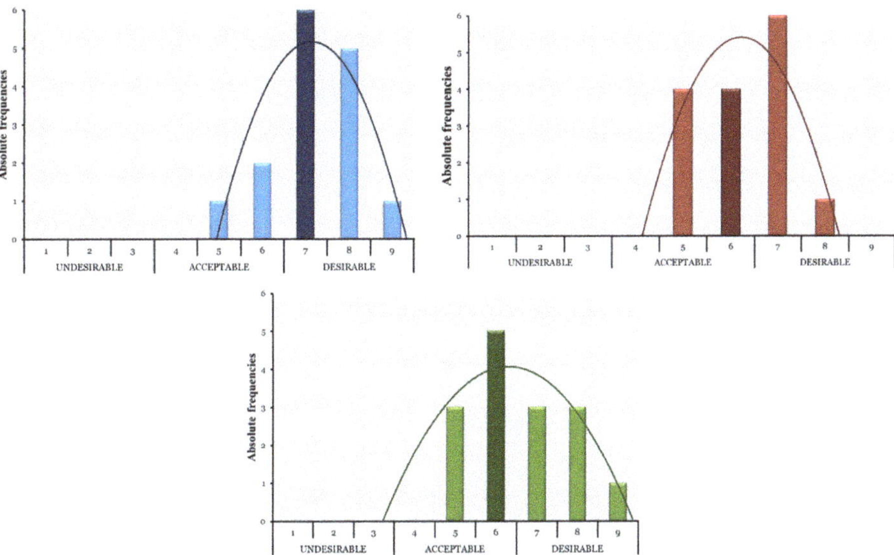

Figure 5. Distribution of absolute frequencies and trend lines obtained in the evaluation of wines obtained at 21 °C by *Saccharomyces cerevisiae* (blue bars); *Torulaspora delbrueckii* (red bars); and *Lachancea thermotolerans* (green bars). Dark bars indicate the median value.

4. Conclusions

L. thermotolerans significantly increases acidity levels due to the high production of lactic acid. Although there is still little knowledge of the routes involved in the production of this acid, in our study, we have been able to correlate its formation with a higher production of isoamyl alcohols, isobutanol, and acetic acid and a lower production of glycerol and 2-phenylethanol.

The principal component analysis showed that the compounds that contribute the most in differentiating between the wines produced by *L. thermotolerans* are lactic acid, ethyl lactate, ethyl acetate, propanol, and acetoin, while isoamyl alcohols, diethyl succinate, and 2,3-butanediol *(meso)* are related to the differentiation of the wines produced by *T. delbrueckii* from those obtained with *S. cerevisiae*. Finally, although the temperature does not have a significant effect on wines produced by *L. thermotolerans*, it does contribute to differentiating between the rest of the wines, with the compounds that contribute the most being glycerol and 2-phenylethanol.

In relation to the organoleptic analysis, the wines preferred by the tasting panel were the control wines, although the wines fermented with *L. thermotolerans* were also well-liked and received a similar score to the control wines. Lastly, the wines fermented with *T. delbrueckii* received the lowest score.

In conclusion, fermentation with *L. thermotolerans* is a great option for the natural acidification of wines from musts with low acidity. However, further research is needed to deepen the understanding of the metabolism of this yeast and its relationship with fermentation conditions. This should be done with the aim of increasing the beneficial effects on the analytical and organoleptic characteristics of wine while reducing the undesirable effects.

Supplementary Materials: The following supporting information can be downloaded at: https://www.mdpi.com/article/10.3390/fermentation10010017/s1, Table S1: Major aroma compounds identified in the wines. Figure S1: Aroma descriptors detected by the tasting panel in the wines produced at 14 °C. The number of times that the tasters detected a given aroma note is shown for *Saccharomyces cerevisiae* (SC); *Torulaspora delbrueckii* (TD); and *Lachancea thermotolerans* (LT). Figure S2: Aroma descriptors detected by the tasting panel in the wines produced at 21 °C. The number of times that the tasters detected a given aroma note is shown for *Saccharomyces cerevisiae* (SC); *Torulaspora delbrueckii* (TD); and *Lachancea thermotolerans* (LT).

Author Contributions: Conceptualization, R.A.P. and F.S.-S.; methodology, R.A.P. and F.S.-S.; formal analysis, R.A.P.; investigation, R.A.P. and F.S.-S.; data curation, R.A.P.; writing—original draft preparation, F.S.-S.; writing—review and editing, R.A.P. and F.S.-S.; funding acquisition, R.A.P. and F.S.-S. All authors have read and agreed to the published version of the manuscript.

Funding: This research was funded by the Ministry of Science and Innovation of Spain (TED2021-129208B-100 project).

Institutional Review Board Statement: Not applicable.

Informed Consent Statement: Not applicable.

Data Availability Statement: The data are available upon request to the authors.

Conflicts of Interest: The authors declare no conflicts of interest.

References

1. Giorgi, F.; Lionello, P. Climate Change Projections for the Mediterranean Region. *Glob. Planet. Chang.* **2008**, *63*, 90–104. [CrossRef]
2. Ficklin, D.L.; Novick, K.A. Historic and Projected Changes in Vapor Pressure Deficit Suggest a Continental-scale Drying of the United States Atmosphere. *J. Geophys. Res. Atmos.* **2017**, *122*, 2061–2079. [CrossRef]
3. Dusenge, M.E.; Duarte, A.G.; Way, D.A. Plant Carbon Metabolism and Climate Change: Elevated CO_2 and Temperature Impacts on Photosynthesis, Photorespiration and Respiration. *New Phytol.* **2019**, *221*, 32–49. [CrossRef] [PubMed]
4. van Leeuwen, C.; Destrac-Irvine, A.; Dubernet, M.; Duchêne, E.; Gowdy, M.; Marguerit, E.; Pieri, P.; Parker, A.; de Rességuier, L.; Ollat, N. An Update on the Impact of Climate Change in Viticulture and Potential Adaptations. *Agronomy* **2019**, *9*, 514. [CrossRef]
5. Van Leeuwen, C.; Destrac-Irvine, A. Modified Grape Composition under Climate Change Conditions Requires Adaptations in the Vineyard. *OENO One* **2017**, *51*, 147–154. [CrossRef]
6. Winefield, C.; Lee, T.-M.; Gambetta, J.M.; Friedel, M.; Holzapfel, B.P.; Stoll, M. Sunburn Grapes: A Review. *Front. Plant Sci.* **2021**, *11*, 604691. [CrossRef]
7. Cataldo, E.; Eichmeier, A.; Mattii, G.B. Effects of Global Warming on Grapevine Berries Phenolic Compounds—A Review. *Agronomy* **2023**, *13*, 2192. [CrossRef]
8. Kliewer, W.M.; Torres, R.E. Effect of Controlled Day and Night Temperatures on Grape Coloration. *Am. J. Enol. Vitic.* **1972**, *23*, 71–77. [CrossRef]
9. Reynolds, A.G.; Vanden Heuvel, J.E. Influence of Grapevine Training Systems on Vine Growth and Fruit Composition: A Review. *Am. J. Enol. Vitic.* **2009**, *60*, 251–268. [CrossRef]
10. Gatti, M.; Frioni, T.; Garavani, A.; Biagioni, A.; Poni, S. Impact of Delayed Winter Pruning on Phenology and Ripening Kinetics of Pinot Noir Grapevines. *BIO Web Conf.* **2019**, *13*, 04002. [CrossRef]
11. Zheng, W.; García, J.; Balda, P.; Martínez de Toda, F. Effects of Late Winter Pruning at Different Phenological Stages on Vine Yield Components and Berry Composition in La Rioja, North-Central Spain. *OENO One* **2017**, *51*, 363. [CrossRef]
12. de Toda, F.; Sancha, J.; Zheng, W.; Balda, P. Leaf Area Reduction by Trimming, a Growing Technique to Restore the Anthocyanins: Sugars Ratio Decoupled by the Warming Climate. *Vitis* **2014**, *53*, 189–192.
13. de Toda, F.; Balda, P. Delaying Berry Ripening through Manipulating Leaf Area to Fruit Ratio. *Vitis J. Grapevine Res.* **2013**, *52*, 171–176.
14. Intrigliolo, D.S.; Lizama, V.; García-Esparza, M.J.; Abrisqueta, I.; Álvarez, I. Effects of Post-Veraison Irrigation Regime on Cabernet Sauvignon Grapevines in Valencia, Spain: Yield and Grape Composition. *Agric. Water Manag.* **2016**, *170*, 110–119. [CrossRef]
15. Dinis, L.T.; Malheiro, A.C.; Luzio, A.; Fraga, H.; Ferreira, H.; Gonçalves, I.; Pinto, G.; Correia, C.M.; Moutinho-Pereira, J. Improvement of Grapevine Physiology and Yield under Summer Stress by Kaolin-Foliar Application: Water Relations, Photosynthesis and Oxidative Damage. *Photosynthetica* **2018**, *56*, 641–651. [CrossRef]
16. Frioni, T.; Tombesi, S.; Luciani, E.; Sabbatini, P.; Berrios, J.G.; Palliotti, A. Kaolin Treatments on Pinot Noir Grapevines for the Control of Heat Stress Damages. *BIO Web Conf.* **2019**, *13*, 04004. [CrossRef]
17. Coniberti, A.; Ferrari, V.; Dellacassa, E.; Boido, E.; Carrau, F.; Gepp, V.; Disegna, E. Kaolin over Sun-Exposed Fruit Affects Berry Temperature, Must Composition and Wine Sensory Attributes of Sauvignon Blanc. *Eur. J. Agron.* **2013**, *50*, 75–81. [CrossRef]

18. van Leeuwen, C.; Roby, J.-P.; Alonso-Villaverde, V.; Gindro, K. Impact of Clonal Variability in *Vitis vinifera* Cabernet Franc on Grape Composition, Wine Quality, Leaf Blade Stilbene Content, and Downy Mildew Resistance. *J. Agric. Food Chem.* **2013**, *61*, 19–24. [CrossRef]
19. Salgado, C.M.; Fernández-Fernández, E.; Palacio, L.; Carmona, F.J.; Hernández, A.; Prádanos, P. Application of Pervaporation and Nanofiltration Membrane Processes for the Elaboration of Full Flavored Low Alcohol White Wines. *Food Bioprod. Process.* **2017**, *101*, 11–21. [CrossRef]
20. Biyela, B.N.E.; du Toit, W.J.; Divol, B.; Malherbe, D.F.; van Rensburg, P. The production of reduced-alcohol wines using Gluzyme Mono®10.000 BG-treated grape juice. *South. Afr. J. Enol. Vitic.* **2016**, *30*, 124–132. [CrossRef]
21. Liguori, L.; Albanese, D.; Crescitelli, A.; Di Matteo, M.; Russo, P. Impact of Dealcoholization on Quality Properties in White Wine at Various Alcohol Content Levels. *J. Food Sci. Technol.* **2019**, *56*, 3707–3720. [CrossRef] [PubMed]
22. Contreras, A.; Hidalgo, C.; Henschke, P.A.; Chambers, P.J.; Curtin, C.; Varela, C. Evaluation of Non-Saccharomyces Yeasts for the Reduction of Alcohol Content in Wine. *Appl. Environ. Microbiol.* **2014**, *80*, 1670–1678. [CrossRef] [PubMed]
23. Peinado, R.A.; Moreno, J.; Bueno, J.E.; Moreno, J.A.; Mauricio, J.C. Comparative Study of Aromatic Compounds in Two Young White Wines Subjected to Pre-Fermentative Cryomaceration. *Food Chem.* **2004**, *84*, 585–590. [CrossRef]
24. Benito, S. The Impact of *Torulaspora delbrueckii* Yeast in Winemaking. *Appl. Microbiol. Biotechnol.* **2018**, *102*, 3081–3094. [CrossRef] [PubMed]
25. Bely, M.; Stoeckle, P.; Masneuf-Pomarède, I.; Dubourdieu, D. Impact of Mixed *Torulaspora delbrueckii–Saccharomyces cerevisiae* Culture on High-Sugar Fermentation. *Int. J. Food Microbiol.* **2008**, *122*, 312–320. [CrossRef] [PubMed]
26. Belda, I.; Navascués, E.; Marquina, D.; Santos, A.; Calderon, F.; Benito, S. Dynamic Analysis of Physiological Properties of *Torulaspora delbrueckii* in Wine Fermentations and Its Incidence on Wine Quality. *Appl. Microbiol. Biotechnol.* **2015**, *99*, 1911–1922. [CrossRef] [PubMed]
27. Azzolini, M.; Tosi, E.; Lorenzini, M.; Finato, F.; Zapparoli, G. Contribution to the Aroma of White Wines by Controlled *Torulaspora delbrueckii* Cultures in Association with *Saccharomyces cerevisiae*. *World J. Microbiol. Biotechnol.* **2015**, *31*, 277–293. [CrossRef]
28. Belda, I.; Ruiz, J.; Beisert, B.; Navascués, E.; Marquina, D.; Calderón, F.; Rauhut, D.; Benito, S.; Santos, A. Influence of *Torulaspora delbrueckii* in Varietal Thiol (3-SH and 4-MSP) Release in Wine Sequential Fermentations. *Int. J. Food Microbiol.* **2017**, *257*, 183–191. [CrossRef]
29. Renault, P.; Coulon, J.; Moine, V.; Thibon, C.; Bely, M. Enhanced 3-Sulfanylhexan-1-Ol Production in Sequential Mixed Fermentation with *Torulaspora delbrueckii / Saccharomyces cerevisiae* Reveals a Situation of Synergistic Interaction between Two Industrial Strains. *Front. Microbiol.* **2016**, *7*, 293. [CrossRef]
30. Ruiz, J.; Belda, I.; Beisert, B.; Navascués, E.; Marquina, D.; Calderón, F.; Rauhut, D.; Santos, A.; Benito, S. Analytical Impact of *Metschnikowia pulcherrima* in the Volatile Profile of Verdejo White Wines. *Appl. Microbiol. Biotechnol.* **2018**, *102*, 8501–8509. [CrossRef]
31. Morata, A.; Bañuelos, M.A.; Vaquero, C.; Loira, I.; Cuerda, R.; Palomero, F.; González, C.; Suárez-Lepe, J.A.; Wang, J.; Han, S.; et al. *Lachancea thermotolerans* as a Tool to Improve PH in Red Wines from Warm Regions. *Eur. Food Res. Technol.* **2019**, *245*, 885–894. [CrossRef]
32. Hranilovic, A.; Gambetta, J.M.; Schmidtke, L.; Boss, P.K.; Grbin, P.R.; Masneuf-Pomarede, I.; Bely, M.; Albertin, W.; Jiranek, V. Oenological Traits of *Lachancea thermotolerans* Show Signs of Domestication and Allopatric Differentiation. *Sci. Rep.* **2018**, *8*, 14812. [CrossRef] [PubMed]
33. Benito, S. The Impacts of *Lachancea thermotolerans* Yeast Strains on Winemaking. *Appl. Microbiol. Biotechnol.* **2018**, *102*, 6775–6790. [CrossRef] [PubMed]
34. Balikci, E.K.; Tanguler, H.; Jolly, N.P.; Erten, H. Influence of *Lachancea thermotolerans* on Cv. Emir. Wine Fermentation. *Yeast* **2016**, *33*, 313–321. [CrossRef] [PubMed]
35. Benito, Á.; Calderón, F.; Palomero, F.; Benito, S. Quality and Composition of Airen Wines Fermented by Sequential Inoculation of *Lachancea thermotolerans* and *Saccharomyces cerevisiae*. *Food Technol. Biotechnol.* **2016**, *54*, 135–144. [CrossRef] [PubMed]
36. Vicente, J.; Kelanne, N.; Navascués, E.; Calderón, F.; Santos, A.; Marquina, D.; Yang, B.; Benito, S. Combined Use of *Schizosaccharomyces pombe* and a *Lachancea thermotolerans* Strain with a High Malic Acid Consumption Ability for Wine Production. *Fermentation* **2023**, *9*, 165. [CrossRef]
37. Benito, Á.; Calderón, F.; Palomero, F.; Benito, S. Combine Use of Selected *Schizosaccharomyces Pombe* and *Lachancea thermotolerans* Yeast Strains as an Alternative to the Traditional Malolactic Fermentation in Red Wine Production. *Molecules* **2015**, *20*, 9510–9523. [CrossRef]
38. Gobbi, M.; Comitini, F.; Domizio, P.; Romani, C.; Lencioni, L.; Mannazzu, I.; Ciani, M. *Lachancea thermotolerans* and *Saccharomyces cerevisiae* in Simultaneous and Sequential Co-Fermentation: A Strategy to Enhance Acidity and Improve the Overall Quality of Wine. *Food Microbiol.* **2013**, *33*, 271–281. [CrossRef]
39. International Organisation of Vine and Wine. *Compendium of International Methods of Wine and Must Analysis*; International Organisation of Vine and Wine: Dijon, France, 2021; ISBN 9782850380686.
40. AENOR. *AENOR. Análisis Sensorial. Tomo I. Alimentación*; AENOR: Madrid, Spain, 1997.
41. Llauradó, J.M.; Rozès, N.; Constantí, M.; Mas, A. Study of Some *Saccharomyces cerevisiae* Strains for Winemaking after Preadaptation at Low Temperatures. *J. Agric. Food Chem.* **2005**, *53*, 1003–1011. [CrossRef]

42. Kapsopoulou, K.; Mourtzini, A.; Anthoulas, M.; Nerantzis, E. Biological Acidification during Grape Must Fermentation Using Mixed Cultures of *Kluyveromyces thermotolerans* and *Saccharomyces cerevisiae*. *World J. Microbiol. Biotechnol.* **2007**, *23*, 735–739. [CrossRef]

43. Benito, S. Combined Use of *S. pombe* and *L. thermotolerans* in Winemaking. Beneficial Effects Determined Through the Study of Wines' Analytical Characteristics. *Molecules* **2016**, *21*, 1744. [CrossRef]

44. Escribano, R.; González-Arenzana, L.; Portu, J.; Garijo, P.; López-Alfaro, I.; López, R.; Santamaría, P.; Gutiérrez, A.R. Wine Aromatic Compound Production and Fermentative Behaviour within Different Non-*Saccharomyces* Species and Clones. *J. Appl. Microbiol.* **2018**, *124*, 1521–1531. [CrossRef]

45. Puertas, B.; Jiménez, M.J.; Cantos-Villar, E.; Cantoral, J.M.; Rodríguez, M.E. Use of *Torulaspora delbrueckii* and *Saccharomyces cerevisiae* in Semi-Industrial Sequential Inoculation to Improve Quality of Palomino and Chardonnay Wines in Warm Climates. *J. Appl. Microbiol.* **2017**, *122*, 733–746. [CrossRef]

46. Moreno, J.J.; Peinado, R.A. *Enological Chemistry*; Academic Press: Cambridge, MA, USA, 2012.

47. Everitt, B.S.; Landau, S.; Leese, M.; Stahl, D. *Cluster Analysis*; Wiley: Hoboken, NJ, USA, 2011; ISBN 9780470749913.

48. Vaquero, C.; Izquierdo-Cañas, P.M.; Mena-Morales, A.; Marchante-Cuevas, L.; Heras, J.M.; Morata, A. Use of *Lachancea thermotolerans* for Biological vs. Chemical Acidification at Pilot-Scale in White Wines from Warm Areas. *Fermentation* **2021**, *7*, 193. [CrossRef]

 **_fermentation_**

Article

Survey on Yeast Assimilable Nitrogen Status of Musts from Native and International Grape Varieties: Effect of Variety and Climate

Elisavet Bouloumpasi [1,*], Adriana Skendi [2] and Evangelos H. Soufleros [1]

[1] Department of Food Science and Technology, Faculty of Agriculture, Aristotle University of Thessaloniki, P.O. Box 235, GR-54124 Thessaloniki, Greece; esoufler@agro.auth.gr

[2] Department of Agricultural Biotechnology and Oenology, International Hellenic University, 1st Km Dramas—Mikrohoriou, GR-66100 Drama, Greece; andrianaskendi@hotmail.com or antrsken@teiemt.gr

* Correspondence: elisboul@abo.ihu.gr

Abstract: Yeast assimilable nitrogen (YAN), besides the oenological parameters (sugar content, titratable acidity, and pH) in grape musts of sixteen native and international varieties of *Vitis vinifera* cultivated in six regions of Northern Greece, was assessed in the frame of the present study. Low levels of YAN are frequently thought to be the cause of problematic fermentations and originate significant changes in the organoleptic aspects of the finished product. The objective of this multivariety study was to assess factors affecting the YAN amount and composition in technologically mature grapes and, therefore, to evaluate the necessity of YAN supplementation with ammonium salts in musts across different native and international grape varieties. Free amino nitrogen was measured colorimetrically, ammoniacal nitrogen was measured enzymatically, and their values for each must sample were summed to obtain the total amount of YAN. Statistical analysis was carried out including principal component analysis (PCA) to discover relationships among must samples and the parameters studied. PCA analysis classified samples depending on grape varieties and region of origin, bringing knowledge about native and international cultivars of great commercial interest. Moreover, these findings could help to understand how commercial varieties can behave in different climates in the climate change context.

Keywords: nitrogen; yeast assimilable nitrogen (YAN); must; grape variety; climate

Citation: Bouloumpasi, E.; Skendi, A.; Soufleros, E.H. Survey on Yeast Assimilable Nitrogen Status of Musts from Native and International Grape Varieties: Effect of Variety and Climate. *Fermentation* **2023**, *9*, 773. https://doi.org/10.3390/fermentation9080773

Academic Editors: Maria Dimopoulou, Spiros Paramithiotis, Yorgos Kotseridis and Jayanta Kumar Patra

Received: 31 July 2023
Revised: 16 August 2023
Accepted: 18 August 2023
Published: 19 August 2023

1. Introduction

Nitrogen is an important parameter to monitor in grapes for wine production. Different factors influence nitrogen quantity in grapes, with the nitrogen fertilization of vines being the most important. Amino acids represent one of the most important groups of nitrogen compounds present in grapes, must, and wine. The quantity of amino acids in grape juice can vary significantly depending on the climate as well as the ripeness of the grapes [1]. Among the amino acids present in the must, only α-amino acids, named free amino nitrogen (FAN), are used by the *Saccharomyces cerevisiae* yeasts during their growth. Together with ammoniacal nitrogen (ammonia NH_3 plus NH_4^+ nitrogen), α-amino acids represent yeast assimilable nitrogen (YAN), also called by some authors "promptly assimilable nitrogen by yeasts" (PAN) [2].

Nitrogen compounds in grapes usually reach maximum values at the technological maturity stage, according to Garde-Cerdan et al. [3], who studied grape ripening in conjunction with the evolution of assimilable nitrogen (amino and ammoniacal nitrogen). The nitrogen content of the grape juice can highly affect the yeast metabolism and different critical aspects of fermentation, i.e., the amount of yeast biomass, the rate of fermentation, the time for complete fermentation, as well as the type and quantity of end products [4].

Lack of assimilable nitrogen sources during alcoholic fermentation can delay fermentation [5,6] or produce high amounts of non-desirable by-products (fusel alcohols, as well as hydrogen sulfide) [1,7]. On the other hand, high levels of assimilable nitrogen increase the rate and decrease the overall time of fermentation [8] and favor the synthesis of high levels of acetate esters [9]. Low ammoniacal nitrogen content could increase the production of higher alcohols as the yeasts have the ability to use amino acids as a nitrogen source. Yeast metabolism allows the transformation of amino acids. The corresponding α-keto acids, produced during the transamination process, are transformed by decarboxylation into aldehydes that are reduced further to higher alcohols [10].

The yeasts prefer nitrogen sources that can be converted quickly into vital elements of their metabolism, but parameters such as the strain, carbon source, and growth conditions can affect the order of utilization of available nitrogen sources in the must [7,11]. Yeasts utilize α-amino acids present in the must for their growth, and their lack is linked with the formation of fusel alcohols and hydrogen sulfide [1]. On the other hand, elevated YAN concentration may result in microbial instability, the production of carcinogens (ethyl carbamate), allergens (biogenic amines), and haze formation [12]. The type of nitrogen also affects the aroma profile. So, higher alcohols are produced when amino acids are added compared to the ammonium form [13].

It is commonly accepted that a minimum of 140 mg of assimilable nitrogen per liter is indispensable for a complete and good-quality fermentation; musts with lower levels have a reduced capacity to produce alcohol, resulting in a slow or stationary fermentation Butzke [14]. The addition of diammonium hydrogen phosphate or ammonium sulphate at levels up to 1 g L^{-1} (expressed in salts) or ammonium bisulphate up to 0.2 g L^{-1} (expressed in salts) is allowed by European legislation [15] in order to help increase the amount of assimilable nitrogen needed for yeast fermentation. If used in excess, it can promote the production of higher concentrations of ethyl carbamate [16]. Generally, nitrogen fertilization in the field during growth represents an alternative for increasing the YAN of the must [17–20]. Nitrogen fertilization increased YAN levels by over 50% in the must of the Greek variety *Vitis vinifera* L. cv. Savvatiano (considered a "neutral variety"), whereas in the respective wine, it increased the intensity of aroma [17]. It seems that the effect of fertilization on YAN levels depends on factors such as timing, irrigation, and climate, besides cultivar [18,19].

Different surveys on YAN content have been conducted for *Vitis vinifera* varieties in diverse grape-growing regions differing in climate and growing practices (i.e., United States [14,21]), South Africa [22] and Europe (Italy [2], Spain [23], France [24], Czech Republic [25], and Germany [26]). Some authors have focused their investigation on the variability of YAN depending on the variety [27,28]) region or district ([27,28]), ripeness [27] and vintage [27,28]), while others have been involved in prediction studies to establish variety-specific regression models based on YAN measurements before harvest [21,29]. Differences in YAN between cultivars, vintage, and vineyard locations were reported in the literature [27]. According to Petropoulos et al. [30], climatic conditions and the cultivar are more likely to be the key parameters affecting amino acid concentration in grapes.

However, due to the variable nature of YAN, the information gathered in these surveys may not necessarily be applicable in a Greek context. Limited research was done to assess YAN in local and foreign grape types cultivated in Greece's major wine-producing regions. Only one study monitored the effect of climatic conditions and vineyard altitude on YAN concentration in a native Greek grape variety (*Vitis vinifera* L. cv. Agiorgitiko) during a three-year period in one single region [30]. Since the nitrogen amount in the must represents a crucial factor for quality wines, monitoring this parameter considering the factors that affect it (i.e., climate and variety) is of help to the wine industry. Therefore, it would be beneficial to gain insight into the nitrogen status of the Greek grapes used for wine production since no study exists about YAN in Greek musts for wine production from international and native grape varieties.

In this context, the aim of the study was (i) to assess the nitrogen status of grape juices currently used to make commercial wines in Northern Greece, and (ii) to examine the impacts of variety and climate on yeast assimilable nitrogen of grape musts. This investigation was conducted with the aim to bring into consideration international grape varieties that are not reported in the literature, besides native Greek varieties. The aim was accomplished by evaluating a large number of grape musts for free amino nitrogen (FAN) and ammoniacal nitrogen and by employing exploratory statistical methods.

2. Materials and Methods

2.1. Samples

A survey of the YAN status of 274 musts (132 white and 142 red), corresponding to 16 different grape varieties from six regions of Greece and 12 regional units from the northern part of Greece (see Figure 1), was conducted during the 2017 harvest period. Grape samples were harvested at a ripeness level suitable for commercial winemaking according to the various producers' (wineries') harvest protocols. Upon collection, samples were coded and analyzed immediately. The grape bunches were pressed by hand to separate the must from the pomace. The distribution of samples according to variety and region appears in Table 1.

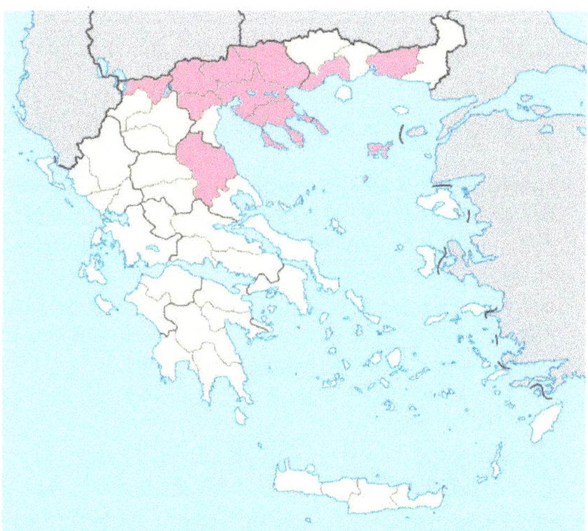

Figure 1. Regional units of Greece involved in the present study (noted in pink).

Table 1. Name, color, origin of grape variety, number of the must samples collected and regional units of cultivation.

Grape Variety	Variety Color	Origin of Variety	Number of Samples	Regional Units
Assyrtiko	1	N	5	Chalkidiki
Cabernet Sauvignon	2	I	2	Chalkidiki
Chardonnay	1	I	11	Pella, Rodopi
Cinsault	2	I	16	Serres
Debina	1	N	8	Larissa
Grenache Rouge	2	I	4	Chalkidiki

Table 1. *Cont.*

Grape Variety	Variety Color	Origin of Variety	Number of Samples	Regional Units
Limnio	2	N	6	Rodopi
Merlot	2	I	74	Chalkidiki, Kavala, Kilkis, Larissa, Thessaloniki, Rodopi
Muscat Hamburg	2	I	4	Larissa
Muscat of Alexandria	1	I	12	Lemnos
Roditis	1/2	N	33	Chalkidiki, Larissa, Mount Athos, Thessaloniki
Sauvignon Blanc	1	I	54	Chalkidiki, Kavala, Mount Athos, Rodopi, Thessaloniki
Syrah	2	I	12	Imathia, Kilkis, Larissa, Rodopi
Ugni Blanc	1	I	6	Larissa
Xinomavro	2	N	23	Chalkidiki, Florina, Kilkis
Zoumiatiko	1	N	4	Serres

Variety color: 1: white; 2: red; origin of variety: I: international; N: native.

2.2. Enological Parameters and Yeast Assimilable Nitrogen Analysis

The fresh must aliquot for chemical analysis was centrifuged (4000 rpm, 10 min) and classical enological parameters, namely Baume, pH, and titratable acidity (expressed as $g\,L^{-1}$ tartaric acid), were determined according to the official methods of OIV [31].

The components of YAN, free amino nitrogen (FAN) and ammoniacal nitrogen were measured separately. Free amino nitrogen was measured by the NOPA (Nitrogen by *o*-Phthaldialdehyde) colorimetric method [32]. Ammoniacal nitrogen (NH_4-N) was measured enzymatically using Ammonia (Rapid) Assay (K-AMIAR 07/14, Megazyme, Ireland). Values for FAN and ammonia were then summed to obtain the total amount of YAN (expressed as $mg\,N\,L^{-1}$) available in the must.

2.3. Meteorological Data

Meteorological data were collected from each grape production area in order to assess the effect of climatic factors on the YAN of grape musts. Therefore, a dataset containing temperature and rainfall data for each region, comprising the period from April to October, was built (see Supplementary Table S1). This frame period was chosen since it covers the grape growth phases.

2.4. Statistical Analysis

Statistical data processing was performed using IBM SPSS Statistics, v.25 statistical software (International Business Machines—IBM Corporation, Chicago, IL, USA). Non-parametric tests for independent samples were performed. Kruskal–Wallis tests (when compared more than two independent variables) were used to detect evidence of differences among the grape varieties and the regional units ($p < 0.05$). Dunn's post hoc tests with Bonferroni adjustment are carried out on each pair of groups. The median test and Mann–Whitney U test were used to test two independent variables differences.

Pearson linear and Spearman's correlation (2-tailed, $p < 0.05$) were employed to recognize any relationship between the parameters studied, depending on the type of data distribution.

The mathematical processing of the standardized data was done by hierarchical cluster analysis (HCA) using Ward's method. The results from the cluster analysis are presented by a dendrogram. HCA analysis was performed using the statistical software package JMP 14 (SAS Institute Inc., Cary, NC, USA).

3. Results

3.1. YAN Status of Grape Samples

A very broad variation was observed in the values of FAN (31–223 mg N L^{-1}) and ammoniacal nitrogen (1–123 mg N L^{-1}) among the studied samples, resulting consequently in an 8-fold variation in the YAN (sum of them) values (Table 2). Nevertheless, the variation observed for YAN (40 to 332 mg N L^{-1}) in this study was lower than those reported by other authors [2,14,22]. Besides the cultivar, identified as the principal factor affecting YAN concentration, vintage and climate are of great importance [22]. Petrovic, Kidd and Buica [22] reported an 11-fold variation for South African musts however, their samples were spread over two vintage years and belong to many other cultivars besides those included in this study. In a similar way, Nicolini, Larcher and Versini [2] attribute the very high (25-fold) variation found in Italian samples to varietal and vintage differences. A 13-fold variation (40 to 559 mg N L^{-1}) observed by Butzke [14] for musts, besides the varietal differences, could be accounted for by the different climates of the three grape-growing regions in the United States (California, Oregon and Washington) from where the samples were collected.

Table 2. Yeast assimilable nitrogen, free amino nitrogen, ammoniacal nitrogen and oenological parameters of the musts according to the grape color [1,2].

Samples	No of Samples	YAN (N L^{-1})	FAN (N L^{-1})	NH$_4$-N Ammonia (N L^{-1})	Baume (Be)	pH	Total Acidity (g L^{-1} Tartaric Acid)
All samples	274	39.79–277.50 (161.20/158.90)	30.58–222.90 (107.77/103.90)	0.70–142.86 (53.43/56.15)	9.10–16.10 (11.85/11.90)	2.87–3.79 (3.42/3.44)	3.51–10.59 (6.41/6.34)
White grapes	132	39.79–277.50 (166.68/172.67)	30.58–222.90 (117.44/117.25) [a]	0.70–142.86 (49.24/47.14) [a]	9.10–16.10 (11.39/11.35) [a]	3.15–3.65 (3.40/3.41) [a]	4.46–8.81 (6.46/6.51)
Red grapes	142	41.73–265.55 (156.11/155.39)	30.77–193.00 (98.77/94.36) [a]	2.30–112.40 (57.33/59.59) [a]	9.96–14.10 (12.28/12.30) [a]	2.87–3.79 (3.44/3.47) [a]	3.51–10.59 (6.36/6.14)

[1] The mean/median values of the measurements are shown in parentheses. [2] Within the same column, medians between white and red groups of musts with the same letter differ significantly from each other ($p < 0.05$), as determined by the independent median test.

Significant statistical differences were observed between the medians of red and white varieties in terms of FAN and ammoniacal nitrogen, while medians of YAN do not differ significantly. White grape varieties exhibited higher FAN but lower ammoniacal nitrogen than the red ones. The literature associates red cultivars with lower YAN and FAN values than white cultivars [22]. An increase in FAN content in grapes is reported during ripening, starting from veraison, with the level reaching in some varieties a maximum before harvest [33]. Similarly, the total nitrogen level of grapes rises during maturation, but can stabilize after an initial increase and, in some situations, may even start to drop as ripening progresses, as reviewed by Bell and Henschke [12]. On the contrary, the concentration of ammoniacal nitrogen varies widely in grapes and declines between veraison and harvest [12].

The collected must samples were characterized by a broad variation in the sugar content and total acidity (expressed as tartaric acid), while they showed less than 1 unit of pH range variation (Table 1). Red and white samples differed in pH and sugar content while exhibiting statistically similar total acidity. The red musts were higher in sugars and exhibited higher pH values than the white samples. This behavior is in agreement with correct oenological practices for producing red wines with higher alcohol content and with the literature that reports pH and TA (as tartaric acid) values in the range of 3.0–3.4, and 6–9 g/L for white and 3.3–3.7, and 5–8 g/L for red wines, respectively [34].

3.2. Grape Variety

The nitrogen content of the samples, as well as basic oenological parameters, categorized based on the grape variety, are presented in Table 3. According to the Kruskal–Wallis test, significant differences in the medians of all the determined must parameters were observed among the varieties. Among the varieties considered, Chardonnay had the highest YAN value (mean 254 mg N L^{-1}), followed by Cinsault, Syrah and Sauvignon Blanc. Average concentrations of YAN in international varieties cultivated in Greece followed the ranking: Chardonnay > Cinsault > Syrah > Sauvignon blanc > Merlot > Ugni blanc > Muscat Hamburg > Cabernet Sauvignon > Grenache rouge > Muscat of Alexandria (also known as Zibibbo in Sicily, Italy). Nevertheless, the differences in the number of samples tested for each variety could affect this classification, especially for the samples present in low numbers (Ugni Blanc, Cabernet Sauvignon and Grenache Rouge). Different surveys conducted in various countries do not concord in ranking specific varieties according to YAN [2,14,21,28]. Butzke [14] found a similar ranking, (i.e., Sauvignon blanc > Merlot > Cabernet Sauvignon) while Hagen, Keller and Edwards [28] and Petrovic, Kidd and Buica [22] reported a slightly different one for grape musts from the Pacific Northwest (U.S.A.) (Chardonnay > Syrah > Cabernet Sauvignon > Merlot) and South Africa (Chardonnay > Sauvignon blanc > Cinsault > Syrah > Cabernet Sauvignon > Merlot), respectively. In all those studies, the average YAN content of the Chardonnay variety exceeds 200 mg L^{-1}.

Overall, despite the fact that the absolute YAN numbers vary amongst the numerous studies, what is remarkable is the proximity in ranking. Our results regarding the red varieties Merlot and Syrah are much higher than those reported by Petrovic, Kidd and Buica [22]. In addition, other studies performed in various world wine areas have consistently indicated Chardonnay to have a high average YAN level [2,14,21]. To our knowledge, no information on the YAN levels of Grenache rouge, Ugni blanc, Muscat of Alexandria, and Muscat Hamburg grapes have been reported in the literature. Yet, in this investigation was discovered that all four varieties had mean YAN levels lower than 140 mg L^{-1}, exhibiting values of 110 mg L^{-1} (n = 4), 126 mg L^{-1} (n = 6), 104 mg L^{-1} (n = 12), and 118 mg L^{-1} (n = 4), for Grenache rouge, Ugni blanc, Muscat of Alexandria, and Muscat Hamburg, respectively. Since the nitrogen content of grapes varies with the vintage [22] but also with viticultural techniques [35], differences among the studies conducted in different countries are expected. Nevertheless, regardless of the study location, they serve to evidence the influence of cultivars on the YAN levels.

The highest mean value of YAN among Greek native varieties found in Roditis (157 mg N L^{-1}) is nearly 80 mg N L^{-1} less than the highest value presented in the international variety Chardonnay. Based on YAN concentration, the native varieties are ranked as follows: Roditis > Xinomavro > Debina > Assyrtiko > Zoumiatiko > Limnio. Although the mean YAN value of international varieties is higher than that of native, only a few differences were observed between the values of single cultivars. No clear trend was observed among native and international varieties regarding the median values of YAN. This fact could be due to a high variation detected among the samples collected. Statistical analysis revealed the existence of only a few Greek varieties that differed from the international ones (Table 3). In a previous study, Bouloumpasi et al. [36], reported that international *Vitis vinifera* varieties (i.e., Chardonnay, Sauvignon blanc, Malvasia aromatica) had a significantly ($p < 0.05$) higher average soluble protein content in comparison to grapes of Greek native *Vitis vinifera* varieties (i.e., Malagousia, Roditis). Possibly genetic factors are those affecting YAN accumulation in grapes. The literature suggests that YAN may be a cultivar-specific trait [22]. In general, it seems that native cultivars have similar contents of FAN but different ammoniacal nitrogen from the international ones.

Table 3. Variation (minimum–maximum value observed) of yeast assimilable nitrogen, free amino nitrogen, ammoniacal nitrogen and oenological parameters in grape musts categorized per grape variety (mg N L^{-1} must) [1,2].

Variety	No of Samples	YAN (N L^{-1})	FAN (N L^{-1})	NH$_4$-N Ammonia (N L^{-1})	Baume (Be)	pH	Total Acidity (g L^{-1} Tartaric Acid)
Sauvignon blanc	54	110.02–271.86 (190.43/187.99) aeksw	61.36–176.90 (119.88/119.10) elm	15.56–142.86 (70.55/81.33) ek	9.10–13.00 (10.80/10.60) abc	3.15–3.58 (3.34/3.36) f	5.02–8.81 (6.87/6.73) af
Chardonnay	11	178.14–277.50 (235.67/228.20) bfiloqtxz	136.80–222.90 (186.55/194.80) bchkl	10.61–79.51 (49.11/47.10)	10.50–12.49 (11.56/11.70) i	3.26–3.61 (3.43/3.40)	6.19–8.44 (7.27/7.16) b
Ugni blanc	6	89.60–176.40 (124.55/111.14) op	54.71–160.60 (95.14/68.58)	10.00–63.18 (29.41/22.21)	9.80–11.67 (10.80/10.65)	3.32–3.45 (3.38/3.37)	5.43–6.64 (6.00/5.95)
Muscat of Alexandria	12	39.79–174.94 (104.01/86.86) klmn	30.58–145.60 (82.40/66.64) cd	9.21–31.67 (21.62/20.92) fghijk	10.90–13.20 (12.55/12.65) aeg	3.33–3.58 (3.49/3.50) a	4.87–7.39 (5.93/5.97)
Zoumiatiko	4	84.00–120.40 (100.10/98.00) efgh	51.53–89.14 (66.34/62.35)	31.26–35.76 (33.76/34.01)	9.11–10.40 (10.00/10.25)	3.34–3.40 (3.38/3.39)	5.17–5.48 (5.39/5.45)
Debina	8	107.80–192.96 (135.20/126.00) qr	46.33–95.32 (67.41/66.13) ab	47.17–97.64 (67.79/61.96) i	11.30–12.50 (11.97/12.00)	3.41–3.55 (3.48/3.50) b	5.17–7.20 (5.81/5.66)
Assyrtiko	5	100.80–154.00 (133.45/133.25)	44.84–76.33 (65.73/67.14)	55.96–77.94 (67.72/68.98)	11.80–13.80 (12.80/12.80)	3.17–3.35 (3.23/3.20)	7.09–8.4 (7.68/7.54)
Roditis	33	70.00–275.41 (156.85/154.00) zα	62.04–212.10 (134.76/133.20) fi	0.70–83.51 (22.09/14.90) abcde	10.20–16.10 (11.79/11.60) j	3.24–3.65 (3.48/3.49) cg	4.46–7.95 (5.86/5.85) f
Merlot	74	89.71–238.86 (152.66/150.05) wxy	50.46–148.90 (89.72/85.49) efgh	18.44–89.96 (62.94/61.53) bh	10.80–14.00 (12.68/12.75) bdhij	3.27–3.68 (3.48/3.50) dh	3.51–8.78 (6.27/6.19) c
Cinsault	16	110.76–265.55 (211.75/217.00) cgjmpruyα	92.72–193.00 (156.67/160.90) dgj	18.04–81.87 (55.08/59.30) cg	9.96–12.70 (11.17/11.10) def	3.52–3.79 (3.60/3.59) fgh	3.83–6.38 (4.93/4.91) abcde
Grenache rouge	4	73.06–124.60 (109.67/120.50) ij	70.76–117.70 (94.06/93.89)	2.30–30.43 (15.61/14.85)	13.70–14.10 (13.83/13.75)	3.45–3.49 (3.46/3.46)	5.00–6.08 (5.54/5.53)
Syrah	12	148.40–252.00 (203.66/209.41) dhnv	82.35–170.60 (130.53/140.65) a	33.30–112.40 (73.14/71.44) dj	11.50–13.70 (12.57/12.50) cf	3.26–3.57 (3.45/3.47) e	6.04–8.66 (7.00/6.73) d
Cabernet Sauvignon	2	105.21–123.20 (114.21/114.21)	59.50–64.88 (62.19/62.19)	45.71–58.32 (52.02/52.02)	11.90–13.80 (12.85/12.85)	3.41–3.52 (3.47/3.47)	5.29–5.95 (5.62/5.62)
Muscat Hamburg	4	95.20–144.20 (118.35/117.00)	87.62–128.10 (108.41/108.95)	3.40–16.10 (9.95/10.14)	12.20–12.60 (12.45/12.50)	3.51–3.6 (3.57/3.58)	3.60–4.56 (4.00/3.92)
Xinomavro	23	63.12–204.65 (142.08/149.33) stuv	31.53–141.00 (86.07/95.25) ijkm	17.08–87.49 (56.01/57.79) af	10.30–12.80 (11.55/11.50) gh	2.87–3.47 (3.19/3.15) abcde	4.61–10.59 (8.00/7.61) e
Limnio	6	41.73–110.88 (72.53/76.55) abcd	30.77–43.66 (39.33/41.15)	4.12–67.59 (33.21/38.00)	10.80–12.20 (11.38/11.30)	3.21–3.43 (3.32/3.33)	5.63–7.91 (6.40/6.19)

[1.] The mean/median values of the measurements are shown in parentheses. [2.] Within one column, medians of groups of grape musts with the same letter differ significantly from each other ($p < 0.05$), as determined by the independent samples Kruskal–Wallis test followed by Dunn's pairwise tests adjusted using the Bonferroni correction.

Concerning the FAN values, the literature reports that Grenache Rouge wines had a high amino acid content, higher than Syrah and Merlot [37]. However, in this study, Syrah grapes had higher mean FAN content, followed by Grenache Rouge and Merlot. As only four samples of Grenache Rouge were examined in the current survey, this amount could not be considered enough to draw conclusions about the variety. The FAN content of Merlot musts ranged from 50 to149 mg L^{-1}. Etiévant et al. [38] reported average Merlot amino acid content of 124 mg L^{-1} for wines originating from France, well within the above range. Muscat of Alexandria wines also have been reported to contain low amounts of FAN [39]. The low levels found could not be explained in terms of grape variety or the location because the samples originated from a single region (Lemnos—Aegean Islands) characterized by semi-arid climate and mainly rocky soils.

Ammoniacal nitrogen in our study ranged from 1–123 mg N L^{-1}, while other authors reported values ranging between 5 and 325 mg N L^{-1} [5,8,14]. The NH$_4$-N contribution to YAN varies highly among varieties (i.e., between 2 and 53% of the YAN as reported by Huang and Ough [40], and it is variety-related, as in varieties of high proline accumulation (e.g., Cabernet Sauvignon) the FAN contribution to YAN is lower, resulting in a higher contribution by NH$_4$-N to YAN compared with varieties that are high arginine accumulators [40].

Significant differences are observed, in general, in oenological parameters between native and international varieties. Native varieties seem to have more sugars, higher pH values and lower acidity than the international varieties. This behavior could be due to genetic factors or viticultural techniques applied by vine growers.

As displayed in Figure 2, for most of the varieties, the mean concentration of YAN does not exceed 140 mg N L^{-1}, which is widely considered to be the minimum requirement for a successful fermentation for *Saccharomyces*, as reviewed by Nicolini, Larcher and Versini [2], Bell and Henschke [12], and Verdenal et al. [41], otherwise, there is a high risk for the fermentation to stop prematurely, while the risk factor is moderate when the YAN content ranges from 140 to 200 mg N L^{-1} [12]. This amount of YAN is necessary, especially for white musts. On the other hand, Australian Wine Research Institute [42] recommends a minimum of 100 mg N L^{-1} YAN for red musts due to extended contact with grape skins, which leads to the extraction of more nitrogen [43]. The international varieties Sauvignon blanc, Chardonnay, Merlot, Cinsault and Syrah, along with the native varieties Xinomavro and Roditis, had more than 140 mg N L^{-1} YAN on average. On the contrary, international varieties Ugni blanc, Muscat of Alexandria, Grenache rouge, Cabernet Sauvignon, Muscat Hamburg, and the natives Zoumiatiko, Debina, Assyrtiko, and Limnio did not exceed the minimum required quantity of YAN. Limnio, followed by Zoumiatiko found to have the lowest average concentration of YAN, at 73 mg N L^{-1} and 100 mg N L^{-1}, respectively. It appears that most native Greek varieties are deficient in YAN and would probably need nitrogen supplementation in order to achieve a successful fermentation. The percentage (36%) of Greek must samples with YAN levels below the threshold level of 140 mg N L^{-1} was lower than that reported for Italian musts (58%) [2] but higher than that of must grapes from the West Coast of United States (13%) [14]. However, as depicted in Figure 2, high variance exists in the values within varieties. This behavior is in line with the findings of Petrovic, Kidd and Buica [22] and also Nicolini, Larcher and Versini [2]. The latter noted that YAN was the most variable factor in grape juice used to evaluate the quality of the grape at harvest. High variation in the YAN values within the same variety could be due to the growing region characteristics and differences in cultivation techniques adopted.

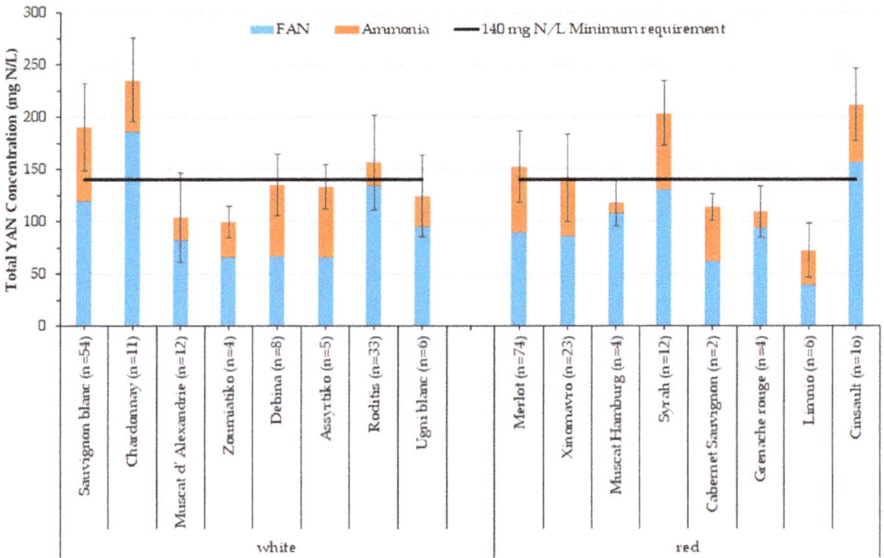

Figure 2. Average YAN concentrations (in mg N L^{-1}) according to variety, represented as a stacked plot of FAN and ammonia concentrations.

3.3. Region and Effect of Climate

The collected samples originated from 12 different regional units belonging to six regions: the Central Macedonia region (Chalkidiki, Imathia, Kilkis, Serres, Thessaloniki, and Mount Athos), Western Macedonia (Florina, Pella), Eastern Macedonia (Kavala), Thrace (Rodopi), Aegean Islands (Lemnos), and Thessaly (Larissa). (Supplementary Table S2). We elected to examine the effect of grape origin based on regional units (smaller areas than regions) as it would reflect better climatic conditions in the respective vineyards. In order to evaluate each region's climate, data regarding precipitation and temperatures were collected from each regional unit for the period between May and September, as most of the varieties are harvested during August and September.

For the explanatory analysis, grape varieties with fewer than eight samples were not included. Merlot and Sauvignon Blanc samples were collected from six and five different regional units, respectively, trying to represent each regional unit. With respect to the climatic data, the general description of each regional unit based on the collected data is as follows:

- Thrace/Rodopi: high temperatures over 25 °C during July–August, rainfall of 5 mm during July–September, and rainfall of 62 mm during May–September.
- Eastern Macedonia/Kavala: high temperatures over 25 °C during July–August, rainfall of 18 mm during July–September, and rainfall of 62 mm in May–September.
- Central Macedonia/Serres: high temperatures over 25 °C during July–August, rainfall of 254 mm during July–September, and high rainfall (357 mm) in May–September.
- Central Macedonia/Kilkis: high temperatures over 25 °C during June–August, 27 mm of rainfall during June–August and >100 mm in September, and a total rainfall of 292 mm during May–September.
- Central Macedonia/Thessaloniki: high temperatures over 25 °C during June–August, rainfall of 47 mm during June–August and >90 mm in September, and a total rainfall 245 mm in May–September.

- Central Macedonia/Chalkidiki: high temperatures > 25 °C July–August, rainfall of 43 mm during June–August and >350 mm in September, and a total rainfall 460 mm in May–September.
- Central Macedonia/Athos: high temperatures > 25 °C July–August, rainfall of 16 mm during June–August and >75 mm in September, and a total rainfall of 233 mm in May–September.
- Central Macedonia/Imathia: high temperatures > 25 °C July–August, rainfall of 42 mm during June–August and >100 mm in September, total rainfall < 250 mm in May–September.
- Central Macedonia/Pella: high temperatures > 25 °C July–August, rainfall of 78 mm during June–September, 10 mm in September, and a total rainfall of 186 mm in May–September.
- Western Macedonia/Florina: low temperatures below 25 °C during all months, rainfall of 300 mm during July–September, 150 mm in September, and a total rainfall of 420 mm in May–September.
- Thessaly/Larissa: high temperatures over 25 °C during June–August and low total rainfall of 150 mm during May and September (none during July–August, and nearly 100 mm in September).
- North Aegean/Lemnos: high temperatures during June–August, no rainfall during June–September, and a total rainfall of 45 mm in May–September.

Cluster analysis was performed to find relationships among the samples collected using nitrogen composition, basic oenological parameters and climate data (rainfall and temperatures), which are different per regional unit. Genetic classification of different plant matrices (e.g., wine, maize, and sweet cherries) has been done previously using cluster analysis, occasionally or in conjunction with principal components analysis [44–47].

3.3.1. White Grape Varieties

In the previous sections, it was noted that the high variation in FAN and YAN content cannot be attributed only to the grape variety. Therefore, cluster analysis was performed to reveal possible relationships between the nitrogen fractions and oenological attributes and climate data (Figure 3), as climate affects both maturity (sugar content, titratable acidity) and nitrogen compounds in grapes [30]. Samples from white varieties were divided into groups of increasing dissimilarity according to the heat map generated. White samples were arranged in seven well-defined groups (noted with different colors: red, blue, green, brown, blue-green, violet, and yellow), suggesting similar composition within the same group. It seems that the arrangement is done mainly according to the grape variety and regional unit of origin.

On the other hand, the variables are organized into two clusters (shown in red, and blue). The red cluster comprises variables FAN, NH_4-N, total acidity, total rainfall (May–September), as well as rainfall for each month from June to September. The rainfall during ripening and total acidity have been correlated. Tartaric and malic acids build up before the veraison. Due to respiration and the loss of malic acid, as well as berry enlargement (and the dilution of both acids), the acid concentration reduces during ripening and sugar accumulation. Because the rate of malic acid breakdown increases as the temperature rises, grapes cultivated in cooler climates or harvested earlier in the season typically have higher amounts of acidity in the must and wine. In the blue cluster are the variables linked with temperatures from May to September, rainfall in May, pH value, and sugar level (Baume) [48].

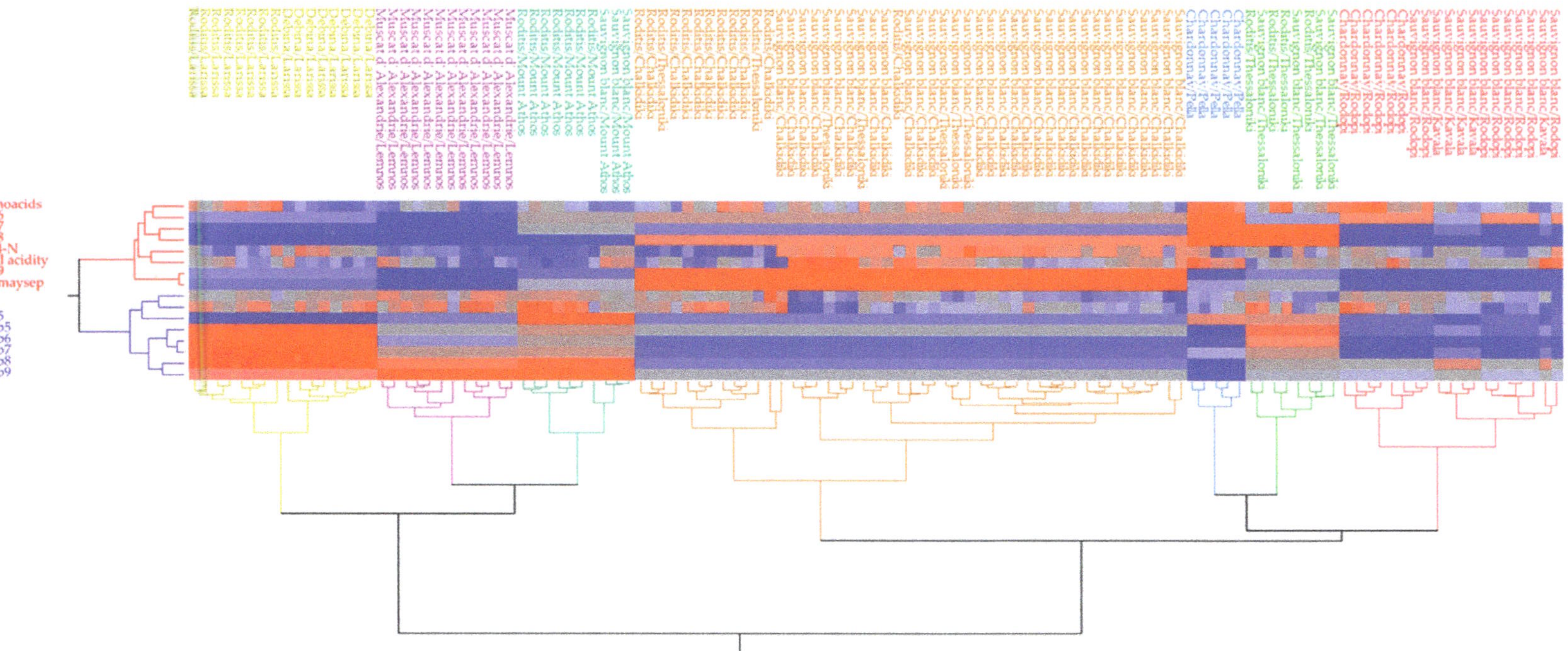

Figure 3. Heatmap of correlations for musts from white grape varieties between the nitrogen composition, basic oenological parameters, and climate data (rainfall and temperatures for months May to September. Rainfall (rain) and temperatures (temp) for months May (5) to September (9); Blue areas in the map dendrogram indicate low values, whereas the red areas indicate high values.

Grouping in clusters of different colors revealed differences among white grape musts. More specifically, the red group consisted of Sauvignon Blanc and Chardonnay samples from Rodopi and Kavala regional units, characterized by medium-to-higher FAN content, low-to-no rainfall (<5 mm) in July–August, and low total rainfall (<62 mm) from May to September. The green cluster grouped Sauvignon Blanc and Roditis varieties from the regional unit Thessaloniki and appeared to be closely related to the blue group consisting of the Chardonnay variety from the Pella regional unit. Those groups were characterized by 67 mm of rain during July and August, higher FAN values, and less than 250 mm of rainfall during May and September. The brown group included various subgroups from two cultivars Sauvignon Blanc and Roditis, from regional units of Thessaloniki and Chalkidiki, which are geographically close to each other. These samples were characterized by intermediate-to-higher FAN levels, rainfall over 90 mm during September and >245 mm during May–September, rainfall during June and August, and lower sugar levels. The samples in the blue-green clusters were of Sauvignon Blanc and Roditis grapes from the Mount Athos regional unit, which were distinguished by low FAN and NH_4-N levels, nearly no rain during the June to August period, high temperatures during August and September, higher-than-average sugars and pH values, and lower total acidity. The purple group contained Muscat of Alexandria grapes from the Lemnos regional unit, with low-to-intermediate FAN values, low NH_4-N, higher pH values, dry thermal climate (a total lack of rainfall from June to September, and high temperatures during August and September). Lastly, the yellow group grouped two Greek native cultivars, Roditis and Debina, from the Larissa regional unit. Those samples were characterized either by high FAN values and lower NH_4-N values (Roditis), or low FAN and higher NH_4-N (Debina), higher pH values, intermediate levels of sugars, high temperatures from June to September, and almost no rainfall during the summer months. Specific clusters were identified based on regions.

It appears that the same white variety was grouped in separate clusters based on the regional units where it was cultivated. This was the case for Sauvignon blanc, Roditis, and Chardonnay grapes, suggesting different compositions for each group. Soufleros, Bouloumpasi, Tsarchopoulos, and Biliaderis [39] noted differences in the a-amino acid content of Roditis wines cultivated in different regions, with samples originating from Macedonia, Central Greece, and Peloponnesos containing on average 496, 428, and 394 mg L^{-1} of α-amino acids. Soufleros et al. [39] and Schrader et al. [49] have previously reported that the region's climate and the grapes' level of maturity could affect the total content of amino acids in the juice. Petropoulos, Metafa, Kotseridis, Paraskevopoulos, and Kallithraka [30] found that the temperature generally increased amino acid content, although rainfall showed the opposite trend. Garde-Cerdan, Lorenzo, Lara, Pardo, Ancin-Azpilicueta, and Salinas [3] also noted that the amount of nitrogen compounds in grapes depended on their maturation stage.

A principal component analysis (PCA) was performed for white grape varieties using FAN (a-amino acids), ammoniacal nitrogen (NH_4-N), basic must analyses (Be, pH, and total acidity), and climate data to examine whether these compounds may lead to any grouping of samples according to variety. The resulting plots for white must are presented in Figure 4. The PCA results showed that the two main components (Component 1 × Component 2) explained only 54.2% of the total variance (Components 3, 4, and 5 explained other 13%, 11%, and 6.56%, explaining about 84.76% of the total variation). The PCA score plot (Figure 4, left) shows a pattern in which the samples from Muscat of Alexandria are placed closely together in the lower right quartile, the samples from Debina are placed together in the upper right quartile, and the samples from other varieties form separate groups. The Chardonnay, Sauvignon blanc, and Roditis samples are spread in two or more groups. Specifically, the Chardonnay samples are all placed in the lower left quartile, forming two separate groups, one of which is located in the left lower part and comprises samples from a single regional unit (Pella), thus suggesting grouping not only based on the cultivar but also on the regional unit.

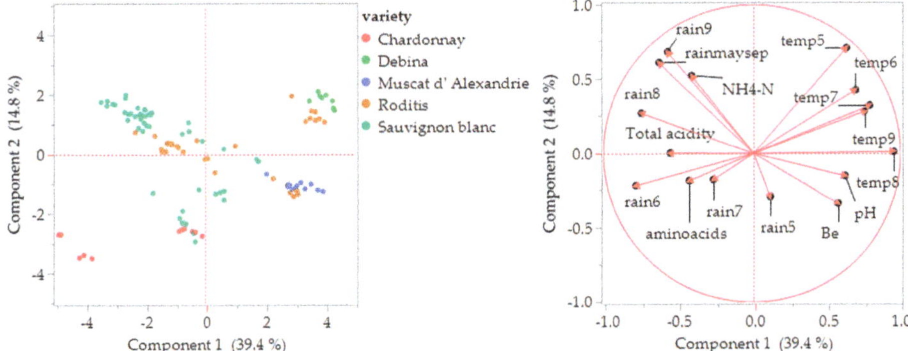

Figure 4. Plots of principal component analysis scores (**left**) and loadings (**right**) for the composition of the white musts. Rainfall (rain) and temperatures (temp) for months May (5) to September (9); rainmaysep is total rainfall from months May to September.

The loading plots (Figure 4, right) show that amino acids were negatively affected by both components, while ammoniacal nitrogen and total acidity were negatively affected by the first component and positively by the second component. Amino acid content (FAN) appears to be positively correlated with rainfall during June and July and negatively to temperature from May to July. Ammoniacal nitrogen positively correlated with total rainfall during May–September, especially rainfall in August and September, and a negative correlation with pH, Baume, the temperature during August (ripening period), and the rainfall during May (growing period). No correlation was found between amino acid and ammoniacal nitrogen levels. Similarly to the heat map, the PCA grouped samples based on both the variety and the region, suggesting that the region, with its unique climate, has a significant input on the content of assimilable nitrogen compounds in the grape.

3.3.2. Red Grape Varieties

Similarly to the white grape samples, a heatmap cluster analysis for the red grape samples based on nitrogen status, oenological parameters, and climate data for rainfall and temperature revealed the presence of seven clusters (Figure 5). On the other hand, the variables are organized into three clusters (shown in red, blue, and green). The red cluster comprises the variables FAN, total acidity, and rainfall for June, July, and August. The blue cluster groups variables such as NH_4-N content, Baume, total rainfall (May–September), as well as rainfall in August and September. Lastly, in the green cluster the rest of the variables (temperatures from May to September and pH value) are included. The grouping of red musts under different colors reveals differences among the samples. More specifically, the red cluster is composed of Merlot and Syrah samples from Rodopi and Kavala regional units, characterized by low FAN content, medium-to-higher NH_4-N content, nearly no rainfall (<5 mm) during July–August, and low total rainfall (<62 mm) from May to September. The green group, which appears to be closely related to the red group, includes Syrah and Merlot samples from the Larissa regional unit and is also characterized by a lack of rainfall (<1 mm) during July–August, low total rainfall from May–September (<140 mm), and higher temperatures from May to September in contrast to the red group. The blue cluster comprises the Merlot variety from the Thessaloniki regional unit and is characterized by intermediate-to-higher FAN levels and high temperatures from May to September. The brown cluster includes two cultivars, Xinomavro and Syrah, from the Kilkis regional unit and is characterized by low-to-intermediate FAN levels, low NH_4-N levels, high temperatures from May to August, and rainfall of 150 mm during May–June. The blue-green group includes the Cinsault variety from the Serres regional unit, and its characteristics include the highest FAN along with low NH_4-N, sugar, total acidity levels, intermediate temperatures, and rainfall of 360 mm from May to September. The purple cluster comprises three cultivars, Merlot, Syrah, and Xinomavro, from three regional units, Thessaloniki, Chalkidiki, and Imathia, and consists of several subgroups. It is characterized by low FAN content, low rainfall from June to August but high in September, intermediate-to-high sugars, and ammoniacal nitrogen values. Lastly, the yellow cluster includes the Xinomavro variety grown in the Florina regional unit and differs from the others due to higher total acidity, lower sugar content, rainfall of 240 mm from May to July, and lower temperatures from May to September in comparison to other groups.

It appears that the heat map grouped samples based primarily on the climate conditions of the regional unit and secondarily on the variety. Regional units such as Florina, Larissa, and Rodopi/Kavala are distinguished based on rainfall and temperatures (low or high), independently of the variety, while the other regional units of Central Macedonia do not discriminate much. Petropoulos, Metafa, Kotseridis, Paraskevopoulos, and Kallithraka [30] reported that rainfall had a detrimental impact on the amino acid content of Agiorgitiko grapes in the viticultural region of Nemea from veraison through harvest, although the increased ambient temperature had the reverse effect.

The principal component analysis performed for red grape varieties explained 62.6% (Component 1 × Component 2) of the total variance (Component 3 and 4 explained the other 12.3% and 7.92%, explaining about 82.82% of the total variation) (Figure 6). The PCA score plot (Figure 6, left) grouped the Xinomavro cultivar originating from the Florina regional unit positioned upon the horizontal axis at the far-left side. The other Xinomavro samples are positioned near the center of the plot, together with the other samples from the Central Macedonia region (the Chalkidiki, Thessaloniki, Imathia, Kilkis, and Serres regional units). All those areas are characterized by a lack of rainfall during ripening and high temperatures during the June–August period. Similarly, the Syrah cultivar is spread into four groups depending on the origin (Rodopi, Kilkis, Imathia, and Larissa). The placement of the samples from the same variety in different quartiles or the same implies that they are arranged based on other parameters, e.g., samples placed in the upper right quartile are affected by temperatures during May–September, the period of growing and ripening of grapes, and Larissa is the area with the highest mean temperatures among the examined regional units for all months, except September (see Supplementary Table S2).

The loading plots (Figure 6, right) reveal that amino acids were positively affected by the second component, while ammoniacal nitrogen was negatively affected by both components. No correlation was found between amino acid and ammoniacal nitrogen levels. In conclusion, the heat map and PCA grouped the red must samples based on both the variety and the region.

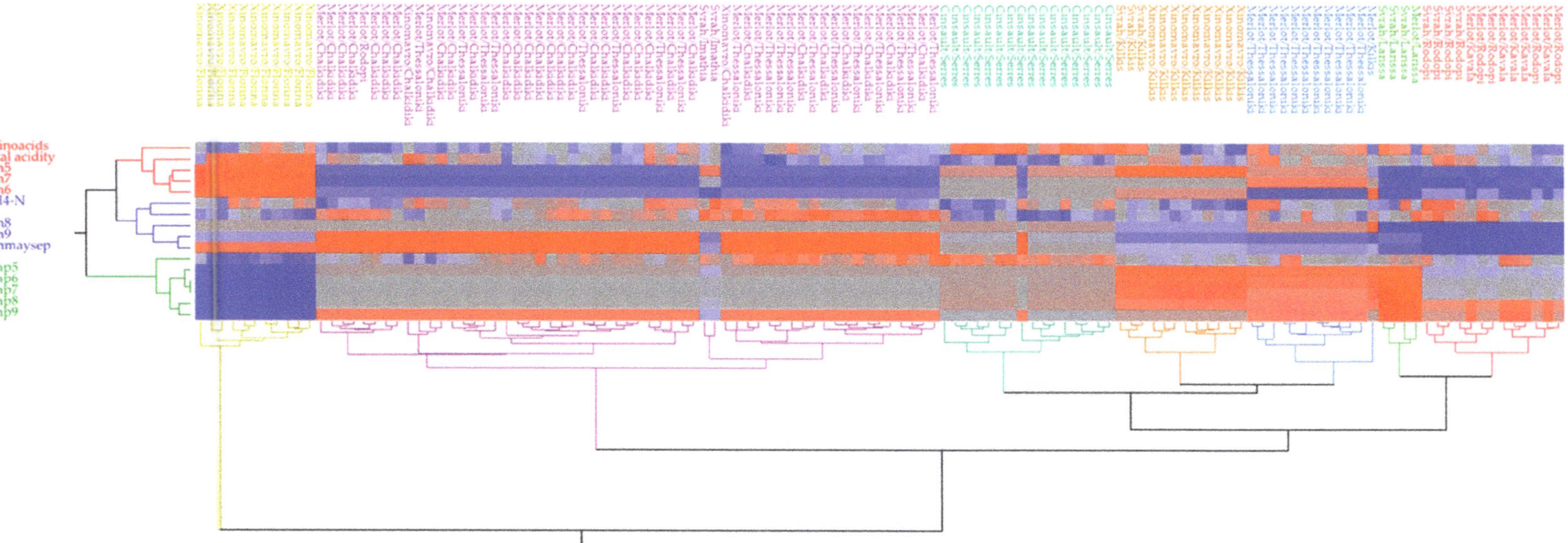

Figure 5. Heatmap of correlations for musts from red grape varieties between the nitrogen composition and basic oenological parameters and climate data rainfall (rain) and temperatures (temp) for months May (5) to September (9)); Blue areas in the map dendrogram indicate low values whereas the red areas indicate high values.

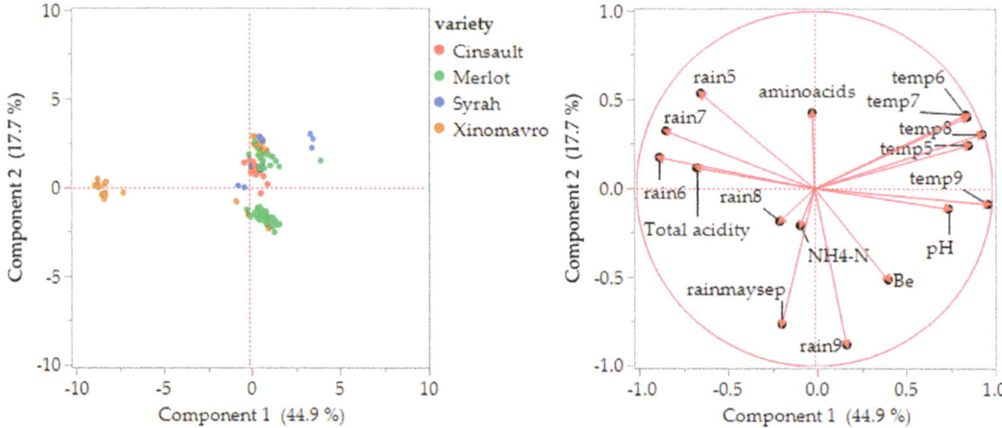

Figure 6. Plots of principal component analysis scores (**left**) and loadings (**right**) for the composition of the red musts. Rainfall (rain) and temperatures (temp) for months May to September; rainmaysep is total rainfall from months May to September.

4. Conclusions

In the present study, the yeast assimilable nitrogen status of the musts of several Greek grape-growing areas and varieties, both native and international, was surveyed. The musts from technologically ripe grapes analyzed varied largely in their assimilable nitrogen content. The content of international varieties cultivated in Greece fell within the range noted by other researchers. About 36% of the Greek samples were below the suggested by some authors deficiency threshold of 140 mg L^{-1}, thus posing a danger of problems with the nutrition of yeasts in the course of fermentation unless it occurs an addition of a nitrogen source. Significant differences were observed among varieties, with international varieties found to be richer in assimilable nitrogen than most of the natives included in this survey. Some international varieties had a high content of YAN, similar to surveys conducted in Italy and the United States. This study also revealed that not only grape variety but also cultivation region is a determinator of the concentration and composition of YAN. This work brings insights into the knowledge about native and international cultivars of great commercial interest. In addition, this work can be useful to understand how commercial varieties included in this study can behave in different climates in the context of climate change.

Supplementary Materials: The following supporting information can be downloaded at: https://www.mdpi.com/article/10.3390/fermentation9080773/s1, Table S1: Meteorological data (temperature and rainfall) during growth period May–September 2017; Table S2: Variation (minimum–maximum value observed) of Yeast Assimilable Nitrogen, Free Amino Nitrogen, ammoniacal Nitrogen in grape musts categorized per grape origin (mg N L^{-1} must).

Author Contributions: Conceptualization, E.B. and E.H.S.; methodology, E.B.; software, E.B. and A.S.; formal analysis, E.B.; investigation, E.B.; resources, E.H.S.; data curation, E.B. and A.S.; writing—original draft preparation, E.B. and A.S.; writing—review and editing, E.B., A.S. and E.H.S.; visualization, E.B. and A.S.; supervision, E.H.S.; project administration, E.H.S. All authors have read and agreed to the published version of the manuscript.

Funding: This research received no external funding.

Data Availability Statement: The data presented in this study are available on request from the corresponding author (pending privacy and ethical considerations).

Conflicts of Interest: The authors declare no conflict of interest.

References

1. Sponholz, W. Nitrogen compounds in grapes, must, and wine. In Proceedings of the International Symposium on Nitrogen in Grapes and Wine, Seattle, WA, USA, 18–19 June 1991; pp. 67–77.
2. Nicolini, G.; Larcher, R.; Versini, G. Status of yeast assimilable nitrogen in Italian grape musts and effects of variety, ripening and vintage. *Vitis* **2004**, *43*, 89–96. [CrossRef]
3. Garde-Cerdan, T.; Lorenzo, C.; Lara, J.F.; Pardo, F.; Ancin-Azpilicueta, C.; Salinas, M.R. Study of the evolution of nitrogen compounds during grape ripening. Application to differentiate grape varieties and cultivated systems. *J. Agric. Food Chem.* **2009**, *57*, 2410–2419. [CrossRef] [PubMed]
4. Bisson, L. Yeast and biochemistry of ethanol formation. In *Principles and Practices of Winemaking*; Boulton, R.B., Singleton, V.L., Bisson, L.F., Kunkee, R.E., Eds.; Chapman & Hall: New York, NY, USA, 1996; pp. 140–172.
5. Henschke, P.A.; Jiranek, V. Yeasts-metabolism of nitrogen compounds. In *Wine Microbiology Biotechnology*; Fleet, G.H., Ed.; Harwood Academic: Chur, Switzerland, 1993; Volume 77, pp. 77–164.
6. Su, Y.; Heras, J.M.; Gamero, A.; Querol, A.; Guillamón, J.M. Impact of nitrogen addition on wine fermentation by *S. cerevisiae* strains with different nitrogen requirements. *J. Agric. Food Chem.* **2021**, *69*, 6022–6031. [CrossRef] [PubMed]
7. Barbosa, C.; Mendes-Faia, A.; Mendes-Ferreira, A. The nitrogen source impacts major volatile compounds released by *Saccharomyces cerevisiae* during alcoholic fermentation. *Int. J. Food Microbiol.* **2012**, *160*, 87–93. [CrossRef]
8. Bely, M.; Sablayrolles, J.-M.; Barre, P. Automatic detection of assimilable nitrogen deficiencies during alcoholic fermentation in oenological conditions. *J. Ferment. Bioeng.* **1990**, *70*, 246–252. [CrossRef]
9. Hernandez-Orte, P.; Bely, M.; Cacho, J.; Ferreira, V. Impact of ammonium additions on volatile acidity, ethanol, and aromatic compound production by different Saccharomyces cerevisiae strains during fermentation in controlled synthetic media. *Aust. J. Grape Wine R.* **2006**, *12*, 150–160. [CrossRef]
10. Barbosa, C.; Falco, V.; Mendes-Faia, A.; Mendes-Ferreira, A. Nitrogen addition influences formation of aroma compounds, volatile acidity and ethanol in nitrogen deficient media fermented by *Saccharomyces cerevisiae* wine strains. *J. Biosci. Bioeng.* **2009**, *108*, 99–104. [CrossRef]
11. Christofi, S.; Papanikolaou, S.; Dimopoulou, M.; Terpou, A.; Cioroiu, I.B.; Cotea, V.; Kallithraka, S. Effect of yeast assimilable nitrogen content on fermentation kinetics, wine chemical composition and sensory character in the production of assyrtiko wines. *Appl. Sci.* **2022**, *12*, 1405. [CrossRef]
12. Bell, S.J.; Henschke, P.A. Implications of nitrogen nutrition for grapes, fermentation and wine. *Aust. J. Grape Wine R.* **2005**, *11*, 242–295. [CrossRef]
13. Webster, D.R.; Edwards, C.G.; Spayd, S.E.; Peterson, J.C.; Seymour, B.J. Influence of vineyard nitrogen fertilization on the concentrations of monoterpenes, higher alcohols, and esters in aged riesling wines. *Am. J. Enol. Vitic.* **1993**, *44*, 275–284. [CrossRef]
14. Butzke, C.E. Survey of yeast assimilable nitrogen status in musts from California, Oregon, and Washington. *Am. J. Enol. Vitic.* **1998**, *49*, 220–224. [CrossRef]
15. EU. European Commission Delegated Regulation (EU) 2019/934 of 12 March 2019 Supplementing Regulation (EU) No 1308/2013 of the European Parliament and of the Council as Regards Wine-Growing Areas Where the Alcoholic Strength May Be Increased, Authorised Oenological Practices and Restrictions Applicable to the Production and Conservation of Grapevine Products, the Minimum Percentage of Alcohol for By-Products and Their Disposal, and Publication. *OJ* **2019**, *149*, 1–52.
16. Henschke, P.A.; Ough, C.S. Urea accumulation in fermenting grape juice. *Am. J. Enol. Vitic.* **1991**, *42*, 317. [CrossRef]
17. Miliordos, D.E.; Kanapitsas, A.; Lola, D.; Goulioti, E.; Kontoudakis, N.; Leventis, G.; Tsiknia, M.; Kotseridis, Y. Effect of nitrogen fertilization on Savvatiano (*Vitis vinifera* L.) grape and wine composition. *Beverages* **2022**, *8*, 29. [CrossRef]
18. Hannam, K.D.; Neilsen, G.H.; Forge, T.; Neilsen, D. The concentration of yeast assimilable nitrogen in Merlot grape juice is increased by N fertilization and reduced irrigation. *Can. J. Plant Sci.* **2013**, *93*, 37–45. [CrossRef]
19. Linsenmeier, A.W.; Loos, U.; Löhnertz, O. Must composition and nitrogen uptake in a long-term trial as affected by timing of nitrogen fertilization in a cool-climate Riesling vineyard. *Am. J. Enol. Vitic.* **2008**, *59*, 255. [CrossRef]
20. Hilbert, G.; Soyer, J.; Molot, C.; Giraudon, J.; Milin, M.; Gaudillere, J. Effects of nitrogen supply on must quality and anthocyanin accumulation in berries of cv. Merlot. *Vitis* **2015**, *42*, 69. [CrossRef]
21. Nisbet, M.A.; Martinson, T.E.; Mansfield, A.K. Accumulation and prediction of yeast assimilable nitrogen in New York winegrape cultivars. *Am. J. Enol. Vitic.* **2014**, *65*, 325. [CrossRef]
22. Petrovic, G.; Kidd, M.; Buica, A. A statistical exploration of data to identify the role of cultivar and origin in the concentration and composition of yeast assimilable nitrogen. *Food Chem.* **2019**, *276*, 528–537. [CrossRef]
23. Garde-Cerdán, T.; Martínez-Gil, A.M.; Lorenzo, C.; Lara, J.F.; Pardo, F.; Salinas, M.R. Implications of nitrogen compounds during alcoholic fermentation from some grape varieties at different maturation stages and cultivation systems. *Food Chem.* **2011**, *124*, 106–116. [CrossRef]
24. van Leeuwen, C.; Friant, P.; Soyer, J.-P.; Molot, C.; Choné, X.; Dubourdieu, D. Measurement of total nitrogen and assimilable nitrogen in grape juice to assess vine nitrogen status. *J. Int. Sci. Vigne Vin.* **2000**, *34*, 75–82. [CrossRef]
25. Baron, M. Yeast assimilable nitrogen in South Moravian grape musts and its effect on acetic acid production during fermentation. *Czech J. Food Sci.* **2011**, *29*, 603–609. [CrossRef]

26. Berger, S.; Schober, V.; Korntheuer, K.U.; Eder, R. Einfluss des hefeverwertbaren Stickstoffes auf die Gärung in Mosten der Sorten Grüner Veltliner, Rheinriesling, Welschriesling und Neuburger. *Mitt. Klosterneubg.* **1999**, *49*, 117–123.
27. Nicolini, G.; Versini, G.; Corradin, L.; Larcher, R.; Beretta, C.; Olivari, A.; Eccli, E. Misura dell'azoto prontamente assimilabile dal lievito nei mosti d'uva ed esempi di applicazione. *Riv. Vitic. Enol.* **2004**, *1–2*, 13–27. Available online: http://hdl.handle.net/1044 9/17619 (accessed on 17 August 2023).
28. Hagen, K.M.; Keller, M.; Edwards, C.G. Survey of biotin, pantothenic acid, and assimilable nitrogen in winegrapes from the Pacific Northwest. *Am. J. Enol. Vitic.* **2008**, *59*, 432. [CrossRef]
29. Nisbet, M.A.; Martinson, T.E.; Mansfield, A.K. Preharvest prediction of yeast assimilable nitrogen in Finger Lakes Riesling using linear and multivariate modeling. *Am. J. Enol. Vitic.* **2013**, *64*, 485. [CrossRef]
30. Petropoulos, S.; Metafa, M.; Kotseridis, Y.; Paraskevopoulos, I.; Kallithraka, S. Amino acid content of Agiorgitiko (*Vitis vinifera* L. cv.) grape cultivar grown in representative regions of Nemea. *Eur. Food Res. Technol.* **2018**, *244*, 2041–2050. [CrossRef]
31. OIV. *Compendium of International Methods of Wine and Must Analysis*; OIV, International Organisation of Vine and Wine: Paris, France, 2018; Volume 1–2.
32. Dukes, B.C.; Butzke, C.E. Rapid determination of primary amino acids in grape juice using an o-Phthaldialdehyde/N-Acetyl-L-Cysteine spectrophotometric assay. *Am. J. Enol. Vitic.* **1998**, *49*, 125. [CrossRef]
33. Hernández-Orte, P.; Guitart, A.; Cacho, J.F. Changes in the concentration of amino acids during the ripening of *Vitis vinifera* Tempranillo variety from the denomination d'origine Somontano (Spain). *Am. J. Enol. Vitic.* **1999**, *20*, 144–154. [CrossRef]
34. Sowalsky, R.A.; Noble, A.C. Comparison of the effects of concentration, pH and anion species on astringency and sourness of organic acids. *Chem. Senses* **1998**, *23*, 343–349. [CrossRef]
35. Panzeri, V.; Ipinge, H.N.; Buica, A. Evaluation of South African Chenin Blanc wines made from six different Trellising systems using a chemical and sensorial approach. *S. Afr. J. Enol. Vitic.* **2020**, *41*, 133–150. [CrossRef]
36. Bouloumpasi, E.; Koutsouridou, V.; Soufleros, E.; Greveniotis, V. Nitrogen status of grapes cultivated under conventional and organic farming systems (PoS2-82). In Proceedings of the XV European Society for Agronomy Congress, Geneva, Switzerland, 27–31 August 2018; p. 164.
37. Bouloumpasi, E.; Soufleros, E.H.; Tsarchopoulos, C.; Biliaderis, C.G. Primary amino acid composition and its use in discrimination of Greek red wines with regard to variety and cultivation region. *Vitis* **2015**, *41*, 195–202. [CrossRef]
38. Etiévant, P.; Schlich, P.; Bouvier, J.-C.; Symonds, P.; Bertrand, A. Varietal and geographic classification of French red wines in terms of elements, amino acids and aromatic alcohols. *J. Sci. Food Agr.* **1988**, *45*, 25–41. [CrossRef]
39. Soufleros, E.H.; Bouloumpasi, E.; Tsarchopoulos, C.; Biliaderis, C.G. Primary amino acid profiles of Greek white wines and their use in classification according to variety, origin and vintage. *Food Chem.* **2003**, *80*, 261–273. [CrossRef]
40. Huang, Z.; Ough, C.S. Effect of vineyard locations, varieties, and rootstocks on the juice amino acid composition of several cultivars. *Am. J. Enol. Vitic.* **1989**, *40*, 135. [CrossRef]
41. Verdenal, T.; Dienes-Nagy, Á.; Spangenberg, J.E.; Zufferey, V.; Spring, J.-L.; Viret, O.; Marin-Carbonne, J.; van Leeuwen, C. Understanding and managing nitrogen nutrition in grapevine: A review. *OENO One* **2021**, *55*, 1–43. [CrossRef]
42. AWRI. Yeast Assimilable Nitrogen. Available online: https://www.awri.com.au/industry_support/winemaking_resources/wine_fermentation/yan/ (accessed on 16 August 2023).
43. Stines, A.P.; Grubb, J.; Gockowiak, H.; Henschke, P.A.; HØJ, P.B.; van Heeswijck, R. Proline and arginine accumulation in developing berries of *Vitis vinifera* L. in Australian vineyards: Influence of vine cultivar, berry maturity and tissue type. *Aust. J. Grape Wine R.* **2000**, *6*, 150–158. [CrossRef]
44. Skendi, A.; Papageorgiou, M.; Stefanou, S. Preliminary study of microelements, phenolics as well as antioxidant activity in local, homemade wines from North-East Greece. *Foods* **2020**, *9*, 1607. [CrossRef]
45. Greveniotis, V.; Bouloumpasi, E.; Zotis, S.; Korkovelos, A.; Ipsilandis, C.G. Stability, the last frontier: Forage yield dynamics of peas under two cultivation systems. *Plants* **2022**, *11*, 892. [CrossRef]
46. Greveniotis, V.; Giourieva, V.; Bouloumpasi, E.; Sioki, E.; Mitlianga, P. Morpho-physiological characteristics and molecular markers of maize crosses under multi-location evaluation. *J. Agric. Sci.* **2018**, *10*, 79–90. [CrossRef]
47. Ganopoulos, I.; Moysiadis, T.; Xanthopoulou, A.; Ganopoulou, M.; Avramidou, E.V.; Aravanopoulos, F.A.; Tani, E.; Madesis, P.; Tsaftaris, A.; Kazantzis, K. Diversity of morpho-physiological traits in worldwide sweet cherry cultivars of GeneBank collection using multivariate analysis. *Sci. Hortic.* **2015**, *197*, 381–391. [CrossRef]
48. Lakso, A.N.; Kliewer, W.M. The influence of temperature on malic acid metabolism in grape berries: I. Enzyme responses. *Plant Physiol.* **1975**, *56*, 370–372. [CrossRef] [PubMed]
49. Schrader, U.; Lemperle, E.; Becker, N.J.; Bergner, K.E. Der Aminosäure-Zucker, Saure-, and Mineralstoffgehalt von Weinbeeren in Abhängigkeit vom Kleinklima des Standortes der Rebe. *Wein-Wissensch* **1976**, *31*, 160–175.

Article

Integrated Metagenomics and Network Analysis of Metabolic Functional Genes in the Microbial Community of Chinese Fermentation Pits

Mingyi Guo [1,2], Yan Deng [1], Junqiu Huang [3], Chuantao Zeng [1], Huachang Wu [1], Hui Qin [2] and Suyi Zhang [2,*]

1   College of Food Science and Technology, Sichuan Tourism University, Chengdu 610100, China; guomy@sctu.edu.cn (M.G.)
2   National Engineering Research Center of Solid-State Brewing, Luzhou Laojiao Group Co., Ltd., Luzhou 646000, China
3   College of Bioengineering, Sichuan University of Science and Engineering, Yibin 644005, China
*   Correspondence: zhangsy@lzlj.com

Abstract: Traditional Chinese strong-aroma baijiu (CSAB) fermentation technology has been used for thousands of years. Microbial communities that are enriched in continuous and uninterrupted fermentation pits (FPs) are important for fermentation. However, changes in the metabolic functional genes in microbial communities of FPs are still under-characterized. High-throughput sequencing technology was applied to comprehensively analyze the diversity, function, and dynamics of the metabolic genes among FPs of different ages, positions, and geographical regions. Approximately 1,375,660 microbial genes derived from 259 Gb metagenomic sequences of FPs were assembled and characterized to understand the impact of FP microorganisms on the quality of CSAB and to assess their genetic potential. The core functional gene catalog of FPs, consisting of 3379 ubiquitously known gene clusters, was established using Venn analysis. The functional profile confirmed that the flavor compounds in CSAB mainly originate from the metabolism of carbohydrates and amino acids. Approximately 17 key gene clusters that determine the yield and quality of CSAB were identified. The potential mechanism was associated with the biosynthesis of host compounds in CSAB, which relies on the abundance of species, such as *Lactobacillus*, *Clostridium*, *Saccharomycetales*, and the abundance of functional genes, such as CoA dehydrogenase, CoA transferase, and NAD dehydrogenase. Furthermore, the detailed metabolic pathways for the production of main flavor compounds of CSAB were revealed. This study provides a theoretical reference for a deeper understanding of substance metabolism during CSAB brewing and may help guide the future exploration of novel gene resources for biotechnological applications.

Keywords: Chinese strong-aroma baijiu; fermentation pit; metagenomics; functional gene; metabolic profile

Citation: Guo, M.; Deng, Y.; Huang, J.; Zeng, C.; Wu, H.; Qin, H.; Zhang, S. Integrated Metagenomics and Network Analysis of Metabolic Functional Genes in the Microbial Community of Chinese Fermentation Pits. *Fermentation* **2023**, *9*, 772. https://doi.org/10.3390/fermentation9080772

Academic Editors: Maria Dimopoulou, Spiros Paramithiotis, Yorgos Kotseridis and Jayanta Kumar Patra

Received: 19 July 2023
Revised: 15 August 2023
Accepted: 17 August 2023
Published: 18 August 2023

1. Introduction

Baijiu is a conventional fermented beverage that has been produced for thousands of years. Chinese strong-aroma baijiu (CSAB), also called Chinese strong-aroma liquor (CSAL), is the most consumed spirit in China (>70%) [1] and is produced using unique and traditional solid-state fermentation. Fermentation pits (FPs) are key fermentative bioreactors that are used for the solid-state fermentation of CSAB and significantly affect the quality of CSAB. These bioreactors are rectangular in shape (2000–3000 mm in length, 1500–2000 mm in width, and 1800–2000 mm in height) and are coated with a mixture of fermented cereals (including sorghum, wheat, corn, rice, and sticky rice) and mud. Studies have suggested that the characteristics of FPs are related to the final characteristics of the CSAB produced [2,3]; however, the mechanisms underlying this relationship between FPs and CSAB still need to be elucidated.

To reveal the fermentation mechanism and improve the quality of CSAB, attention should be paid to the analysis of flavor compounds and microorganisms. On the one hand, the application of high-throughput screening and confirmation of flavor compounds has attracted a large amount of attention and interest from researchers. The key flavor compounds of CSAB have been identified using hyphenated chromatographic techniques based on various treatment approaches [4,5]. Gas chromatography-mass spectrometry (GC-MS) and liquid chromatography–mass spectrometry (LC-MS) are two widely used approaches [6,7]. GC-MS is suitable for the detection of semipolar and nonpolar flavor compounds of CSAB, whereas LC-MS specifically targets the semipolar and polar flavor compounds of CSAB [8,9]. Therefore, these two methods are typically combined. More than 400 flavor compounds, such as alcohols, acids, esters, aldehydes, and ketones [2,10], have been identified in CSAB through the unremitting efforts of baijiu researchers. Ethyl caproate, ethyl lactate, ethyl acetate, and ethyl butyrate have been confirmed as the four critical flavor compounds of CSAB. This laid the foundation for the exploration of the fermentation mechanisms associated with the manufacturing of CSAB. On the other hand, a vast number of research studies have focused on microbial community analysis of the CSAB ecosystem, specifically using culture-independent analysis methods [11]. Various sources of CSAB microbial diversity have been studied, including workshop environments [12], daqu [13–15], zaopei [16,17], and pit mud [18–21]. These studies reported that *Lactobacillus*, *Bacillus*, *Clostridium*, *Syntrophomonas*, *Methanobacterium*, *Methanoculleus*, *Galactomyces*, *Sedimentibacte*, *Candida*, *Pichia*, *Aspergillus*, *Penicillium*, and so on significantly affect the yield and quality of CSAB [2]. FPs are vital sources of microorganisms during CSAB fermentation. During brewing, multiple microorganisms coexist in FPs [21], and these microorganisms produce CSAB flavor compounds [22]. Therefore, the quality of CSAB partly depends on the metabolism of microorganisms in FPs. Furthermore, the possible application of systems biology approaches to elucidate the molecular mechanisms underlying CSAB production has been gradually explored [11,23]. To comprehensively understand the complex microbial communities of FPs, links between flavor compounds and functional genes must be established. However, changes in metabolic functional genes in the FPs of CSAB are still poorly understood. Therefore, it is difficult to comprehensively understand the processes and mechanisms underlying CSAB fermentation. A metagenomic approach based on total DNA analysis provides reliable information on the functional genes in microbial communities [24,25]. Whole metagenome sequencing is a powerful method used to analyze complex gene diversities [26]. This method has been extensively applied to characterize microbial gene catalogs in a large variety of environments [27–29]. Thus, this efficient approach could also be employed to characterize the diversity of metabolic functional genes in the FPs for CSAB production.

In the present study, a metagenomic approach using Illumina high-throughput sequencing was applied to reveal the functional gene profiles of FPs. A core functional gene catalog of FPs was established. Variations in the functional genes of FPs of different ages, spatial distributions, and geographical locations were analyzed. The overall metabolic pathway profile and biosynthesis of secondary metabolites was constructed. Key gene clusters and metabolic pathways affecting CSAB yield and quality were identified. The metabolic relationship between the functional gene profile and the flavor components of CSAB was determined.

2. Materials and Methods

2.1. Sampling and Metagenomic DNA Extraction

Samples were collected from representative famous traditional FPs for CSAB production [2] in Chengdu City (30.6586° N, 104.0647° E), Luzhou City (28.8833° N, 105.4500° E), and Mianyang City (31.4667° N, 104.6833° E), China. Sample selection was based on the statistical methods described by Carter [30] and the authors' previous work [2]. Twenty FPs of four different ages were sampled from Luzhou and 10 FPs of the same age were sampled from Chengdu and Mianyang. The FPs in Luzhou were 440, 220, 140, and 50 years old. The FPs in Chengdu and Mianyang were approximately 50 years old. Samples were collected from three depths (top, middle, and bottom) of the FP. The top layer was 0–50 cm from the top of the FP, the bottom layer was 0–50 cm from the bottom of the FP, and the remaining layer was termed the middle layer. Sampling was performed at the end of each fermentation cycle after emptying the fermented cereals. Ten representative pit mud samples (10 g each) were collected from each sampling sites and mixed (total 100 g) to provide a single sample. For each sample, five samples with the same FP characteristics (age, position, and geographical region) were collected as parallel samples. The samples were kept at −80 °C until genomic DNA was extracted. Total genomic DNA was extracted from the pit mud of the FPs using a Power Max® Soil DNA Isolation Kit (MO BIO Laboratories, Inc., Carlsbad, CA, USA). Quality checks and quantification of metagenomic DNA were performed with the Qubit™ 1X dsDNA HS/BR Assay Kit using a Qubit 4 fluorometer (Thermo Fisher Scientific, Inc., Waltham, MA, USA) to obtain a DNA concentration of $\geq$30 ng/μL and a weight of $\geq$3 μg. Thereafter, the metagenomic DNA bands were detected using 1% agarose gel electrophoresis yielding clear $\geq$23 kb bands with no obvious degradation.

2.2. Paired-End (PE) Library Generation and Illumina Hiseq Sequencing

The PE library generation and the Illumina Hiseq sequencing of metagenomic DNA was achieved in the group's previous work [2]. Finally, ~3.4 billion sequence reads, including more than 259 gigabases (Gb) of metagenomic sequence data, were obtained from the FPs. The further bioinformatic analysis was carried out in the current study.

2.3. Data Processing of Metagenomic Sequence Reads

The metagenomic data of the FPs were processed using the Perl utility module of Velvet-shuffleSeqences_fastq.pl from the Velvet toolset [31] to match and pair the paired-end sequences generated by the Illumina HiSeq platform. Adapter sequences and low-quality reads were removed using LUCY2 [32] and DynamicTrim pl. [33]. Adapter sequences were defined based on the information from the header sequences added during the PE library generation experiment. Reads with a Phred quality score below 38 and lengths exceeding 40 bp were discarded as low-quality reads. Reads with more than 5 bp N bases and an overlap with an adapter exceeding 15 bp were removed. Furthermore, to identify and remove potential nonmicrobial sequence contaminants present in the FPs, the SOAP Aligner [34] tool was employed with the following parameter settings: identity $\geq$99%, -l 30, -v 7, -M 4, -m 100, -x1000 to align and filter out highly specific sequences related to brewing raw materials.

2.4. Metagenome Assembly of Fps

The preprocessed clean data were assembled using a de Bruijn-type algorithm with the Meta-IDBA [35] tool. The parameter settings for Kmer were as follows: min 70, max 100, and step 10 for the number of iterations. First, overlapping regions were assembled into contiguous sequences without gaps to generate a set of contigs. The order of the contigs was then determined based on paired-end relationships to further assemble them into scaffolds. The obtained scaffolds were fragmented at potential N base locations to obtain scaffold sequences without N bases. These sequences were used as the assembly results for subsequent statistical analyses and gene predictions. The assembly process was performed in the second round to maximize the utilization of the sequences and improve the assembly efficiency. The SOAP Align [34] tool was used to align the obtained scaffolds in the first assembly step from each categorized sample with the quality-controlled clean metagenomic data of all FPs samples, which facilitated the identification of unused reads. Thereafter, the unused reads were combined and subjected to a new round of assembly using the Meta-IDBA [35] tool with the same parameter settings, resulting in supplementary scaffolds for potential mixed samples. The completeness of the two rounds of assembly was evaluated using the important indicator of N50, which represents the minimum sequence length required to cover 50% of the total length of all scaffolds. To improve the accuracy of the conclusions of the downstream analysis, a Perl process was implemented using the BioPerl tool (https://bioperl.org/ (accessed on 1 June 2019)) to remove scaffolds smaller than 10 kb and eliminate the probability of misassembly or confusion in shorter scaffolds.

2.5. Gene Prediction and Abundance Analysis in FPs

Using the assembled scaffolds (length $\geq$ 10 Kb), the MetaGene-ETP [36] tool was employed based on the hidden Markov model (HMM) with parameter settings of -a, -d, -f G, -k, and -r to predict open reading frames (ORFs). The predicted results with lengths smaller than 100 nt were filtered to obtain a high-quality set of genes. The predicted ORFs were subjected to redundancy removal using the CD-HIT tool [37]. The parameter settings for redundancy removal were c 0.95, aS 0.9, aL 0.9, similarity greater than 95%, and coverage >90%. The nonredundant genes were subjected to abundance calculation through the SOAP Align [34] tool with parameter settings of -m 100, -x500, and identity $\geq$95% to align the clean metagenomic data of all FPs samples. Genes with a number of aligned reads supporting <1 in each sample were further filtered out as another part of quality control. Additionally, the MG-RAST [38] tool was used to assist in gene prediction and abundance analysis of the FPs metagenomic data. Finally, nonredundant gene sets (unigenes) obtained from each sample were used for fundamental characteristic statistics, intersample correlation analysis, and Venn diagram analysis.

2.6. Gene Functional Annotation of FPs Aligned to the KEGG/eggNOG/CAZy Databases

First, the predicted FP unigenes were annotated in the Kyoto Encyclopedia of Genes and Genomes (KEGG) database [39] using BLAST with an E-value threshold of 1×10^{-5}. The best alignment results of the supported KEGG Orthology (KO) clusters for the FPs were obtained based on similarity and alignment scores. The corresponding KO abundance of FPs represented the sum of the gene abundances annotated to all genes in the KEGG database. The relative abundances of the five major functional levels in the KEGG database for FPs were calculated. The relative abundances of FPs genes at each subhierarchical functional level were also summed. The diversity in the function of the obtained KO results was analyzed and compared among FPs of different ages, positions, and geographical regions. Secondly, the predicted FPs unigenes were annotated based on the eggnog 6.0 database [40] using the DIAMOND alignment tool [41]. The annotated results with the highest alignment scores were selected as the gene functions of orthologous groups (OGs) for FPs. The relative abundances of annotated genes at each functional level of the eggNOG OGs were also calculated. The functional diversity of the obtained OG results was deciphered among FPs of different ages, positions, and geographical regions. Genes not annotated in eggNOG were considered to have unknown or novel functions. These genes were clustered using BLAST and the Markov cluster algorithm (MCL) [42], with the definition of a gene family as a cluster containing 20 new functional genes. Finally, predicted FP unigenes were annotated in the CAZy database using the dbCAN tool [43]. This provided functional information on the carbohydrate-active enzymes in FPs, including GHs, AAs, GTs, CEs, PLs, and CBMs. The aligned functional results were subjected to statistical analysis, along with the other two annotations.

2.7. Drawing the Overall Metabolic Pathway Profile of FPs

The annotated functional genes were categorized, statistically summarized, and aligned to public pathways [39] to generate metabolic pathway maps of the FPs. In these maps, nodes correspond to various chemical compounds, edges represent a series of enzymatic reactions or protein complexes, and edge thickness represents the relative abundance of functional genes or enzymes. The global metabolic pathway profile and secondary metabolite biosynthesis profiles were generated to depict the overall biological system of the FPs. The key genes that constitute the main metabolic characteristics of FPs are explained in the metabolic network diagram.

2.8. Construction of the Generative Pathways for the Main Flavor Compounds of CSAB

Starting from the functional gene sets of the FP metagenome, the focus was on the main flavor compounds of CSAB previously detected [2] to construct the degradation pathways of raw materials such as starch and cellulose, the generation and utilization of compounds such as glucose and pyruvic acid, as well as the formation pathways of key characteristic compounds of alcohols (including ethanol, butanol), acids (including acetic acid, propionic acid, lactic acid, butyric acid, caproic acid), and esters (including ethyl acetate, ethyl caproate, ethyl lactate, ethyl butyrate, ethyl propionate, and ethyl valerate). Possible metabolic pathways for the formation of the main flavor compounds were identified. The key catalytic enzyme gene information involved in the metabolic pathways was extracted from the annotated functional gene sets of the FP metagenome. Finally, the specific network pathways for the formation of the main flavor compounds in CSAB were identified, summarized, and illustrated.

2.9. Statistical Analysis

Statistical analyses were conducted using R (v 4.3.1) (https://www.r-project.org/ (accessed on 10 June 2023)) with custom scripts under the available packages in the project. The ggplot2 package in R was used to construct a hierarchical heat map. Venn analysis was performed using ImageGP software (v 1) (http://www.ehbio.com/ImageGP/ (accessed on 10 June 2023)). The statistical significance of the difference between the means of the samples was tested using a two-way analysis of variance (ANOVA) with Duncan's test ($p < 0.05$).

3. Results and Discussion

3.1. Statistical Analysis of Fundamental Information for FPs Metagenomic Data

Metagenomic sequencing of the FPs from the Luzhou production area (Figure 1) generated 1,648,547,602 reads, amounting to 167 Gb of data. For the Chengdu production area, 515,952,346 reads, amounting to 51 Gb of data, were obtained. Similarly, for the Mianyang production area, 410,809,062 reads, amounting to 41 Gb of data, were obtained. Overall, 2,575,309,010 reads, equivalent to 259 Gb of data, were obtained from the FP metagenomes. Therefore, the obtained data volume was sufficiently large to meet the common standard requirements for the analysis of metagenomic data. The GC content of the metagenomic sequences was extracted and statistically analyzed for FPs based on the classification of ages, spatial layers, and geographical regions. The GC content, which represents the overall DNA characteristics of the metagenome, was used as an evaluation parameter to compare different environmental metagenomes. The relationship between the GC content distribution and relative abundance of FPs is depicted in Figure 2. The analysis of the relative abundance of the GC content revealed slight differences among the different ages of FPs, layers within the FPs, and geographical regions, with a peak of approximately 42%. This indicates the presence of consistent characteristics in the microbial community composition within the same category. In comparison, typical agricultural soil reference metagenomes (available public data: NCBI ID 13699/MG-RAST ID 4441091.3) had a GC content peak of approximately 63%, indicating a significant difference between these two environments. This suggests that there are notable differences in the microbial composition between FPs and typical agricultural soils, suggesting unique new resource characteristics of FP metagenomes and microbial communities.

3.2. Assembly Results of FPs Metagenome

A total of 2,575,309,010 sequences were generated from the metagenomics of FPs. The assembly results of this metagenomic study revealed 49,322 scaftigs with lengths greater than 10 kb, amounting to a total size of 1,465,978,515 bp (Table 1). The longest single sequence had a length of 938,471 bp, which was almost equivalent to the genome length of a typical single microorganism. The average N50 value was 39,156 base pairs (bp). These results indicated that the assembled scaftigs possessed completeness, and the assembly outcome was satisfactory.

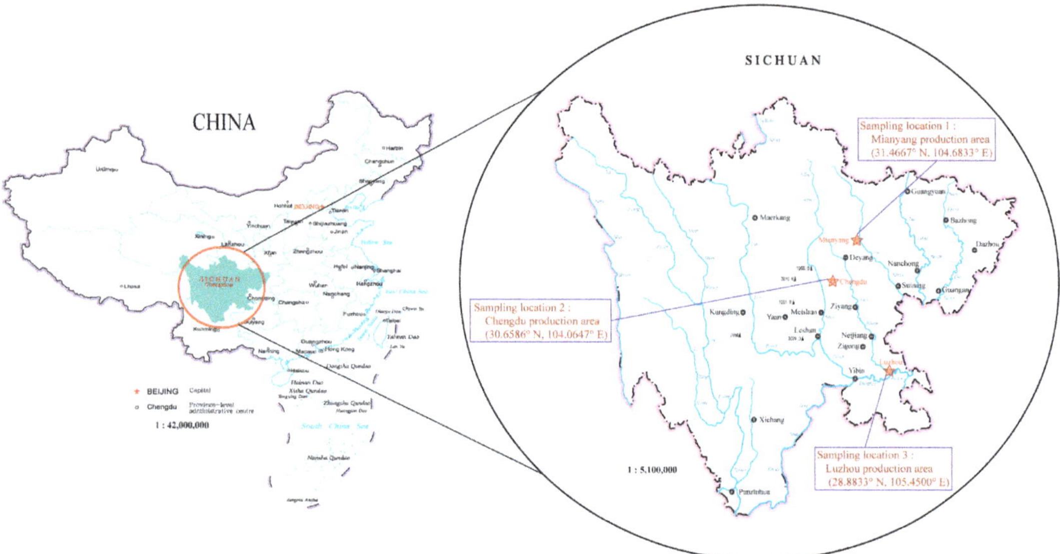

Figure 1. The sampling locations on the geographic map of China and Sichuan province.

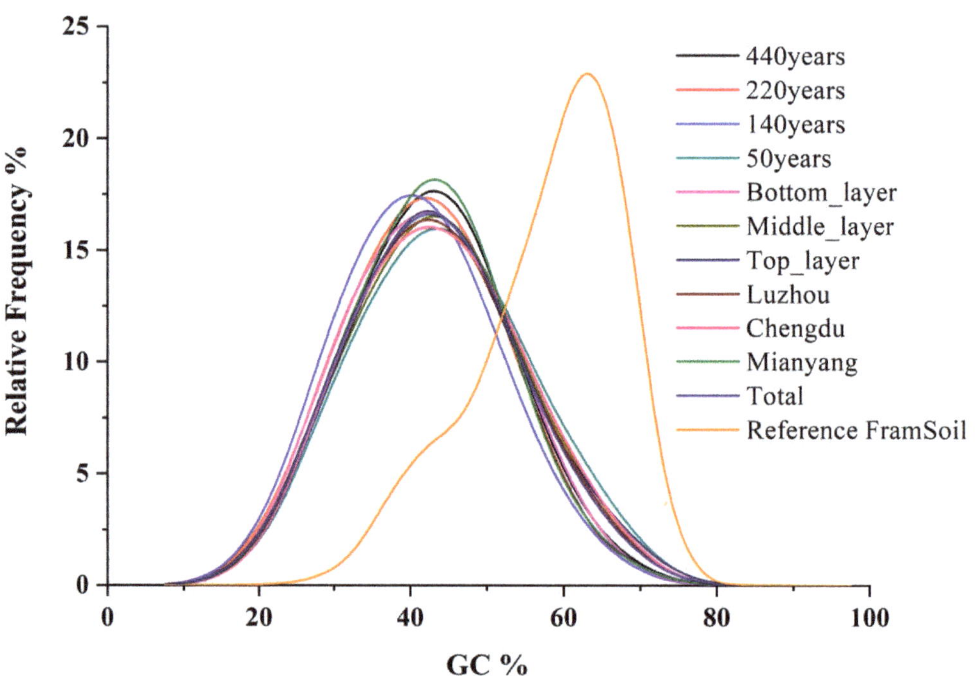

Figure 2. The GC% content for the metagenomic data of the FPs. Note: Reference FrameSoil (Available public data. NCDI ID13699/MG RAST ID 4441091 3).

Table 1. Assembly results of the scaftigs for the samples ($\geq$10 Kb).

Sample Classification			Total Len. (bp)	Num.	Average Len. (bp)	Max Len. (bp)	N50 Len. (bp)
Region	Age	Position					
Luzhou	440 years	All-Bottom_layer	90,762,126	2882	31,492	684,286	43,776
Luzhou	440 years	All-Middle_layer	90,425,571	3338	27,089	383,109	34,736
Luzhou	440 years	All-Top_layer	66,059,291	2301	28,708	383,189	36,358
Luzhou	220 years	All-Bottom_layer	67,906,746	2383	28,496	938,471	34,156
Luzhou	220 years	All-Middle_layer	56,765,588	1953	29,065	454,806	36,701
Luzhou	220 years	All-Top_layer	92,198,789	3334	27,654	362,757	34,395
Luzhou	140 years	All-Bottom_layer	65,720,000	2266	29,002	465,956	36,112
Luzhou	140 years	All-Middle_layer	82,135,058	2861	28,708	544,381	37,429
Luzhou	140 years	All-Top_layer	78,202,504	2789	28,039	353,104	36,019
Luzhou	50 years	All-Bottom_layer	70,386,979	2204	31,936	750,639	43,744
Luzhou	50 years	All-Middle_layer	65,224,763	2307	28,272	294,874	34,635
Luzhou	50 years	All-Top_layer	82,475,229	2717	30,355	426,276	41,806
Chengdu	50 years	All-Bottom_layer	93,242,880	3151	29,591	365,438	38,234
Chengdu	50 years	All-Middle_layer	116,609,574	3931	29,664	793,643	38,912
Chengdu	50 years	All-Top_layer	84,719,849	2791	30,354	411,342	40,105
Mianyang	50 years	All-Bottom_layer	97,967,763	3005	32,601	879,267	44,720
Mianyang	50 years	All-Middle_layer	72,020,509	1928	37,355	689,896	56,266
Mianyang	50 years	All-Top_layer	93,155,296	3181	29,284	644,239	36,710

3.3. Gene Prediction Characteristics of FPs

In total, 1,392,928 ORFs were predicted from the metagenomic data of the FPs. The specific results for the number of genes obtained for each FP classification are listed in Table 2. The total length of the predicted open ORFs was 1,292,881,866 bp. After removing redundant genes, a total of 1,375,660 nonredundant genes (unigenes) with a cumulative length of 1,292,669,970 bp were obtained. These nonredundant genes were defined as the gene set used to establish a foundational database of the gene resource catalog for CSAB FPs.

Table 2. The results of the scaftigs for the samples.

Sample Classification			Number of Genes	Number of Unigenes
Region	Age	Position		
Luzhou	440 years	All-Bottom_layer	86,449	85,431
Luzhou	440 years	All-Middle_layer	86,338	85,295
Luzhou	440 years	All-Top_layer	63,079	62,273
Luzhou	220 years	All-Bottom_layer	64,754	63,952
Luzhou	220 years	All-Middle_layer	54,240	53,577
Luzhou	220 years	All-Top_layer	88,622	87,547
Luzhou	140 years	All-Bottom_layer	62,356	61,522
Luzhou	140 years	All-Middle_layer	78,894	77,925
Luzhou	140 years	All-Top_layer	74,567	73,631
Luzhou	50 years	All-Bottom_layer	68,442	67,632
Luzhou	50 years	All-Middle_layer	62,357	61,567
Luzhou	50 years	All-Top_layer	78,485	77,561
Chengdu	50 years	All-Bottom_layer	88,101	86,959
Chengdu	50 years	All-Middle_layer	109,015	107,659
Chengdu	50 years	All-Top_layer	80,540	79,564
Mianyang	50 years	All-Bottom_layer	92,053	90,946
Mianyang	50 years	All-Middle_layer	67,106	66,184
Mianyang	50 years	All-Top_layer	87,530	86,435

3.4. Comparative and Differential Analysis of Gene Sets in FPs

Venn diagrams were constructed based on the confirmed shared and unique genes and their classifications (KO classification) to display the diversity among the FPs with different ages, layers, and geographical regions (Figure 3). It was observed that the number of shared functional genes among different ages of FPs was 3498, accounting for 67.2%. Unique genes in the 50-year, 140-year, 220-year, and 440-year FPs accounted for 14.2%, 0.7%, 1.8%, and 1.7%, respectively. Functional genes exhibited a gradual stabilization pattern, which may correspond to the evolution of the FP community. During the long-term, uninterrupted production and use of FPs, there is a subtle process of microbial domestication. Over time, microorganisms suitable for the FP environment and those that are closely related to brewing were gradually preserved, whereas unsuitable microorganisms were gradually eliminated. Therefore, the microbial community in the FPs also exhibit a gradual stabilization pattern over time. In an analysis of microbial communities [2], it was reported that the number of microbial species in 50-year FPs was higher than that in 440-year FPs. The presence of widely distributed "miscellaneous" microorganisms contributed to the higher number of unique functional genes in the 50-year FPs. Venn analysis of the functional genes from different layers of the FPs (top, middle, and bottom layers) revealed that there were 3847 shared functional genes, accounting for 72.1% of the total. The top layer of FPs contained 1030 unique genes, accounting for 19.3% of the total. The top layer of the FP represents a relatively exposed space and has more opportunities to be influenced by microorganisms from the external environment. Microorganisms from the external environment can infiltrate the microbial community structure of the top layer, thereby altering the genetic diversity. Analysis of the microbial taxonomy of contigs [2] revealed that Bacteroidetes, Gammaproteobacteria, Opisthokonta (fungi), and Chloroflexi were more abundant in the top layer of the FPs. This indicated that these microbial categories potentially contribute to the diversity of functional genes in the top layer. After comparing the functional genes of FPs from different geographical regions, it was concluded that Luzhou, Chengdu, and Mianyang had 3603 shared functional genes, accounting for 67.5% of the total. Luzhou shared more functional genes with Chengdu than with Mianyang. Luzhou also had a significantly higher proportion of unique genes, reaching 21.9%. Venn analysis of the microbial community structure [2] indicated that the diversity of CSAB-related microbial species in Luzhou FPs was higher than those in Chengdu and Mianyang. GC-MS analysis of CSAB produced in different geographical regions has confirmed that the diversity of microbial communities in Luzhou is beneficial for CSAB quality [2]. Therefore, the increased diversity of beneficial microbial communities contributed to the diversity of functional genes in Luzhou. This explains the superior quality of CSAB produced in Luzhou compared to Chengdu and Mianyang, providing a preliminary theoretical explanation for the widely recognized superior quality of Luzhou CSAB.

By comprehensively analyzing the comparative results of functional genes from FPs of different ages, layers, and geographical regions, it was observed that the 50-year FPs had a higher number of unique genes than the 140-year, 220-year, and 440-year FPs. However, older FPs produced higher-quality CSAB. Additionally, the top layer of FPs had a greater number of unique genes than the middle and bottom layers; however, the CSAB quality of the top layer was not as good as that of the middle and bottom layers. Surprisingly, the functional gene sets of FPs in Luzhou had a higher number of unique genes, and the quality of CSAB in Luzhou was superior to that in Chengdu and Mianyang. This indicates that to ensure the quality of fine CSAB, a certain number of functional genes is required. The particular characteristics of the functional gene structure of 440-year-old FPs contributed to the formation of a metabolic network for high-quality CSAB production. Subsequently, the metabolic tendencies of microbial communities progressed to specialization, with beneficial functional genes being preserved and nonbeneficial genes gradually diminishing. Therefore, the presence of other types of nonbeneficial genes may cause deviations in the overall community metabolism, potentially affecting the quality of CSAB. Moreover, by combining the comprehensive results of the Venn analysis of functional gene sets from all the different characteristics of FPs, we concluded that approximately 3379 currently known functional gene clusters were defined as the core functional gene catalog for FPs.

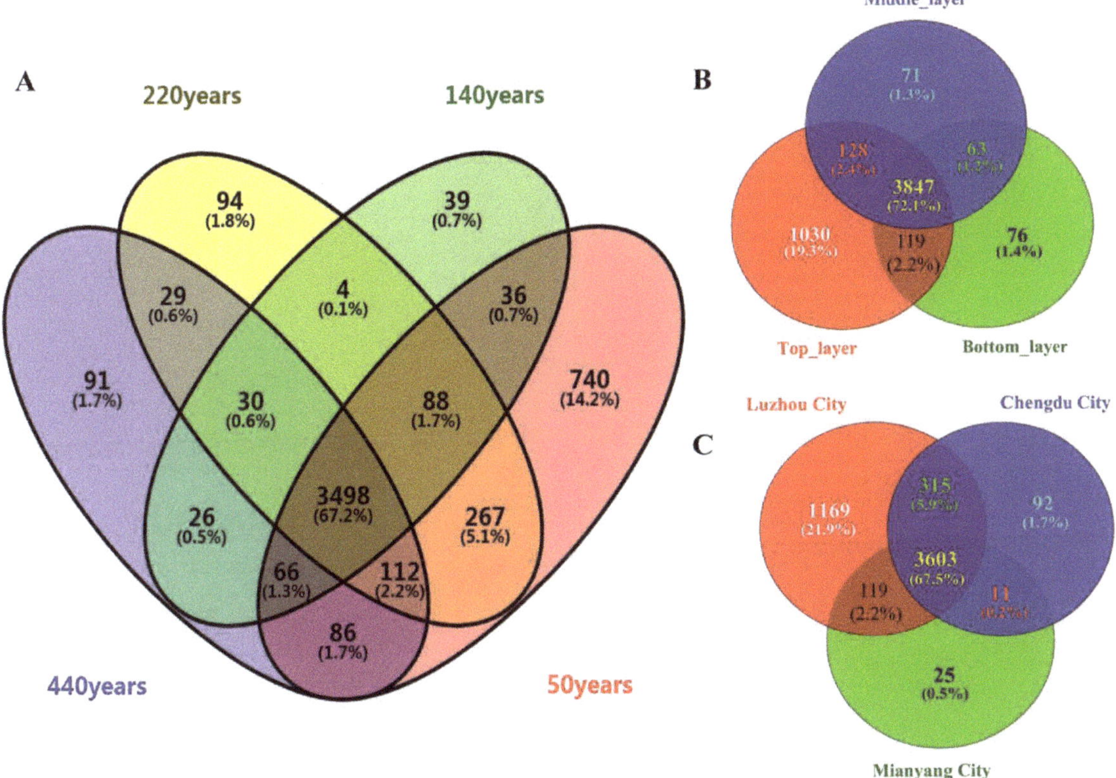

Figure 3. Venn analysis for cross-comparison of the functional genes in FPs with different ages (**A**), layers (**B**), and geographical regions (**C**).

3.5. Classificatory and Differential Analysis of Functional Genes in FPs

The analysis results of functional annotation and classification revealed that the functional microbial genes in the FPs were involved in multiple metabolic pathways (Figure 4). Among them, functional genes related to energy production and conversion, as well as amino acid transport and metabolism were the most abundant, accounting for 6.59% and 6.83%, respectively. Additionally, a relatively high proportion of functional genes in the FPs were associated with carbohydrate transport and metabolism (6.00%), cell wall/membrane/envelope biogenesis (6.02%), translation, ribosomal structure and biogenesis (6.35%), transcription (6.26%), replication, recombination and repair (6.24%), inorganic ion transport and metabolism (5.15%), and lipid transport and metabolism (4.78%). The main components comprising the raw materials used in the fermentation process, such as sorghum, corn, wheat, rice, and glutinous rice, are carbohydrates and proteins, which account for approximately 60–80% and 8–10%, respectively. The major distribution of the functional genes associated with carbohydrate, energy, and amino acid metabolism in microbial communities of FPs has been verified as the fundamental source of flavor compounds in CSAB at the genetic level.

Furthermore, a substantial proportion of unknown functional genes accounted for 23.78% of the total genes. These abundant unknown genes reflect the uniqueness of the FP ecosystem compared with other publicly known environmental microbial ecosystems. These genes represent valuable tools for the exploration of new functional genes for industrial fermentation; however, further research is required.

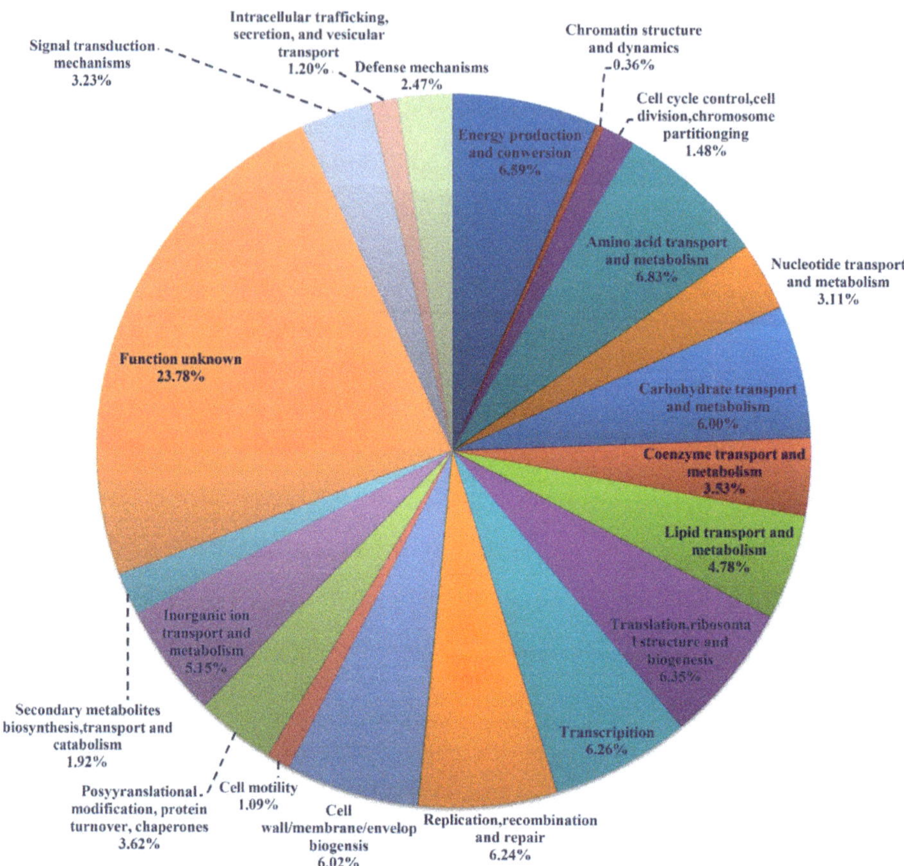

Figure 4. The relative abundance of functional genes in the microbiota of FPs.

Further analysis was conducted to investigate the detailed distinguishable characteristics and metabolic roles of the functional genes in different FPs. These results are displayed in Figure 5. Overall, the differences in the functional gene distribution of microbial ecosystems in FPs across all observed ages were not significant; however, the characteristics were only suitable for the requirement of metabolic flavor compounds in the fermentation production of CSAB. However, subtle differences were observed in key metabolic pathways. As shown in Figure 5, the functional categories of C (energy production and conversion), E (amino acid transport and metabolism), G (carbohydrate transport and metabolism), and I (lipid transport and metabolism) increased in abundance with an increase in age. [G] Carbohydrate metabolism, C (energy metabolism), E (amino acid metabolism), and I (lipid metabolism) were the main pathways involved in microbial utilization, decomposition, and transformation of raw materials used in CSAB production. This could explain why the old FPs produced CSAB with higher quality as opposed to new FPs. With an increase in age, FPs contribute beneficial functional genes that facilitate the metabolism of CSAB raw materials, thereby improving the quality of CSAB. These subtle adjustments in the metabolism of the abovementioned substance categories led to differences in CSAB quality. However, it was also observed that the proportion of unknown functional genes (S, function unknown) slightly decreased with an increase in age. This could be because the functional metabolism of the FP microbial community tends to develop in a more specialized direction during CSAB production as age increases.

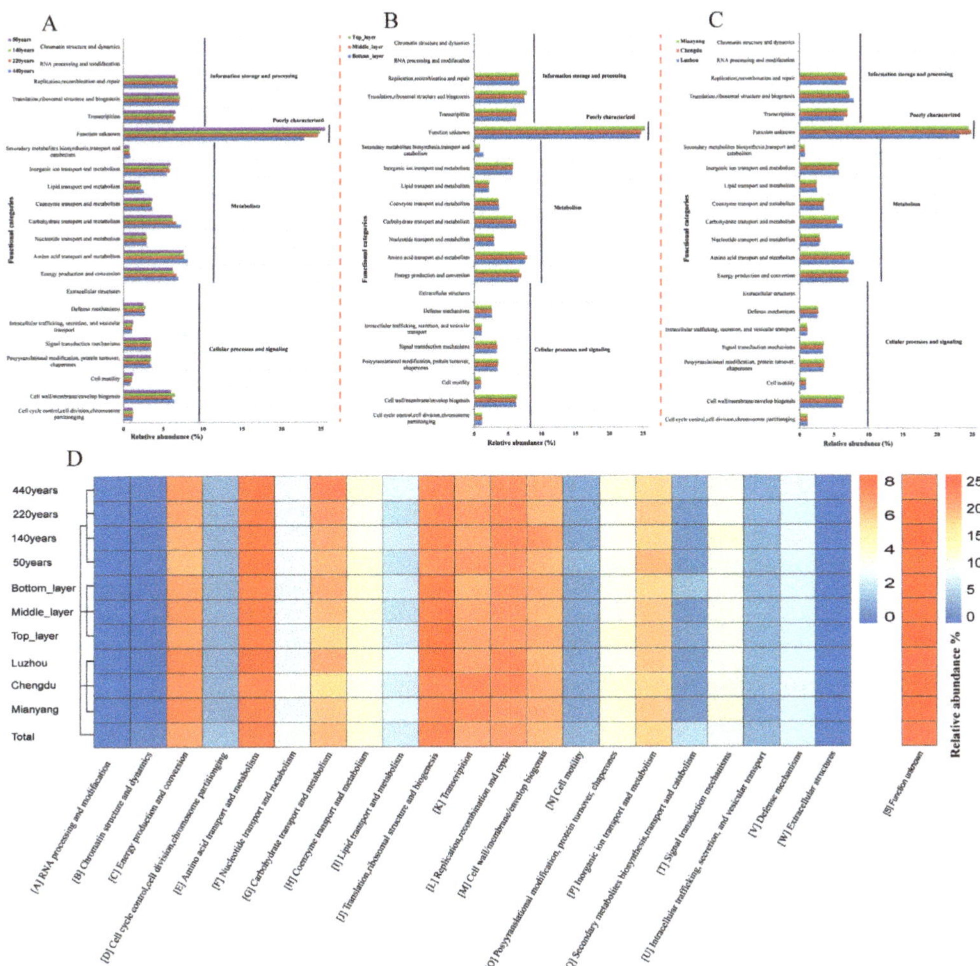

Figure 5. Functional categories of genes in FPs with different ages (**A**), layers (**B**), and geographical regions (**C**); hierarchical clustering analysis heat map of the differential functional categories of genes in FPs (**D**).

Regarding the functional differences in genes in different layers of the FP, the middle layer exhibited a slight superiority in C (energy production and conversion) and E (amino acid transport and metabolism) compared to the bottom and top layers. The bottom layer exhibited a slight advantage in terms of Q (secondary metabolite biosynthesis, transport, and catabolism) compared to the middle and top layers. The top layer had a slightly higher distribution of functional genes associated with J (translation, ribosomal structure, and biogenesis) than the other two layers. By combining the characteristics of CSAB quality and the microbial diversity of FPs [2], we concluded that the CSAB quality of the middle layer was superior to that of the bottom and top layers, due to the slight enhancement in energy metabolism and amino acid metabolism. The CSAB quality of the top layer was inferior to that of the middle layer, possibly because G (carbohydrate transport and metabolism) was slightly reduced in the top layer, resulting in lower levels of fermentation byproducts, such as acetic acid and hexanoic acid, which affected the quality. As the fermentation process progressed, a large amount of fermentation liquid gradually accumulated in the bottom layer of the FPs. Consequently, these environmental conditions may have enhanced Q (secondary metabolite biosynthesis, transport, and catabolism) in the bottom layer when compared to the middle and top layers. Regarding microbial diversity [2], the top layer had more species abundance than the middle and bottom layers, and the abundance of functional genes associated with J (translation, ribosomal structure, and biogenesis) was slightly higher in the top layer, which is consistent with the observations relating to the CSAB quality differences.

In terms of the functional genes of the microbial community in FPs from different geographical regions, a previous analysis revealed that CSAB from the Luzhou region differs from that of the Chengdu and Mianyang regions. This could be attributed to the higher E (amino acid transport and metabolism) and G (carbohydrate transport and metabolism) activities in the FP microbial community in the Luzhou region than in the other two regions. Another possible reason is that the microbial community abundance was higher in the Luzhou region in terms of functional genes related to J (translation, ribosomal structure, and biogenesis).

In conclusion, functional gene distribution exhibited general consistency with regard to the different FP characteristics. Metabolic pathways influence quantity and quality of CSAB. However, the current research mainly focused on the ethanol metabolic pathway derived from carbohydrate metabolism. Therefore, there is still a limited understanding of the metabolic pathways associated with major flavor compounds, such as ethyl caproate, ethyl lactate, ethyl acetate, and ethyl butyrate, which have a significant impact on the quality of CSAB. Hence, it is challenging to analyze differences in the functional gene metabolic classification of FPs. At the same time, this also indicates that the construction and analysis of metabolic pathways, such as ethyl caproate, ethyl lactate, ethyl acetate, and ethyl butyrate, in brewing fermentation would be a meaningful direction. Although it was difficult to perform an in-depth and precise analysis, a general trend can be summarized: the abundance of functional genes constituting the FP microbial community was slightly enhanced in the key main characteristic metabolic classifications of C (energy production), E (amino acid metabolism), G (carbohydrate metabolism), and I (lipid metabolism). This enhancement was beneficial for improving the quality of CSAB, thus exhibiting a positive correlation.

3.6. Overall Characteristics of the Constructed Metabolic Pathway Profile of FPs

The global metabolic network (Figure 6A) and secondary metabolic network diagram (Figure 6B) of the FP microbial community was plotted based on the annotated categorization and abundance of functional genes. The red lines represent the metabolic pathways associated with FP microorganisms, and the thickness of the lines indicates the abundance of the corresponding functional genes. In the FP microbial community, the summarized metabolic support for CSAB was mainly observed in the highly active functional genes in the G (carbohydrate metabolism), I (lipid metabolism), F (nucleotide metabolism), E (amino acid transport and metabolism), and C (energy metabolism) pathways. These metabolic pathways form the life framework of the FP microbial communities. Further analysis revealed specific genes involved in starch and sucrose metabolism, including a glycosyltransferase gene (COG0438) and transaldolase gene (COG0176); in lipid conversion and metabolism, including 3-oxoacyl-(acyl carrier protein) synthase gene (COG0304), phosphopantetheinyl transferase (holo-ACP synthase) gene (COG0736), and (acyl-carrier-protein) s-malonyltransferase gene (COG0331); in starch and sucrose metabolism and energy production and conversion, including a NAD-dependent aldehyde dehydrogenases gene (COG1012) (NADH plays a functional role in bacterial alcohol fermentation to catalyze the conversion of the pyruvic acid produced acetaldehyde to ethanol), and a predicted hydrolases or acyltransferases (alpha/beta hydrolase superfamily) gene (COG0596); in starch and sucrose metabolism and lipid transport and metabolism, including acetyl-CoA acetyltransferase gene (COG0183), acyl dehydratase gene (COG2030), enoyl-CoA hydratase gene (COG1024), dehydrogenases with different specificities gene (related to alcohol dehydrogenases) (COG1028), 3-hydroxyacyl-CoA dehydrogenase gene (COG1250), and acyl-CoA dehydrogenases gene (COG1960, catalyzing the conversion of acetaldehyde to ethanol); in nucleotide metabolism, including tRNA-dihydrouridine synthase gene (COG0042) and intein/homing endonuclease gene (COG1372); and in amino acid transport and metabolism, including acetylornithine deacetylase gene (COG0624) and aspartate/tyrosine/aromatic aminotransferase gene (COG0436). These exhibited highly abundant and active gene clusters that contributed to the characteristic microbial metabolic networks of the FPs. Summary analysis indicated the presence of highly abundant and active genes predominantly associated with CoA hydrolase, CoA dehydrogenase, CoA transferase, and NAD dehydrogenase in the FP microbial ecosystem, as well as the existence of abundant microbial species [2], such as *Lactobacillus*, *Clostridium*, and *Saccharomycetales* (Supporting Table S1). This further helped elucidate the source mechanism of ethanol and the major ester flavor compounds present in CSAB. The brewing materials, namely sorghum, corn, wheat, and glutinous rice, mainly provide starch-derived sugars for fermentation. The starch was metabolized and further converted to pyruvate, which, unlike in aerobic conditions, was not directly oxidized to produce energy and CO_2, and was recruited in the microbial community. Instead, it underwent various fermentation metabolic conversions via different high-abundance microorganisms in the FP, entering the heterolactic fermentation metabolic pathway of *Lactobacillus*, the butyric acid fermentation metabolic pathway of *Clostridium*, and the yeast-type alcoholic fermentation metabolic pathway of *Saccharomycetales* (Supporting Table S2). Under metabolic catalysis by the main enzymes CoA hydrolase, CoA dehydrogenase, CoA transferase, and NAD dehydrogenase, various substrates of flavor compounds, such as lactic acid, ethanol, acetic acid, butyric acid, and hexanoic acid, were generated. Consequently, the main flavor compounds representing the strong aromatic characteristics of CSAB, including ethyl acetate, butyl acetate, hexyl acetate, and ethyl lactate, were produced. This theoretically demonstrates the importance of FPs in the production of CSAB, which provides abundant specific beneficial microorganisms and functional genes for fermentation.

3.7. The Deciphered Generative Pathway Profile for the Main Flavor Compounds of CSAB

The detailed metabolic network profile for the formation of the main flavor compounds detected using GC-MS in CSAB was constructed by progressively aligning the annotated functional genes of the FPs with the metabolic pathways plotted in Figure 6. The results are displayed in Figure 7. A total of 159 catalytic enzymes related to the formation of the main flavor compounds in CSAB were identified. Specifically, fermentative substrates, such as starch, cellulose, sucrose, and trehalose, provided by the raw materials were metabolized into glucose by ko00500, which was further transferred into the cells of the microorganisms for metabolism. The transferred glucose was converted to phosphoenolpyruvate by ko00010, and pyruvic acid was then formed. Pyruvic acid is an important nodal substance in the metabolism and formation of flavor compounds in CSAB. Pyruvic acid can be converted into various flavor compounds through different metabolic pathways. First, acetaldehyde was formed by ko00620, which was then transformed into ethanol by ko00010. Ethanol is the most abundant compound in CSAB and determines the main taste and production yield during fermentation. The acetaldehyde formed in this step can be further metabolized to acetic acid. The esterification reaction between acetic acid and ethanol formed ethyl acetate, which is one of the four pivotal aromatic compounds in CSAB. Pyruvic acid can also be converted into lactic acid, propionic acid, and acetyl-CoA by ko00620. When esterified with ethanol, lactic acid and propionic acid form ethyl lactate and ethyl propionate, respectively. Ethyl lactate is one of the four pivotal aromatic compounds in CSAB. Propionic acid can also be metabolized to valeric acid by ko00290, and valeric acid can be further esterified with ethanol to form ethyl valerate, which also plays a role in the aromatic profile of CSAB. Acetyl-CoA is converted to butyric acid by ko00650, which is then directly esterified with ethanol to form ethyl butyrate. One butyric acid molecule reacts with one acetyl-CoA molecule to generate caproic acid. When esterified with ethanol, the caproic acid forms ethyl caproate. Ethyl butyrate and ethyl caproate are the two remaining pivotal aromatic compounds in CSAB. Additionally, acetic, butyric, and caproic acids are abundant acidic flavor compounds in CSAB. Butyric acid can also be metabolized by ko00650 to form butanol and by ko00290 to form propanol. Both butanol and propanol are abundant alcohols in CSAB. Another abundant alcohol, 2,3-butanediol, is formed from pyruvic acid through ko00290 by metabolizing it to 2-acetolactic acid and 3-hydroxybutanone, which are further transformed through ko00650. Furthermore, the proteins and amino acids provided by brewing raw materials can be metabolized through ko00250 to ko00400, resulting in the production of various higher alcohols in CSAB, thus obtaining the flavor characterized by the sensation of being "heady" after consuming CSAB. Therefore, the metabolic network for the formation of the main flavor compounds in CSAB is well understood.

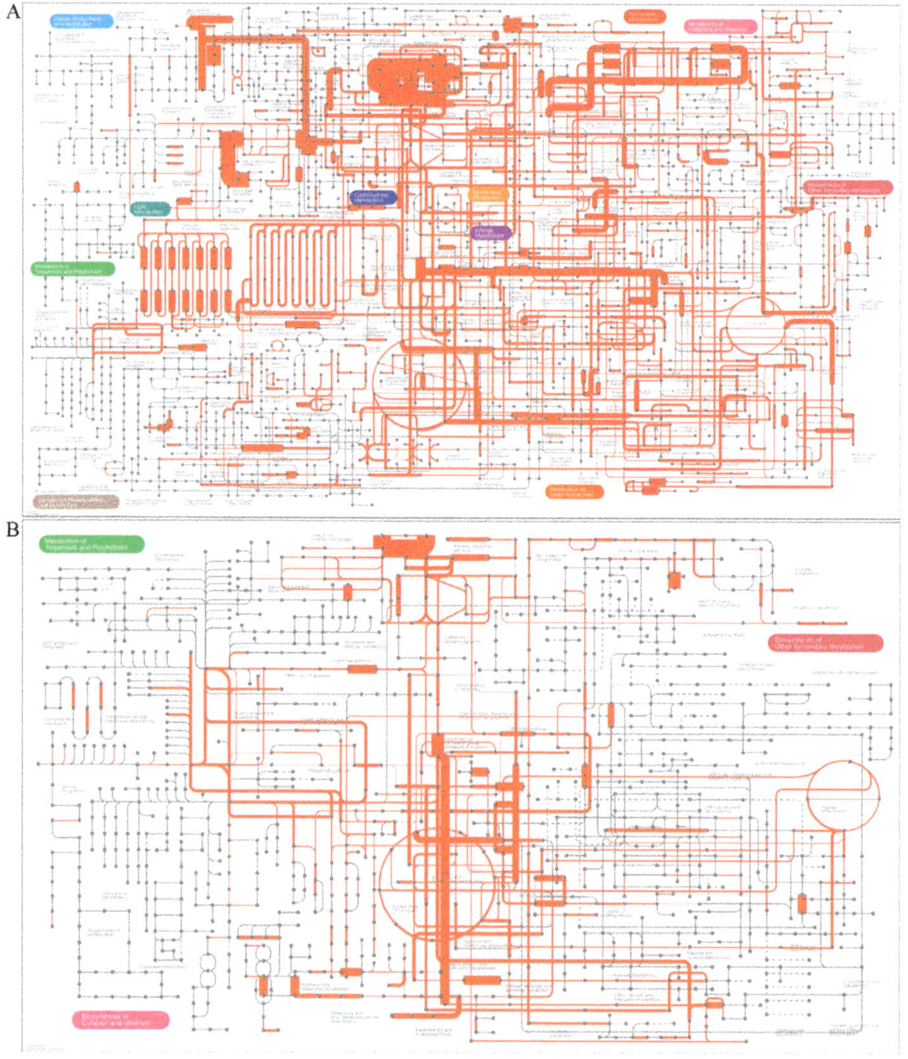

Figure 6. The microbial metabolic pathways of FP microbiotas (**A**); the biosynthesis of secondary metabolites in the FP microbiotas (**B**).

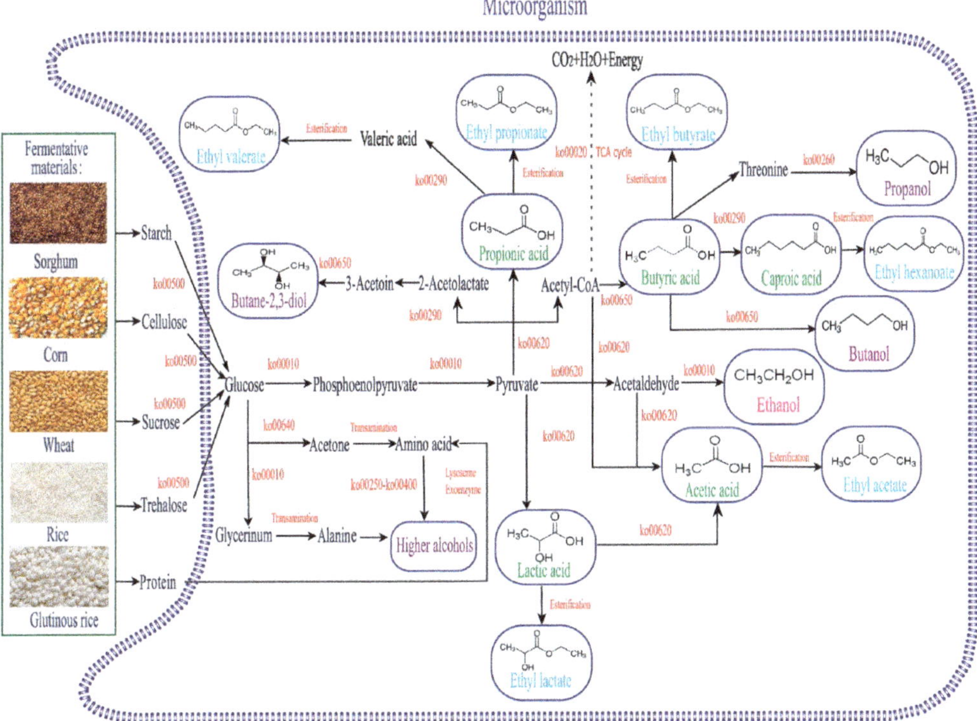

Figure 7. Integrated analysis of the main metabolic network for flavor compounds in industrial CSAB fermentation by FPs.

4. Conclusions

A functional gene catalogue of traditional industrial FPs for CSAB production was established using a metagenomic analysis approach. A total of 259 Gb of metagenomic data were obtained from FPs. After assembly, 1,392,928 ORFs were identified with a cumulative length of 1,292,881,866 bp. Through homology comparisons, 1,375,660 nonredundant genes (unigenes) were identified. Further annotation based on known public functional databases and Venn analysis revealed 3379 known functional gene clusters as the core functional gene set of FPs. Based on the KO annotation using the KEGG database and COG annotation from the eggNOG database, the functional classification features of the microbial genes in the microbial community of FPs were identified. The main functional genes were as follows: carbohydrate metabolism, 6.00%; energy metabolism, 6.59%; amino acid metabolism, 6.83%; cell wall/membrane formation, 6.02%; translation and genetic processes, 6.35%; transcriptional processes, 6.26%; replication/recombination/repair processes, 6.24%; ion transport processes, 5.15%; and lipid metabolism, 4.78%. Additionally, newly discovered genes with unknown functions accounted for a significant proportion (23.78%). These functional compositional features indicate that the microbial ecosystem of FPs is a community system primarily driven by the utilization of carbohydrates and amino acids as the main sources for metabolic consumption. This revealed that the flavor compounds formed during CSAB were mainly derived from the conversion of carbohydrates and amino acid metabolism. Global and secondary metabolic network profiles of the microbial community in FPs were constructed based on known functional gene metabolic pathways. Approximately 17 clusters of key genes affecting CSAB yield and quality were identified. The underlying mechanisms that influence the production (ethanol) and quality (main flavor compounds, such as ethyl caproate, ethyl lactate, ethyl acetate, and ethyl butyrate) of CSAB by FPs were elucidated. These mechanisms are formed through the specific metabolic activities of highly abundant microbial species, such as *Lactobacillus*, *Clostridium*, and yeast (*Saccharomycetales*), and through specific pathways involving highly active specific species of functional genes, including CoA hydrolases, CoA dehydrogenases, CoA transferases, and NAD dehydrogenases. The relationship between flavor compound formation and microbial metabolism was studied based on the main flavor compounds. In total, 159 catalytic enzymes related to the formation of main flavor compounds in CSAB were annotated. The pathways involved in the formation of flavor compounds were elucidated.

However, the number of new homologous gene clusters in the microbial ecosystem of FPs was much larger than that in the known homologous groups. There are still a significant number of unknown microbial genes in the FP ecosystem, indicating that the functionality of FPs is still not fully understood. Further research is required to explore these unknown genes, and continuous updates to functional gene databases are required.

Supplementary Materials: The following supporting information can be downloaded at: https://www.mdpi.com/article/10.3390/fermentation9080772/s1, Table S1: The relative abundance of the top 100 microorganisms in the microbial communities from FPs for the production of CSAB; Table S2: The phylogenetic profile of the top 20 Eukaryotes in the FPs.

Author Contributions: M.G. and Y.D.: conceptualization; M.G., Y.D. and J.H.: methodology, software and visualization; J.H., C.Z. and H.Q.: formal analysis; H.W., H.Q. and S.Z.: investigation and resources; M.G., J.H., C.Z. and H.Q.: writing—original draft preparation; M.G., Y.D., H.W. and S.Z.: writing—review and editing. All authors have read and agreed to the published version of the manuscript.

Funding: This work was supported by the Natural Science Foundation of Sichuan Province (No. 2022NSFSC1676).

Institutional Review Board Statement: Not applicable.

Informed Consent Statement: Not applicable.

Data Availability Statement: Supporting Information may be found in the online version of this article. The datasets generated during the current study are available from the corresponding author upon reasonable request.

Acknowledgments: We gratefully acknowledge the financial support of this research provided by the Natural Science Foundation of Sichuan Province.

Conflicts of Interest: The authors declare no conflict of interest. HQ and SYZ were employed by the company of Luzhou Laojiao Group Co. Ltd. The remaining authors declare that the research was conducted in the absence of any commercial or financial relationships that could be construed as a potential conflict of interest.

Abbreviations

AAs	Auxiliary activities
ANOVA	Analysis of variance
BLAST	Basic local alignment search tool
CAZy	Carbohydrate active enzyme
CBMs	Carbohydrate-binding modules
CEs	Carbohydrate esterases
CoA	Coenzyme A
COG	Clusters of orthologous genes
CSAB	Chinese strong-aroma baijiu
eggNOG	Evolutionary gene genealogy non-supervised orthologous groups
FP	Fermentation pit
GC-MS	Gas chromatography-mass spectrometry
GHs	Glycoside hydrolases
GTs	Glycosyl transferases
HMM	Hidden Markov model

KEGG	Kyoto Encyclopedia of Genes and Genomes
KO	KEGG orthology
LC-MS	Liquid chromatography-mass spectrometry
MCL	Markov cluster algorithm
NAD	Nicotinamide adenine dinucleotide
OGs	Orthologous groups
ORFs	Open reading frames
PE	Paired-end
PLs	Polysaccharide lyases

References

1. Ye, H.; Wang, J.; Shi, J.; Du, J.; Zhou, Y.; Huang, M.; Sun, B. Automatic and Intelligent Technologies of Solid-State Fermentation Process of Baijiu Production: Applications, Challenges, and Prospects. *Foods* **2021**, *10*, 680. [CrossRef] [PubMed]
2. Guo, M.Y.; Hou, C.J.; Bian, M.H.; Shen, C.H.; Zhang, S.Y.; Huo, D.Q.; Ma, Y. Characterization of microbial community profiles associated with quality of Chinese strong-aromatic liquor through metagenomics. *J. Appl. Microbiol.* **2019**, *127*, 750–762. [CrossRef]
3. Chai, L.-J.; Xu, P.-X.; Qian, W.; Zhang, X.-J.; Ma, J.; Lu, Z.-M.; Wang, S.-T.; Shen, C.-H.; Shi, J.-S.; Xu, Z.-H. Profiling the Clostridia with butyrate-producing potential in the mud of Chinese liquor fermentation cellar. *Int. J. Food Microbiol.* **2019**, *297*, 41–50. [CrossRef] [PubMed]
4. Jia, W.; Fan, Z.; Du, A.; Li, Y.; Zhang, R.; Shi, Q.; Shi, L.; Chu, X. Recent advances in Baijiu analysis by chromatography based technology-A review. *Food Chem.* **2020**, *324*, 126899. [CrossRef] [PubMed]
5. Liu, H.; Sun, B. Effect of Fermentation Processing on the Flavor of Baijiu. *J. Agric. Food Chem.* **2018**, *66*, 5425–5432. [CrossRef]
6. Xie, X.; Chen, L.; Chen, T.; Yang, F.; Wang, Z.; Hu, Y.; Lu, J.; Lu, X.; Li, Q.; Zhang, X.; et al. Profiling and annotation of carbonyl compounds in Baijiu Daqu by chlorine isotope labeling-assisted ultrahigh-performance liquid chromatography-high resolution mass spectrometry. *J. Chromatogr. A* **2023**, *1703*, 464110. [CrossRef]
7. Yu, Y.; Nie, Y.; Chen, S.; Xu, Y. Characterization of the dynamic retronasal aroma perception and oral aroma release of Baijiu by progressive profiling and an intra-oral SPME combined with GC × GC-TOFMS method. *Food Chem.* **2023**, *405*, 134854. [CrossRef]
8. Sharma, A.; Rai, P.K.; Prasad, S. GC–MS detection and determination of major volatile compounds in Brassica juncea L. leaves and seeds. *Microchem. J.* **2018**, *138*, 488–493. [CrossRef]
9. Xu, Y.; Sun, L.; Wang, X.; Zhu, S.; You, J.; Zhao, X.-E.; Bai, Y.; Liu, H. Integration of stable isotope labeling derivatization and magnetic dispersive solid phase extraction for measurement of neurosteroids by in vivo microdialysis and UHPLC-MS/MS. *Talanta* **2019**, *199*, 97–106. [CrossRef]
10. Mu, Y.; Huang, J.; Zhou, R.; Zhang, S.; Qin, H.; Tang, H.; Pan, Q.; Tang, H. Characterization of the differences in aroma-active compounds in strong-flavor Baijiu induced by bioaugmented Daqu using metabolomics and sensomics approaches. *Food Chem.* **2023**, *424*, 136429. [CrossRef]
11. Zou, W.; Zhao, C.; Luo, H. Diversity and Function of Microbial Community in Chinese Strong-Flavor Baijiu Ecosystem: A Review. *Front. Microbiol.* **2018**, *9*, 671. [CrossRef] [PubMed]
12. Tu, W.; Cao, X.; Cheng, J.; Li, L.; Zhang, T.; Wu, Q.; Xiang, P.; Shen, C.; Li, Q. Chinese Baijiu: The Perfect Works of Microorganisms. *Front. Microbiol.* **2022**, *13*, 919044. [CrossRef] [PubMed]
13. Li, H.; Liu, S.; Liu, Y.; Hui, M.; Pan, C. Functional microorganisms in Baijiu Daqu: Research progress and fortification strategy for application. *Front. Microbiol.* **2023**, *14*, 1119675. [CrossRef]
14. Huang, Y.; Yi, Z.; Jin, Y.; Zhao, Y.; He, K.; Liu, D.; Zhao, D.; He, H.; Luo, H.; Zhang, W.; et al. New microbial resource: Microbial diversity, function and dynamics in Chinese liquor starter. *Sci. Rep.* **2017**, *7*, 14577. [CrossRef] [PubMed]
15. Liu, W.H.; Chai, L.J.; Wang, H.M.; Lu, Z.M.; Zhang, X.J.; Xiao, C.; Wang, S.T.; Shen, C.H.; Shi, J.S.; Xu, Z.H. Bacteria and filamentous fungi running a relay race in Daqu fermentation enable macromolecular degradation and flavor substance formation. *Int. J. Food Microbiol.* **2023**, *390*, 110118. [CrossRef]
16. Tan, Y.; Zhong, H.; Zhao, D.; Du, H.; Xu, Y. Succession rate of microbial community causes flavor difference in strong-aroma Baijiu making process. *Int. J. Food Microbiol.* **2019**, *311*, 108350. [CrossRef]
17. Xu, S.; Zhang, M.; Xu, B.; Liu, L.; Sun, W.; Mu, D.; Wu, X.; Li, X. Microbial communities and flavor formation in the fermentation of Chinese strong-flavor Baijiu produced from old and new Zaopei. *Food Res. Int.* **2022**, *156*, 111162. [CrossRef]
18. Chai, L.J.; Qian, W.; Zhong, X.Z.; Zhang, X.J.; Lu, Z.M.; Zhang, S.Y.; Wang, S.T.; Shen, C.H.; Shi, J.S.; Xu, Z.H. Mining the Factors Driving the Evolution of the Pit Mud Microbiome under the Impact of Long-Term Production of Strong-Flavor Baijiu. *Appl. Environ. Microbiol.* **2021**, *87*, 88521. [CrossRef]
19. Shoubao, Y.; Yonglei, J.; Qi, Z.; Shunchang, P.; Cuie, S. Bacterial diversity associated with volatile compound accumulation in pit mud of Chinese strong-flavor baijiu pit. *AMB Express* **2023**, *13*, 3. [CrossRef]
20. Wu, L.; Fan, J.; Chen, J.; Fang, F. Chemotaxis of Clostridium Strains Isolated from Pit Mud and Its Application in Baijiu Fermentation. *Foods* **2022**, *11*, 3639. [CrossRef]
21. Wang, X.J.; Zhu, H.M.; Ren, Z.Q.; Huang, Z.G.; Wei, C.H.; Deng, J. Characterization of Microbial Diversity and Community Structure in Fermentation Pit Mud of Different Ages for Production of Strong-Aroma Baijiu. *Pol. J. Microbiol.* **2020**, *69*, 151–164. [CrossRef] [PubMed]

22. Xu, M.L.; Yu, Y.; Ramaswamy, H.S.; Zhu, S.M. Characterization of Chinese liquor aroma components during aging process and liquor age discrimination using gas chromatography combined with multivariable statistics. *Sci. Rep.* **2017**, *7*, 39671. [CrossRef] [PubMed]

23. Sun, H.; Chai, L.J.; Fang, G.Y.; Lu, Z.M.; Zhang, X.J.; Wang, S.T.; Shen, C.H.; Shi, J.S.; Xu, Z.H. Metabolite-Based Mutualistic Interaction between Two Novel Clostridial Species from Pit Mud Enhances Butyrate and Caproate Production. *Appl. Environ. Microbiol.* **2022**, *88*, 48422. [CrossRef]

24. Ustick, L.J.; Larkin, A.A.; Garcia, C.A.; Garcia, N.S.; Brock, M.L.; Lee, J.A.; Wiseman, N.A.; Moore, J.K.; Martiny, A.C. Metagenomic analysis reveals global-scale patterns of ocean nutrient limitation. *Science* **2021**, *372*, 287–291. [CrossRef]

25. Luo, T.; He, J.; Shi, Z.; Shi, Y.; Zhang, S.; Liu, Y.; Luo, G. Metagenomic Binning Revealed Microbial Shifts in Anaerobic Degradation of Phenol with Hydrochar and Pyrochar. *Fermentation* **2023**, *9*, 387. [CrossRef]

26. Paoli, L.; Ruscheweyh, H.-J.; Forneris, C.C.; Hubrich, F.; Kautsar, S.; Bhushan, A.; Lotti, A.; Clayssen, Q.; Salazar, G.; Milanese, A.; et al. Biosynthetic potential of the global ocean microbiome. *Nature* **2022**, *607*, 111–118. [CrossRef] [PubMed]

27. Shalon, D.; Culver, R.N.; Grembi, J.A.; Folz, J.; Treit, P.V.; Shi, H.; Rosenberger, F.A.; Dethlefsen, L.; Meng, X.; Yaffe, E.; et al. Profiling the human intestinal environment under physiological conditions. *Nature* **2023**, *617*, 581–591. [CrossRef]

28. Mardanov, A.V.; Gruzdev, E.V.; Beletsky, A.V.; Ivanova, E.V.; Shalamitskiy, M.Y.; Tanashchuk, T.N.; Ravin, N.V. Microbial Communities of Flor Velums and the Genetic Stability of Flor Yeasts Used for a Long Time for the Industrial Production of Sherry-like Wines. *Fermentation* **2023**, *9*, 367. [CrossRef]

29. Levin, D.; Raab, N.; Pinto, Y.; Rothschild, D.; Zanir, G.; Godneva, A.; Mellul, N.; Futorian, D.; Gal, D.; Leviatan, S.; et al. Diversity and functional landscapes in the microbiota of animals in the wild. *Science* **2021**, *372*, 5352. [CrossRef]

30. Carter, M.R.; Gregorich, E.G. *Soil Sampling and Methods of Analysis*, 2nd ed.; CRC Press: Boca Raton, FL, USA, 2007; pp. 1–14. [CrossRef]

31. Afiahayati; Sato, K.; Sakakibara, Y. MetaVelvet-SL: An extension of the Velvet assembler to a de novo metagenomic assembler utilizing supervised learning. *DNA Res.* **2015**, *22*, 69–77. [CrossRef]

32. Li, S.; Chou, H.H. LUCY2: An interactive DNA sequence quality trimming and vector removal tool. *Bioinformatics* **2004**, *20*, 2865–2866. [CrossRef] [PubMed]

33. Cox, M.P.; Peterson, D.A.; Biggs, P.J. SolexaQA: At-a-glance quality assessment of Illumina second-generation sequencing data. *BMC Bioinform.* **2010**, *11*, 485. [CrossRef]

34. Luo, R.; Wong, T.; Zhu, J.; Liu, C.M.; Zhu, X.; Wu, E.; Lee, L.K.; Lin, H.; Zhu, W.; Cheung, D.W.; et al. SOAP3-dp: Fast, accurate and sensitive GPU-based short read aligner. *PLoS ONE* **2013**, *8*, 65632. [CrossRef] [PubMed]

35. Peng, Y.; Leung, H.C.; Yiu, S.M.; Chin, F.Y. IDBA-UD: A de novo assembler for single-cell and metagenomic sequencing data with highly uneven depth. *Bioinformatics* **2012**, *28*, 1420–1428. [CrossRef] [PubMed]

36. Bruna, T.; Lomsadze, A.; Borodovsky, M. GeneMark-ETP: Automatic Gene Finding in Eukaryotic Genomes in Consistence with Extrinsic Data. *bioRxiv* **2023**, *1*, 13. [CrossRef]

37. Wei, Z.G.; Chen, X.; Zhang, X.D.; Zhang, H.; Fan, X.G.; Gao, H.Y.; Liu, F.; Qian, Y. Comparison of methods for biological sequence clustering. *IEEE/ACM Trans. Comput. Biol. Bioinform.* **2023**, *3*, 13. [CrossRef]

38. Abdelsalam, N.A.; Elshora, H.; El-Hadidi, M. Interactive Web-Based Services for Metagenomic Data Analysis and Comparisons. *Methods Mol. Biol.* **2023**, *2649*, 133–174. [CrossRef]

39. Kanehisa, M.; Furumichi, M.; Sato, Y.; Kawashima, M.; Ishiguro-Watanabe, M. KEGG for taxonomy-based analysis of pathways and genomes. *Nucleic Acids Res.* **2023**, *51*, 587–592. [CrossRef]

40. Hernández-Plaza, A.; Szklarczyk, D.; Botas, J.; Cantalapiedra, C.P.; Giner-Lamia, J.; Mende, D.R.; Kirsch, R.; Rattei, T.; Letunic, I.; Jensen, L.J.; et al. eggNOG 6.0: Enabling comparative genomics across 12,535 organisms. *Nucleic Acids Res.* **2023**, *51*, 389–394. [CrossRef]

41. Gautam, A.; Zeng, W.; Huson, D.H. DIAMOND + MEGAN Microbiome Analysis. *Methods Mol. Biol.* **2023**, *2649*, 107–131. [CrossRef]

42. Dong, C.; Zeng, Z.; Pu, D.K.; Wen, Q.F.; Liu, S.; Du, M.Z.; Sun, Y.; Gao, Y.Z.; Rao, N.; Huang, J.; et al. CasLocusAnno: A web-based server for annotating cas loci and their corresponding (sub)types. *FEBS Lett.* **2019**, *593*, 2646–2654. [CrossRef] [PubMed]

43. Zheng, J.; Hu, B.; Zhang, X.; Ge, Q.; Yan, Y.; Akresi, J.; Piyush, V.; Huang, L.; Yin, Y. dbCAN-seq update: CAZyme gene clusters and substrates in microbiomes. *Nucleic Acids Res.* **2023**, *51*, 557–563. [CrossRef] [PubMed]

fermentation

Article

Effect of Microbial Reinforcement on Polyphenols in the Acetic Acid Fermentation of Shanxi-Aged Vinegar

Peng Du [1], Yingqi Li [1], Chenrui Zhen [1], Jia Song [1,*], Jiayi Hou [1], Jia Gou [1], Xinyue Li [1], Sankuan Xie [2], Jingli Zhou [3], Yufeng Yan [3], Yu Zheng [1] and Min Wang [1,*]

1 State Key Laboratory of Food Nutrition and Safety, Tianjin Engineering Research Center of Microbial Metabolism and Fermentation Process Control, College of Biotechnology, Tianjin University of Science & Technology, Tianjin 300457, China; dp9298@163.com (P.D.); liyingqi9710@163.com (Y.L.); zhenchenrui@mail.tust.edu.cn (C.Z.); houjiayi0618@126.com (J.H.); gj20020424@126.com (J.G.); lixinyue020205@163.com (X.L.); yuzheng@tust.edu.cn (Y.Z.)
2 China National Research Institute of Food & Fermentation Industries Co., Ltd., Beijing 100015, China; xskuan@mail.tust.edu.cn
3 Shanxi Province Key Laboratory of Vinegar Fermentation Science and Engineering, Shanxi Zilin Vinegar Industry Co., Ltd., Taiyuan 030400, China; zhoujingli1110@126.com (J.Z.); yanyf8@163.com (Y.Y.)
* Correspondence: tjsongjia@tust.edu.cn (J.S.); minw@tust.edu.cn (M.W.)

Abstract: Polyphenols are important functional substances produced in the acetic acid fermentation (AAF) of Shanxi aged vinegar (SAV). Previous studies have shown that the metabolic activity of microorganisms is closely related to polyphenol production and accumulation. In this study, microorganisms in the AAF of SAV were analyzed to explore how to increase the polyphenol yield by changing the microorganisms and reveal the potential mechanism of the microbial influence on the polyphenol yield. Macrotranscriptome analysis showed that acetic and lactic acid bacteria dominated the AAF fermentation process and initially increased and decreased. Spearman correlation analysis and verification experiments showed that the co-addition of *Acetobacter pasteurianus* and *Lactobacillus helveticus* promoted the accumulation of polyphenols, and the total polyphenol content increased by 72% after strengthening.

Keywords: Shanxi aged vinegar; polyphenols; acetic acid fermentation; microbial enhancement; *Acetobacter pasteurianus*

1. Introduction

Shanxi aged vinegar (SAV) has a long history of use as a condiment and a popular health food in China [1]. Sorghum, daqu, and bran are used for brewing SAV. SAV production includes the following stages. In semi-solid alcohol fermentation (AF), starch and proteins are hydrolyzed into glucose, ethanol, and other small-molecule compounds under the action of *Daqu* and environmental microorganisms. During the acetic acid fermentation (AAF), organic acids, such as acetic and lactic acid, are produced. In the Smoking *pei* (SP) stage (solid form), the mash after acetic acid fermentation is continuously heated and stirred daily. During the aging stage (in the liquid form), the vinegar is exposed to sunlight. Usually, AF, AAF, and SP take more than 20 days, while the aging stage may take decades or even centuries [2]. Due to its unique brewing process, SAV can produce many flavor substances and other functional compounds during fermentation. Polyphenols (PPs) are one of the main bioactive substances in plants and can be classified as flavonoids, non-flavonoids, and phenolic acids [3,4]. SAV is rich in PPs such as gallic acid, tannins, ferulic acid, catechin, quercetin, anthocyanins, p-coumaric acid, and resveratrol [3]. Polyphenols have a variety of biological activities and can prevent diseases such as cardiovascular diseases, liver injury, nerve degenerative diseases, cancer, and hyperlipidemia [5–9]. Various methods exist to obtain PPs from plant-based foods such as vegetables and grains [10,11]. They can also be produced via chemical synthesis or biotransformation [12]. In addition, PPs can be biosynthesized de novo using microorganisms, such as fungi and bacteria, and microbial metabolites as precursors or catalysts [13].

Many types of microorganisms are involved in the AAF of SAV, including yeast, mold, acetic acid bacteria, lactic acid bacteria, and spore bacteria. During SAV fermentation, the species and abundances of microorganisms change constantly. Under the interaction and influence of these microorganisms, various compounds are metabolized and produced during fermentation [14]. The composition of microbial communities affects the composition and concentration of compounds in the AAF process and vice versa, as compounds also have essential effects on the growth and metabolism of microorganisms. Numerous studies have shown that various volatile and non-volatile compounds in SAV are produced by microbial activity [15,16]. By constructing high-yield microbial chassis cells, de novo microbial fermentation using a cheap carbon source as a substrate can effectively improve the efficiency of PPs synthesis [17]. Polyphenols are fermented using microbially modified synthases to yield high polyphenols [18]. However, our understanding of the distribution of PPs in the AAF process is limited, and research on the relationships between microorganisms and PPs in complex systems is scarce.

This study studied changes in the PPs in the AAF of the SAV phase using gas chromatography-mass spectrometry (GC-MS). Correlations between PPs, environmental factors, and microorganisms were analyzed and verified. There are few studies on SAV PPs, especially regarding the effect of microbial enhancement on key PPs in the AAF process. Therefore, this research aimed to study the main PPs and the effects of microbial enhancement on the AAF process using GC-MS to increase the accumulation of PPs in the SAV AAF process. The results of this study provide significant guidance for the production, operation, and quality optimization of SAV.

2. Materials and Methods

2.1. Microculture

2.1.1. Strains

Regarding the strains used in the experiment, *Limosilactobacillus fermentum* SMR-360 (*Lac fer*) was identified as SMR-3601, SMR-3602, and SMR-3603, *Lactiplantibacillus plantarum* SMR-360 (*Lac pla*) was identified as SMR-3604 and SMR-3605, and *Lactobacillus helveticus* SMR-360 (*Lac hel*) was identified as SMR-3606, SMR-3607, SMR-3608, SMR-3609, SMR-3610, SMR-3611, SMR-3612, SMR-3613, SMR-3614, and SMR-3615. All strains were screened and stored at the Microbiology Laboratory of the School of Biological Engineering, Tianjin University of Science and Technology.

2.1.2. Media

The lactic acid bacteria fermentation medium (MRS, w/v) consisted of glucose 2%, peptone 1%, beef extract 1%, yeast extract 0.5%, anhydrous sodium acetate 0.5%, Tween 80 0.1% (v/v), ammonium citrate 0.2%, dipotassium hydrogen phosphate 0.2%, magnesium sulfate 0.058%, and manganese sulfate 0.025%, with a pH of 6.2−6.8.

The acetic acid bacteria fermentation medium (GP, w/v) consisted of glucose 2%, peptone 2% and ethanol 5.0% (v/v). All chemical reagents used in this section were obtained from Beijing Boxbio Science & Technology Co., Ltd. (Beijing, China).

2.1.3. Strain Culture

The strains were inoculated in the fermentation liquid medium. Lactobacilli were incubated in MRS medium at 37 °C for 24 h, and *Acetobacter* was incubated in GP medium at 30 °C with shaking at 180 rpm for 24 h. When the cell concentration reached 1×10^8 CFU/mL, it was used for subsequent experiments.

2.2. Sample Collection

Samples were collected from Shanxi Aged Vinegar Group Co., Ltd. In the AAF process, a uniform sample was obtained using methods described in the literature [19]. Vinegar-fermented grain samples were collected on the 1st, 3rd, 5th, 7th, and 9th day of AAF of SAV. Samples of acetamide were taken from a distance of approximately 30 cm from the surface of the sauce. All samples were collected using a five-point sampling method, with three copies of samples collected in parallel at each time point (i.e., one from three different tanks with the same brewing time), and each sample of the vinegar-fermented grains was approximately 200 g. Samples from the five-time points were mixed and placed in sterile Ziplock bags. The collected samples were immediately refrigerated in an icebox and stored in a refrigerator at $-80\ ^\circ$C for subsequent use.

2.3. Analysis of PPs Contents and Composition during AAF

The PPs contents and composition analyses were carried out in the laboratory before microbial enhancement and at the Shanxi Aged Vinegar Group Co., Ltd. after microbial enhancement.

Before the assay, the solid vinegar paste sample was pretreated to obtain the extract. The solid sample (5 g) was added to 45 mL of water, shaken well for 3 h at room temperature, and centrifuged at 5000 rpm for 10 min (centrifuge TG16-WS, Xiangyi Co., Ltd., Changsha, China). Supernatants were collected for the analysis. The optimized forinophenol method was used to determine the PPs contents of the acetate samples.

The composition of PPs in the AAF phase samples was determined using GC-MS. First, 10 mL of the supernatant obtained in Section 2.2 was collected, and the PPs were extracted using the method described above [20]. Then, 2,4,5-trihydroxybenzoic acid was used as the internal standard, and 0.002 g was added to 10 mL of the extract during pretreatment. One milliliter of bis (trimethylsilyl) trifluoroacetamide (Supelco, Bellefonte, PA, USA) and 1% trimethylchlorosilane (Aladdin Biotechnology Co., Ltd., Shanghai, China) was added to the polyphenol extract, which was then reacted in a water bath at $70\ ^\circ$C for 3 h (DK-8D, Qiaofeng Co., Ltd., Shanghai, China). Compounds were identified by comparison with the standard's retention times and mass spectrometry results. Their concentrations were calculated by comparing the peak areas of the internal standard compounds.

2.4. Macrotranscriptome Sequencing of Microbial Communities during AAF

2.4.1. Extraction of RNA

Total RNA was extracted using the Total RNA Extraction Kit (Mobio, document number: 12866-25). Ribosomal RNA was removed using the Ribo-Zero Magnetic Gold Kit (Epidemiology, article number: MRZE724). The TruSeq Chain mRNA LT Sample Preparation Kit (Illumina, San Diego, CA, USA) was used for reverse transcription and library construction.

2.4.2. Library Quality Checks and Sorting

One microliter of the library was accurately obtained, and 2100 quality checks were performed on an Agilent Bioanalyzer using the Agilent Highly Sensitive Kit (Agilent Technologies Inc., Santa Clara, CA, USA). A qualified library should have a single peak and no adapters. Libraries were quantified on a Promega QuantiFluor using the Quant-it PicoGreen double-stranded DNA Analysis Kit. For qualified libraries, we performed 2×150 bp double-ended sequencing on a NextSeq machine (Personal Bio, Shanghai, China) using the NextSeq 500 High-Yield Kit (300 cycles). Quality control of the raw data was performed by sequencing using FASTQC (http://www.bioinformatics.babraham.ac.uk/projects/fastqc/, accessed on 23 June 2023). CutAdapt (v1.2.1) [21] was used to screen and filter raw data from the sequencing machines. SortMeRNA (http://ioinfo.lifl.fr/rna/sortmerna/, accessed on 23 June 2023) [22,23] was used to eliminate rRNA, and Trinity (http://trinityrnaseq.github.io/, accessed on 23 June 2023) [24] was used for sequence assembly and splicing. The redundancies with a similarity of 0.95 and minimum coverage of 0.9 were merged and removed using a high-consistency tolerance cluster database, and the longest sequence was used as the representative sequence of UniGene to construct the UniGene set. The databases included NR, GO, KEGG, eggnog, CAXY and Swiss-Prot for Unigene feature annotation [23]. The metatranscriptome sequencing data used in this study were uploaded to the NCBI database under the registration numbers SRX7581352-SRX7581356.

2.5. Effects of Microorganisms on PPs

In the single-factor experiment carried out in the laboratory, the exogenous addition of microorganisms was used to control the microbial composition of vinegar-fermented grains. Exogenous microorganisms were added to a 500 mL conical flask containing 300 g of vinegar-fermented grains. The vinegar-fermented grains in the control group (Control) did not contain exogenous microorganisms. *Lactobacillus helveticus*, *Lactiplantibacillus plantarum*, and *Acetobacter pasteurianus* were added to the vinegar-fermented grains in the experimental groups. The addition amount and time were the same as those in the control group, and superior strains were selected. The additional amount of the superior strains was set to 10^8 CFU/100 g of vinegar-fermented grains. On this basis, a suitable fermentation time was selected to evaluate the additional effect of microorganisms.

2.6. Determination of Physical and Chemical Indices during AAF

The pretreated samples were used to determine the physical and chemical indices. The pH was measured using a pH meter S20P (Mettler Toledo Company, Shanghai, China). The total acid content was determined by titration with a standard solution (0.1 mol/L sodium hydroxide) with phenol red as an indicator. The amino nitrogen content was determined using the ninhydrin method of the European winemaking convention. Changes in the pH, total acid content, and amino nitrogen content during AAF were detected using the following methods [25]. The total reduced sugar content was determined using a film test. A fully automatic Sykam amino acid analyzer S433D (Sykam Co., Ltd., Eresing, Germany) was employed for the qualitative and quantitative detection of amino acids in accordance with the manufacturer's instructions. The temperature at the sampling point was measured using a thermometer. The reduced sugar content of the sample was determined using high-performance liquid chromatography (Agilent 1260, Agilent Technologies Inc., Santa Clara, CA, USA) [26].

2.7. Statistical Analysis

R software (version 3.6.3)was used to draw a bar chart of the composition of the dominant species in each sample (the species with the top 20 overall expressions) at each classification level.

Based on the species-level composition spectrum of each sample annotated in the database, R software was used to calculate the number of common groups, and the number of common and unique species in each sample was visually presented using a Venn diagram (https://en.wikipedia.org/wiki/Venn_diagram, accessed on 23 June 2023).

The data for each sample were measured at least three times, and the experimental data were expressed as the mean ± standard deviation (mean ± SD). Spearman's correlations were analyzed using SPSS software (19.0). The correlations between microorganisms, environmental factors, and PPs were analyzed, and a heat map was drawn using Multiple Experiment Viewer software (version 4.9.0).

3. Results

3.1. Correlation Analysis between PPs and Microorganisms during AAF of SAV

3.1.1. Change in the PPs Contents during AAF

The total ion strength chromatogram is shown in Figure 1B,C. From the 1st to 9th day of AAF, the number of substances detected was 509 ± 77, 580 ± 104, 541 ± 53, 546 ± 49, and 483 ± 11, respectively (Figure 1B,C). The number of PPs species gradually increased from seven on the 1st day to nine on the 7th day and then decreased to seven on the 9th day (Figure 1A,D). The identity of the compounds was determined (Table 1). By searching the NIST20 database, 10 types of PPs were found in the AAF of SAV: (1) p-coumaric acid, (2) 4-hydroxy-3-methoxyphenylethylene Glycol, (3) vanillic acid, (4) ferulic acid, (5) L-epicatechin, (6) gallic acid, (7) 3-(3-hydroxy-4-methoxyphenyl)propanoic acid, (8) caffeic acid, (9) (E)-3-(3-hydroxyphenyl)acrylic acid ethyl ester, and (10) vanillin. The type and contents of PPs in the AAF samples differed with time. The PPs types gradually changed from the 1st to the 9th day. Polyphenols 1–4 were detected throughout the AAF process, while PP6 and PP10 were observed in the middle and later AAF stages, respectively. Polyphenol 8 and PP9 appeared on days 1 and 5 and 3 and 7 of the AAF process, respectively (Table 1). The PPs contents changed during AAF. The PP6 and PP10 contents increased gradually with increasing AAF time, whereas the PP1 contents first decreased and then increased. The PP2, 3, 4, 5, 7, 8, and 9 contents showed no apparent trends; however, the PP2, 3, and 5 contents at the end of AAF were higher than at the beginning.

During the AAF process, the total PPs contents gradually increased, reaching a maximum of 47.04 ± 2.74 mg/g of vinegar-fermented grains on the 9th day. Compared with the first day of AAF, the total PPs contents increased by 37.42% (Figure 1A).

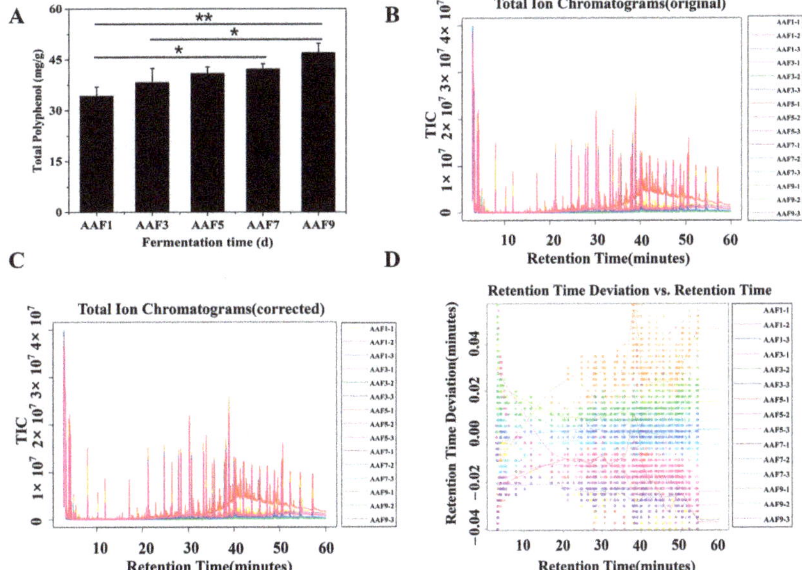

Figure 1. Analysis of polyphenol changes in the acetic acid fermentation (AAF) of Shanxi aged vinegar (SAV). (**A**): Change in the polyphenol contents during AAF; (**B–D**): gas chromatography–mass spectroscopy total ion diagram during the AAF of SAV; * $p < 0.05$; ** $p < 0.01$. TIC, total ion chromatograms.

Table 1. Distribution of polyphenols in the acetic acid fermentation (AAF) of Shanxi aged vinegar (SAV).

Compound Identification	CAS No.	Relative Content (%)				
		AAF 1d	**AAF 3d**	**AAF 5d**	**AAF 7d**	**AAF 9d**
p-Coumaric acid	501-98-4	0.102 ± 0.068 [ab]	0.053 ± 0.037 [a]	0.07 ± 0.06 [ab]	0.127 ± 0.053 [ab]	0.193 ± 0.017 [b]
4-Hydroxy-3-methoxyphenylethylene Glycol	534-82-7	0.156 ± 0.156 [a]	0.074 ± 0.106 [a]	0.013 ± 0.008 [a]	0.067 ± 0.153 [a]	0.0425 ± 0.047 [a]
Vanillic acid	121-34-6	0.127 ± 0.213 [a]	0.013 ± 0.017 [a]	0.05 ± 0.02 [a]	0.037 ± 0.017 [a]	0.047 ± 0.028 [a]
Ferulic Acid	1135-24-6	0.157 ± 0.077 [a]	0.077 ± 0.073 [a]	0.3 ± 0.08 [a]	0.263 ± 0.147 [a]	0.307 ± 0.083 [a]
L-Epicatechin	490-46-0	0.301 ± 0.419 [a]	0	0.075 ± 0.015 [a]	0.11 ± 0.05 [a]	0.053 ± 0.017 [a]
Gallic acid	149-91-7	0	1.727 ± 0.3347 [a]	2.59 ± 0.69 [a]	2.383 ± 0.483 [a]	2.865 ± 2.775 [a]
3-(3-Hydroxy-4-methoxyphenyl)propanoic acid	1135-15-5	0.163 ± 0.137 [a]	0.053 ± 0.037 [a]	0	0.107 ± 0.067 [a]	0
Caffeic acid	331-39-5	0.063 ± 0.097 [a]	0	0.05 ± 0.05 [a]	0	0
2-Propenoic acid, 3-(3-hydroxyphenyl)-, ethyl ester, (2E)-	96251-92-2	0	0.018 ± 0.022 [a]	0	0.14 ± 0.11 [b]	0
Vanillin	121-33-5	0	0	0	0.007 ± 0.007 [a]	0.0175 ± 0.017 [a]

Data are expressed as means ± standard deviations; different letters indicate significant differences among the groups ($p < 0.05$).

3.1.2. Changes in the Physical and Chemical Indices during AAF

Temperature, pH, acidity, reducing sugar, amino nitrogen, and alcohol contents were the key physical and chemical indices monitored during AAF. Figure 2 summarizes the analysis of the changes in the physical and chemical characteristics of AAF after the intensive experiments. The temperature increased at first, then decreased slightly, reaching a maximum of 44.73 °C on the third day (Figure 2A). The amino nitrogen and total acid contents increased gradually and reached a maximum on the 9th day, reaching the maximum values of 0.06 ± 0.012 and 3.94 ± 0.17 g/100 g of vinegar-fermented grains, respectively (Figure 2C,D). The reduced sugar content first increased and then decreased and reached the maximum on the 7th day, which was 2.45 ± 0.09 g/100 g of vinegar-fermented grains (Figure 2B). The alcohol content gradually decreased during the AAF process (Figure 2E). The pH fluctuated slightly in the range of 3.7–4.0, but there was no apparent trend (Figure 2F).

Sixteen amino acids were detected in the AAF culture samples: glutamic acid, glycine, alanine, valine, methionine, isoleucine, leucine, phenylalanine, threonine, aspartic acid, serine, proline, tyrosine, histidine, lysine, and arginine. Regarding the total amino acid concentration, there were significant differences among the different AAF stages. Changes in the contents of various amino acids during AAF are shown in the Supplementary Table S1, and with the exception of threonine, proline, histidine, and lysine, most increased after AAF. Glutamic acid, arginine, and aspartic acid levels significantly increased. From the 1st to the 9th day of AAF, the alanine content decreased slightly and then increased gradually. The tyrosine content increased rapidly from the 1st to the 3rd day, and then its increase slowed.

Six reducing sugars (Supplementary Table S2) were detected during AAF: xylose, fructose, mannose, glucose, maltose, and trehalose. The xylose and mannose contents increased gradually, while the fructose content first increased and then decreased before increasing again, while the trehalose content first increased and then decreased. Changes in the glucose and maltose contents were not obvious.

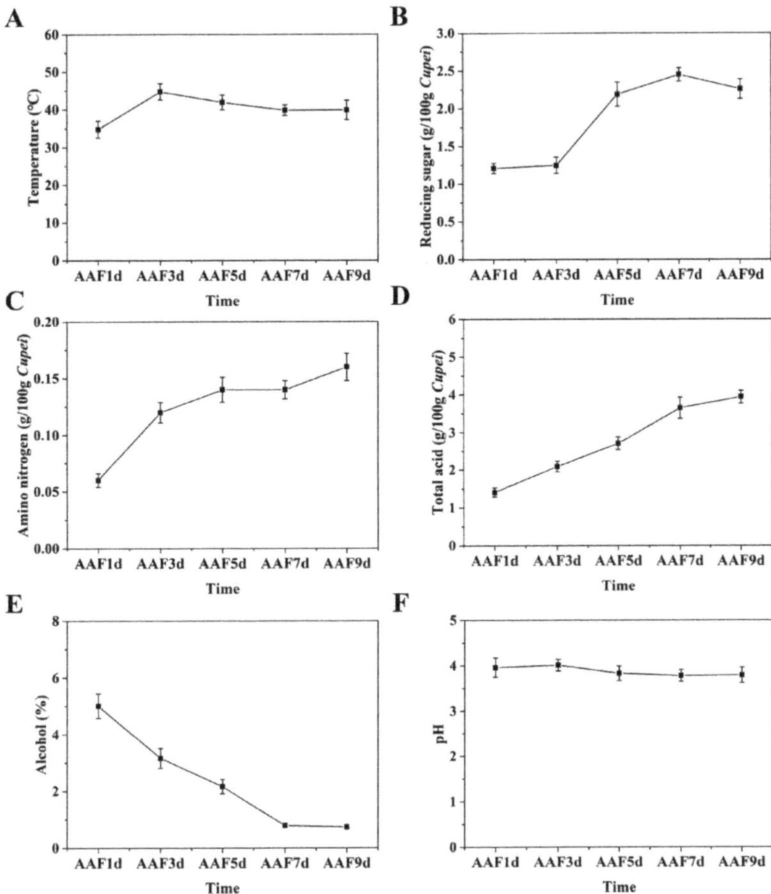

Figure 2. Changes in physicochemical indexes during the acetic acid fermentation (AAF) of Shanxi aged vinegar (SAV). (**A–F**): Changes in the temperature, reducing sugar, amino nitrogen, total acid, alcohol, and pH, respectively, over time. Values are means ± standard deviations.

3.1.3. Succession Law of the Microbial Community during AAF

During the AAF process, microorganisms play an essential role in accumulating PPs. Changes in microbial succession during AAF were analyzed (Figure 3). The number of microbial species constantly changed throughout the AAF process, with 133 species in total and 255, 22, 375, 28, and 62 species unique to AAF1d, 3d, 5d, 7d and 9d, respectively. Differences in the microbial species at different fermentation times were caused by the succession of open-fermentation microorganisms. The sum of the abundance of the top 20 most abundant species accounted for more than 98% of the total microbial abundance. Among these species, the abundance ratios of *Acetobacter bustiensis* in AAF1d-9d samples were: 0.49% (AAF1d), 0.04% (AAF3d), 12.23% (AAF5d), 13.02% (AAF7d), and 18.74% (AAF9d). The abundance ratios of *Lactobacillus* acid fast in AAF1d-9d samples were: 30.74% (AAF1d), 94.55% (AAF3d), 36.22% (AAF5d), 26.42% (AAF7d), and 16.49% (AAF9d), showing an initial increase followed by a decrease. *Lactobacillus acetotolerans* and *Acetobacter pasteurianus* were the dominant flora in the AAF process. Microorganisms with higher abundances included *Lactobacillus helveticus*, Unclassified *Lactobacillales*, and Unclassified *Lactobacillus*.

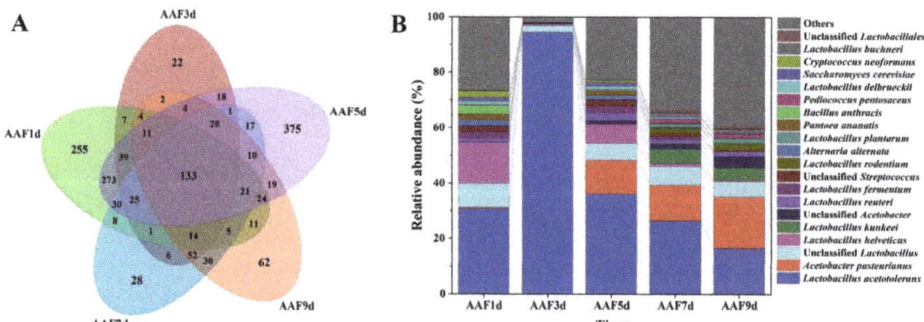

Figure 3. Microbial succession during the acetic acid fermentation (AAF) of Shanxi aged vinegar (SAV). (**A**): Microbial diversity during AAF in SAV; (**B**): changes in microbial abundance during AAF in SAV.

3.1.4. Correlation Analysis between PPs and Microorganisms during AAF

Microorganisms are essential factors affecting PPs in AAF. Using microorganisms as key environmental factors, Spearman's correlation analysis was performed, and correlations between microorganisms and PPs were analyzed (Figure 4). There were significant correlations between many types of microorganisms and PPs. Among these, *Lactobacillus helveticus* and *Cryptococcus neoformans* had significant positive correlations with PP5, PP3, and PP8, *Bacillus anthracis* had a significant positive correlation with PP5 and PP3, and a negative correlation with PP6, acetic acid bacteria had a significant positive correlation with PP4, and *Lactiplantibacillus plantarum*, *Lactobacillus rodentium*, and *Lentilactobacillus buchneri* had a significant positive correlation with PP1 and PP10. Therefore, *Lactobacilli* and acetate bacteria may be essential microorganisms affecting the formation of PPs by altering the production of PP1, 3, 4, 5, 8, and 10. We speculated that these may be the main microorganisms affecting PPs (Figure 4).

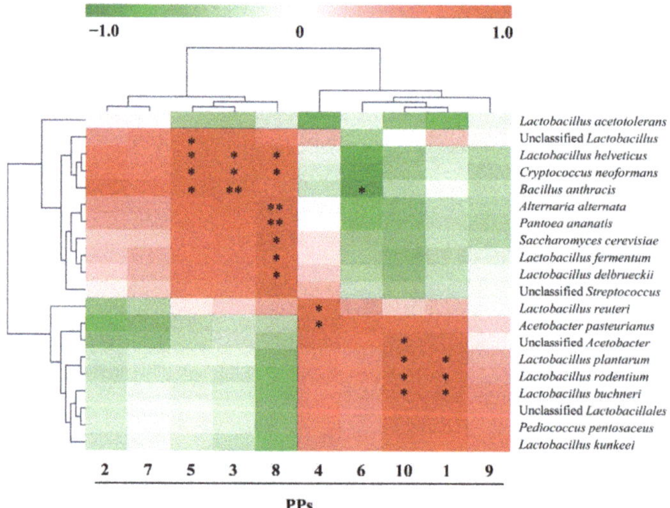

Figure 4. Spearman correlation analysis heat map between polyphenols (PPs) and microorganisms during the acetic acid fermentation (AAF) of Shanxi aged vinegar (SAV). Red and green indicate positive and negative correlations, respectively, and the shade suggests the size of the correlation coefficient; * $p < 0.05$; ** $p < 0.01$. Please refer to Section 3.1.1. for the PPs associated with each PP number.

3.2. Single Factor Verification of Microorganisms Affecting PPs Contents

The above analysis showed that microorganisms may be an essential factor affecting the total PPs content. The effect of microorganisms on the total PPs content was verified by a microbial single-factor screening experiment (Figure 5).

First, the strains producing high levels of PPs were screened experimentally, showing that the total PPs content was higher when *adding Lactobacillus helveticus*. Compared with the control group, exogenous microorganism addition promoted the accumulation of total PPs. On this basis, the microbial addition was screened, and the results showed that with the increase in microbial addition, the total PPs content gradually increased. The total PPs content was highest when *Lactobacillus helveticus* was added at 10^9 CFU/100 g of vinegar-fermented grains or when *Acetobacter pasteurianus* was added at 10^7 CFU/100 g of vinegar-fermented grains. Finally, the addition time was screened, and the total PPs content reached a maximum on the 3rd day of the AAF process. These results showed that adding exogenous microorganisms and increasing the number of microorganisms can increase the total PPs content in the AAF process. In summary, on the 3rd day of AAF, a mixture of *Lactobacillus helveticus* (10^9 CFU/100 g of vinegar-fermented grains) and *Acetobacter pasteurianus* (10^7 CFU/100 g of vinegar-fermented grains) maximized the accumulation of total PPs.

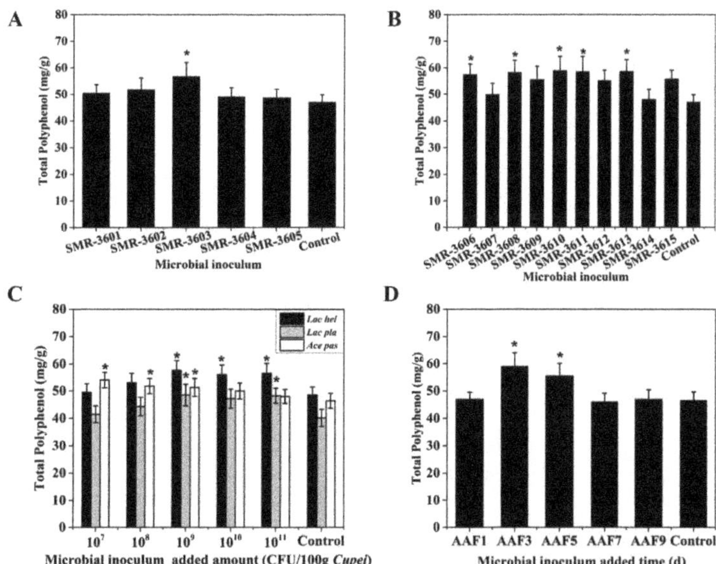

Figure 5. Screening microorganisms related to PPs during the acetic acid fermentation (AAF) of Shanxi aged vinegar (SAV). (**A**,**B**): Screening of microbial species; (**C**): screening of microbial addition amount; (**D**): screening of microbial addition time; * $p < 0.05$. *Lac hel, Lactobacillus helveticus*; *Lac pla, Lactiplantibacillus plantarum*; *Ace pas, Acetobacter pasteurianus*.

3.3. In Situ Regulation of PPs during AAF

Based on the effects of microbial factors on PPs accumulation, exogenous microbial enhancement experiments were performed. A schematic diagram of the GC × MS total ion current and a comparison of the PPs content before and after enhancement are shown in Table 2 and Figure 5. The results showed that the total number of detected substances was higher in the experimental group than in the control group.

After the intensive experiment, the PPs type and contents, reducing sugars and amino acid contents, and utilization rate of raw materials changed significantly during the AAF process (Figure 6, Supplementary Table S1). Twelve types of PPs were detected in the experimental group, with (11) isoferulic acid and (12) salicylic acid detected in addition to the 10 PPs detected in the control group (Table 1). The distribution of PPs in the AAF process was analyzed after an intensive experiment. Like the control group, PP1, 4, and 6 were the main compounds in the experimental group. Compared with the control group, the contents of five PPs in the test group increased: PP1, 2, 4, 6, and 10. Among these, the largest increases were for PP4, 6, and 10. Therefore, the accumulation of PPs mainly increased through the promoted production of PP4, 6, 10, 11, and 12.

Table 2. Comparison of the polyphenol contents during the acetic acid fermentation (AAF) of Shanxi aged vinegar (SAV) before and after the enhancement experiment.

	Compound Identification	CAS No.	Relative Content (%)	
			Before Optimization	**After Optimization**
1	p-Coumaric acid	501-98-4	0.193 ± 0.017	0.256 ± 0.019
2	4-Hydroxy-3-methoxyphenylethylene Glycol	534-82-7	0.0425 ± 0.0047	0.07 ± 0.005
3	Vanillic acid	121-34-6	0.047 ± 0.028	0.052 ± 0.012
4	Ferulic Acid	1135-24-6	0.307 ± 0.083	0.412 ± 0.075 *
5	L-Epicatechin	490-46-0	0.053 ± 0.017	0.043 ± 0.016
6	Gallic acid	149-91-7	2.865 ± 0.175	4.1 ± 0.21 **
7	3-(3-Hydroxy-4-methoxyphenyl)propanoic acid	1135-15-5	0	0
8	Caffeic acid	331-39-5	0	0
9	2-Propenoic acid, 3-(3-hydroxyphenyl)-, ethyl ester, (2E)-	96251-92-2	0	0
10	Vanillin	121-33-5	0.0175 ± 0.007	0.23 ± 0.009 ***
11	Isoferulic acid	1135-16-6	0	0.5 ± 0.05 ***
12	Salicylic acid	69-72-7	0	0.4 ± 0.05 ***

The data are expressed as means ± standard deviations, and the different symbols indicate significant differences between groups (* $p < 0.05$, ** $p < 0.01$, *** $p < 0.001$).

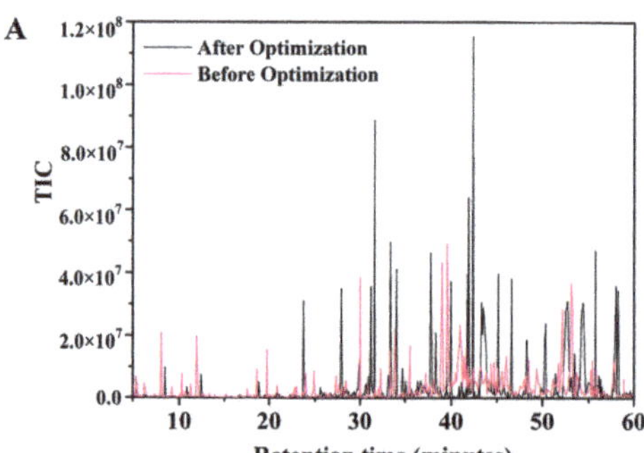

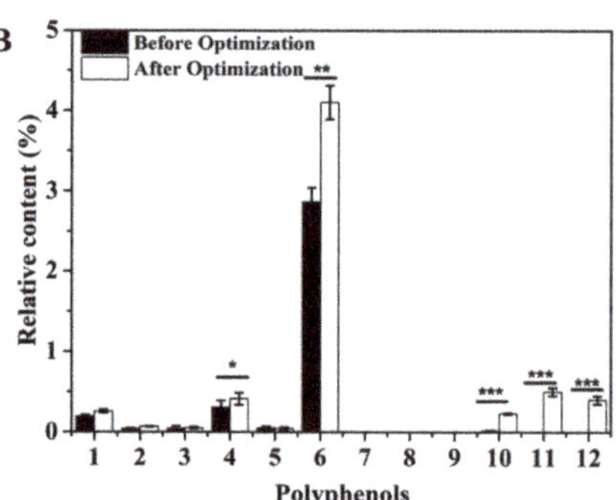

Figure 6. Gas chromatography–mass spectroscopy total ion diagram and polyphenols content during the acetic acid fermentation (AAF) of Shanxi aged vinegar (SAV) before and after the enhancement experiment. (**A**): Gas chromatography–mass spectroscopy total ion diagram. (**B**): Polyphenol contents. * $p < 0.05$, ** $p < 0.01$, *** $p < 0.001$. TIC, total ion chromatogram. Please refer to Sections 3.1.1 and 3.3. for the polyphenols associated with each number.

4. Discussion

Solid-state AAF is the primary process involved in SAV production and brewing. The SAV samples were analyzed during AAF. SAV is rich in amino acids and functional components, such as PPs [19]. The SAV brewing process involves natural fermentation, which leads to the growth of various microorganisms. These microorganisms have an important influence on vinegar quality. Polyphenols accumulate during the AAF process and are affected by various environmental factors, such as microorganisms. In this study, environmental factors affecting PPs, including microorganisms, were analyzed, providing new research data for increasing the yield of PPs.

Increased microbial diversity or abundance can enrich enzyme lineages, which may facilitate the synthesis of functional substances, such as PP4 and PP10 [27,28]. Therefore, adding microorganisms may be an effective means of increasing the polyphenol content. As our results of in situ regulation showed, microorganisms can increase the total PPs content, with the PP4 and PP10 contents increasing by 95.54% and 100%, respectively. PP8 disappeared at the end of the AAF. Simultaneously, the Spearman correlation analysis also showed that PP8 was significantly related to *Lactobacillus helveticus*, *Cryptococcus neoformans*, and other microorganisms. Previous studies have shown that the addition of microorganisms can enhance the activity of caffeic acid 3-O-methyltransferase in catalytic reactions, increase the methylation of PP8, and promote the transformation of PP8 to PP4, resulting in the accumulation of PP4. In addition, microorganisms can produce decarboxylases to decarboxylate PP4 into PP10 [29], which may be an essential reason for the increased accumulation of PP10 (and total PPs).

Furthermore, we speculated that microorganisms may indirectly promote the synthesis of PPs in the AAF process by affecting environmental, physical, and chemical conditions, such as the contents of reducing sugars and amino acids. As shown in Figure 6, the PP4, PP6, and PP10 contents increased significantly. Previous studies have shown that amino acids, particularly aromatic amino acids, are essential precursors for the synthesis of PPs. Therefore, the addition of exogenous microorganisms can promote the accumulation of PPs precursors. For example, free amino acids can be produced by degradation by lactic acid bacteria [30,31], such as tyrosine, which acts as a precursor that promotes PP6 [32]. Recent studies have shown that microorganisms promote PPs production by participating in free amino acid biosynthesis during AAF.

Moreover, the addition of exogenous microorganisms indirectly affects synthetic reactions by promoting the transformation of reducing sugars to PPs. For example, mannose can be a precursor to form PP4 under microbial catalysis [33]. Additionally, glucose and phenylalanine can be precursors for PP10 biosynthesis [34,35].

Some studies have shown that temperature increases in the early stage of AAF contribute to cell wall destruction, which promotes the release of intracellular or bound PPs and increases the content of PPs such as PP4. Simultaneously, high temperatures promote the transformation of PP4 into its corresponding isomers, which may also cause PP11 formation. In addition, some amino acids are non-neutral, which may cause changes in environmental factors such as pH, affecting the release of PPs. The Spearman correlation between PPs and environmental, physical, and chemical factors also showed that amino acids and reducing sugars were significantly correlated with the yield of PPs (PP4, 6, and 10; Supplementary Figure S1).

Macrotranscriptomes were used to analyze the distribution of microorganisms and their effects on vinegar community succession during AAF. As shown in Figure 2, bacterial diversity increased at the initial fermentation stage and decreased gradually in the later stages, consistent with previous community studies. The Spearman correlation analysis between PPs and microorganisms also showed that acetic acid and lactic acid bacteria were significantly correlated with PPs accumulation. Although PP8 was significantly correlated with various microorganisms, it has little influence on vinegar's nutritional and health functions; therefore, it was not considered. In addition, *Lactobacillus rodentium*, *Lactobacillus buchneri*, and other strains were not detected in the fermented grains of SAV vinegar. Therefore, *Lactobacillus helveticus*, *Lactiplantibacillus plantarum* and *Acetobacter pasteurianus* were selected as the dominant strains for the strengthening experiments to improve the PPs yield. The results showed a strong correlation between the microorganisms and PPs during AAF. The metabolic activities of microorganisms play key roles during the AAF stage. Microbial diversity and flavor formation in SAV during fermentation is very different from those of other vinegars [36]. SAV has a unique microbial community structure, phenolic acid content, and physicochemical properties [37].

This study used Spearman correlation analysis and single-factor experiments to screen and verify microbial factors. Based on the effects of microorganisms on PPs, regulatory strategies to promote the accumulation of PPs are proposed from the perspective of microbial enhancement. First, the distribution of the PPs was analyzed. Ten PPs were detected during the AAF of SAV. The effect of microorganisms on PPs yield was observed. By adding exogenous microorganisms, it was verified that a mixture of lactic and acetic acid bacteria could increase the accumulation of PPs. Although microbial factors promote the accumulation of PPs, the mechanism of action of PPs in AAF requires further study. In the future, we will simulate AAF conditions, build a reaction model, and comprehensively study the formation mechanism of PPs in the AAF process.

5. Conclusions

Microorganisms are essential factors that affect the accumulation of PPs during AAF. Metatranscriptomic analysis revealed that acetic and lactic acid bacteria were the dominant bacteria affecting the PPs yield and participating in regulating the AAF process. During fermentation, acetic acid bacteria gradually increase, whereas lactic acid bacteria increase and decrease. Validation experiments with the exogenous addition of *Acetobacter pasteurianus* and *Lactobacillus helveticus* also showed that the total PPs content increased by 72% after their addition. Therefore, in the SAV AAF process, the production of PPs can be increased by the exogenous addition of a mixture of *Acetobacter pasteurianus* and *Lactobacillus helveticus*.

Supplementary Materials: The following supporting information can be downloaded at: https://www.mdpi.com/article/10.3390/fermentation9080756/s1, Figure S1: Spearman correlation between physicochemical and polyphenols. Red and green indicate positive correlation and negative correlation, respectively, and the shadow of color indicates the correlation coefficient between environmental factors and polyphenols; Table S1: Changes of amino acid contents in the AAF of SAV; Table S2: Changes of reducing sugar content in the AAF of SAV.

Author Contributions: Conceptualization, P.D. and Y.L.; methodology, P.D. and C.Z.; software, J.S.; validation, J.H., Y.Y. and J.G.; formal analysis, X.L.; investigation, S.X.; resources, J.Z.; data curation, Y.Z.; writing—original draft preparation, P.D. and Y.L.; writing—review and editing, P.D. and J.S.; visualization, S.X.; supervision, P.D.; project administration, P.D.; funding acquisition, J.S. and M.W. All authors have read and agreed to the published version of the manuscript.

Funding: This research was funded by the National Natural Science Foundation of China (32072203), the Innovation Fund of Haihe Laboratory of Synthetic Biology (22HHSWSS00013), Shanxi Provincial Department of Science and Technology (202204010931002, 2022ZDYF124), Key Research and Development Program of Shanxi, and Key Research and Development Program of Ningxia (2022BBF02010), Project of Tianjin Science and Technology Plan (21ZYJDJC00030).

Institutional Review Board Statement: Not applicable.

Informed Consent Statement: Not applicable.

Data Availability Statement: The data presented in this study are available in article and supplementary.

Acknowledgments: Thanks to Beijing Boxbio Science & Technology Co., Ltd. for providing chemical reagents and technical support.

Conflicts of Interest: The authors declare no conflict of interest.

References

1. Nie, Z.; Zheng, Y.; Xie, S.; Zhang, X.; Song, J.; Xia, M.; Wang, M. Unraveling the correlation between microbiota succession and metabolite changes in traditional Shanxi aged vinegar. *Sci. Rep.* **2017**, *7*, 9240. [CrossRef] [PubMed]
2. Xie, X.; Zheng, Y.; Liu, X.; Cheng, C.; Zhang, X.; Xia, T.; Yu, S.; Wang, M. Antioxidant Activity of Chinese Shanxi Aged Vinegar and Its Correlation with Polyphenols and Flavonoids During the Brewing Process. *J. Food Sci.* **2017**, *82*, 2479–2486. [CrossRef] [PubMed]
3. Del Rio, D.; Rodriguez-Mateos, A.; Spencer, J.P.E.; Tognolini, M.; Borges, G.; Crozier, A. Dietary (Poly)phenolics in Human Health: Structures, Bioavailability, and Evidence of Protective Effects Against Chronic Diseases. *Antioxid. Redox Signal.* **2013**, *18*, 1818–1892. [CrossRef] [PubMed]

4. Zhu, M.J. Dietary polyphenols, gut microbiota, and intestinal epithelial health. In *Nutritional and Therapeutic Interventions for Diabetes and Metabolic Syndrome*; Academic Press: Cambridge, MA, USA, 2018; pp. 295–314.

5. Fan, J.; Zhang, Y.; Chang, X.; Zhang, B.; Jiang, D.; Saito, M.; Li, Z. Antithrombotic and fibrinolytic activities of methanolic extract of aged sorghum vinegar. *J. Agric. Food Chem.* **2009**, *57*, 8683–8687. [CrossRef] [PubMed]

6. Gu, X.; Zhao, H.-L.; Sui, Y.; Guan, J.; Chan, J.C.N.; Tong, P.C.Y. White rice vinegar improves pancreatic beta-cell function and fatty liver in streptozotocin-induced diabetic rats. *Acta Diabetol.* **2010**, *49*, 185–191. [CrossRef] [PubMed]

7. Kahraman, N.K. The effect of vinegar on postprandial glycemia: Does the amount matter? *Acta Endocrinol.* **2011**, *7*, 577–584. [CrossRef]

8. Kondo, S.; Tayama, K.; Tsukamoto, Y.; Ikeda, K.; Yamori, Y. Antihypertensive effects of acetic acid and vinegar on spontaneously hypertensive rats. *Biosci. Biotechnol. Biochem.* **2001**, *65*, 2690–2694. [CrossRef]

9. Sugiyama, A.; Saitoh, M.; Takahara, A.; Satoh, Y.; Hashimoto, K. Acute cardiovascular effects of a new beverage made of wine vinegar and grape juice, assessed using an in vivo rat. *Nutr. Res.* **2003**, *23*, 1291–1296. [CrossRef]

10. Tsao, R. Chemistry and biochemistry of dietary polyphenols. *Nutrients* **2010**, *2*, 1231–1246. [CrossRef]

11. Pastor-Villaescusa, B.; Sanchez Rodriguez, E.; Rangel-Huerta, O.D. Polyphenols in obesity and metabolic syndrome. In *Obesity*; Academic Press: Cambridge, MA, USA, 2018; pp. 213–239. [CrossRef]

12. Rosazza, J.P.N.; Huang, Z.; Dostal, L.; Volm, T.; Rousseau, B. Review: Biocatalytic transformations of ferulic acid: An abundant aromatic natural product. *J. Ind. Microbiol. Biotechnol.* **1995**, *15*, 457–471. [CrossRef]

13. Hansen, E.H.; Møller, B.L.; Kock, G.R.; Bünner, C.M.; Kristensen, C.; Jensen, O.R.; Okkels, F.T.; Olsen, C.E.; Motawia, M.S.; Hansen, J. De novo biosynthesis of vanillin in fission yeast (*Schizosaccharomyces pombe*) and baker's yeast (*Saccharomyces cerevisiae*). *Appl. Environ. Microbiol.* **2009**, *75*, 2765–2774. [CrossRef] [PubMed]

14. Wu, J.J.; Ma, Y.K.; Zhang, F.F.; Chen, F.S. Biodiversity of yeasts, lactic acid bacteria and acetic acid bacteria in the fermentation of "Shanxi aged vinegar", a traditional Chinese vinegar. *Food Microbiol.* **2012**, *30*, 289–297. [CrossRef]

15. Wang, Z.-M.; Lu, Z.-M.; Shi, J.-S.; Xu, Z.-H. Exploring flavour-producing core microbiota in multispecies solid-state fermentation of traditional Chinese vinegar. *Sci. Rep.* **2016**, *6*, 26818. [CrossRef] [PubMed]

16. Wang, Z.-M.; Lu, Z.-M.; Yu, Y.-J.; Li, G.-Q.; Shi, J.-S.; Xu, Z.-H. Batch-to-batch uniformity of bacterial community succession and flavour formation in the fermentation of Zhenjiang aromatic vinegar. *Food Microbiol.* **2015**, *50*, 64–69. [CrossRef] [PubMed]

17. Xu, Y.; Wang, X.; Zhang, C.; Zhou, X.; Xu, X.; Han, L.; Lv, X.; Liu, Y.; Liu, S.; Li, J.; et al. De novo biosynthesis of rubusoside and rebaudiosides in engineered yeasts. *Nat. Commun.* **2022**, *13*, 3040. [CrossRef] [PubMed]

18. Zhou, R.; Bhuiya, M.W.; Cai, X.; Yu, X.; Eilerman, R.G. Methods of Making Vanillin via Microbial Fermentation Utilizing Ferulic Acid Provided by a Modified Caffeic Acid 3-o-Methyltransferase. Francia. Patent WO 106189, 3 July 2014.

19. Sankuan, X.; Cuimei, Z.; Bingqian, F.; Yu, Z.; Menglei, X.; Linna, T.; Jia, S.; Xinyi, Z.; Min, W. Metabolic network of ammonium in cereal vinegar solid-state fermentation and its response to acid stress. *Food Microbiol.* **2020**, *95*, 103684. [CrossRef]

20. Du, P.; Zhou, J.; Zhang, L.; Zhang, J.; Li, N.; Zhao, C.; Tu, L.; Zheng, Y.; Xia, T.; Luo, J.; et al. GC × GC-MS analysis and hypolipidemic effects of polyphenol extracts from Shanxi-aged vinegar in rats under a high-fat diet. *Food Funct.* **2020**, *11*, 7468–7480. [CrossRef]

21. Martin, M. Cutadapt removes adapter sequences from high-throughput sequencing reads. *EMBnet. J.* **2011**, *17*, 10–12. [CrossRef]

22. Kopylova, E.; Noé, L.; Touzet, H. SortMeRNA: Fast and accurate filtering of ribosomal RNAs in metatranscriptomic data. *Bioinformatics* **2012**, *28*, 3211–3217. [CrossRef]

23. Robert, S.; Yan, L.; Robert, E. Identification and removal of ribosomal RNA sequences from metatranscriptomes. *Bioinformatics* **2012**, *28*, 433–435. [CrossRef]

24. Grabherr, M.G.; Haas, B.J.; Yassour, M.; Levin, J.Z.; Thompson, D.A.; Amit, I.; Adiconis, X.; Fan, L.; Raychowdhury, R.; Zeng, Q.D.; et al. Full-length transcriptome assembly from RNA-Seq data without a reference genome. *Nat. Biotechnol.* **2011**, *29*, 644–652. [CrossRef] [PubMed]

25. Nie, Z.; Zheng, Y.; Wang, M.; Han, Y.; Wang, Y.; Luo, J.; Niu, D. Exploring microbial succession and diversity during solid-state fermentation of Tianjin duliu mature vinegar. *Bioresour. Technol.* **2013**, *148*, 325–333. [CrossRef] [PubMed]

26. Song, J.; Jia, Y.-X.; Su, Y.; Zhang, X.-Y.; Tu, L.-N.; Nie, Z.-Q.; Zheng, Y.; Wang, M. Initial Analysis on the Characteristics and Synthesis of Exopolysaccharides from *Sclerotium rolfsii* with Different Sugars as Carbon Sources. *Polymers* **2020**, *12*, 348. [CrossRef] [PubMed]

27. Li, P.; Li, S.; Cheng, L.; Luo, L. Analyzing the relation between the microbial diversity of DaQu and the turbidity spoilage of traditional Chinese vinegar. *Appl. Microbiol. Biotechnol.* **2014**, *98*, 6073–6084. [CrossRef] [PubMed]

28. Xu, W.; Huang, Z.; Zhang, X.; Li, Q.; Lu, Z.; Shi, J.; Xu, Z.; Ma, Y. Monitoring the microbial community during solid-state acetic acid fermentation of Zhenjiang aromatic vinegar. *Food Microbiol.* **2011**, *28*, 1175–1181. [CrossRef] [PubMed]

29. Priefert, H.; Rabenhorst, J.; Steinbüchel, A. Biotechnological production of vanillin. *Appl. Microbiol. Biotechnol.* **2001**, *56*, 296–314. [CrossRef] [PubMed]

30. Fadda, S.; Oliver, G.; Vignolo, G. Protein Degradation by *Lactobacillus plantarum* and *Lactobacillus casei* in a Sausage Model System. *J. Food Sci.* **2002**, *67*, 1179–1183. [CrossRef]

31. Pfeiler, E.A.; Klaenhammer, T.R. The genomics of lactic acid bacteria. *Trends Microbiol.* **2007**, *15*, 546–553. [CrossRef]

32. Rao, B.V.; Ramanjaneyulu, K.; Rambabu, T.; Devi, C.H.B.T.S. Synthesis and antioxidant activity of galloyl tyrosine derivatives from young leaves of *Inga laurina*. *Int. J. Pharma Bio Sci.* **2011**, *2*, 39–44.

33. Yan, Y.; Kohli, A.; Koffas, M.A.G. Biosynthesis of natural flavanones in *Saccharomyces cerevisiae*. *Appl. Environ. Microbiol.* **2005**, *71*, 5610–5613. [CrossRef]
34. Li, K.; Frost, J.W. Synthesis of Vanillin from Glucose. *J. Am. Chem. Soc.* **1998**, *120*, 10545–10546. [CrossRef]
35. Nethaji, J.G.; Birger, L.M. Vanillin–Bioconversion and Bioengineering of the Most Popular Plant Flavor and Its *De Novo* Biosynthesis in the Vanilla Orchid. *Mol. Plant* **2015**, *8*, 40–57. [CrossRef]
36. Lu, Z.-M.; Xu, W.; Yu, N.-H.; Zhou, T.; Li, G.-Q.; Shi, J.-S.; Xu, Z.-H. Recovery of aroma substances from Zhenjiang aromatic vinegar by supercritical fluid extraction. *Int. J. Food Sci. Technol.* **2011**, *46*, 1508–1514. [CrossRef]
37. Chen, J.-C.; Chen, Q.-H.; Guo, Q.; Ruan, S.; Ruan, H.; He, G.-Q.; Gu, Q. Simultaneous determination of acetoin and tetramethylpyrazine in traditional vinegars by HPLC method. *Food Chem.* **2010**, *122*, 1247–1252. [CrossRef]

fermentation

Article

The Effect of Yeast Inoculation Methods on the Metabolite Composition of Sauvignon Blanc Wines

Farhana R. Pinu [1,*], Lily Stuart [2], Taylan Topal [2,3], Abby Albright [2], Damian Martin [2] and Claire Grose [2]

1 Biological Chemistry and Bioactives Group, The New Zealand Institute for Plant and Food Research Limited, Auckland 1025, New Zealand
2 Viticulture and Oenology Group, The New Zealand Institute for Plant and Food Research Limited, Blenheim 7201, New Zealand; lily.stuart@plantandfood.co.nz (L.S.); taylan.topal@plantandfood.co.nz (T.T.); abby.albright@plantandfood.co.nz (A.A.); damian.martin@plantandfood.co.nz (D.M.); claire.grose@plantandfood.co.nz (C.G.)
3 Marine Products Team, The New Zealand Institute for Plant and Food Research Limited, Nelson 7010, New Zealand
* Correspondence: farhana.pinu@plantandfood.co.nz; Tel.: +64-9-926-3565

Abstract: Evidence from the literature suggests that different inoculation strategies using either active dry yeast (ADY) or freshly prepared yeast cultures affect wine yeast performance, thus altering biomass and many primary and secondary metabolites produced during fermentation. Here, we investigated how different inoculation methods changed the fermentation behaviour and metabolism of a commercial wine yeast. Using a commercial Sauvignon blanc (SB) grape juice, fermentation was carried out with two different inoculum preparation protocols using *Saccharomyces cerevisiae* X5: rehydration of commercial ADY and preparation of pre-inoculum in a rich laboratory medium. We also determined the effect of different numbers of yeast cells inoculation (varying from 1×10^6 to 1×10^{12}) and successive inoculation on fermentation and end-product formation. The yeast inoculation method and number of cells significantly affected the fermentation time. Principal component analysis (PCA) using 60 wine metabolites showed a separation pattern between wines produced from the two inoculation methods. Inoculation methods influenced the production of amino acids and different aroma compounds, including ethyl and acetate esters. Varietal thiols, 3-mercaptohexanol (3MH), and 4-methyl-4-mercaptopentan-2-one (4MMP) in the wines were affected by the inoculation methods and numbers of inoculated cells, while little impact was observed on 3-mercaptohexyl acetate (3MHA) production. Pathway analysis using these quantified metabolites allowed us to identify the most significant pathways, most of which were related to central carbon metabolism, particularly metabolic pathways involving nitrogen and sulphur metabolism. Altogether, these results suggest that inoculation method and number of inoculated cells should be considered in the production of different wine styles.

Keywords: grape juice; gas chromatography and mass spectrometry; wine yeast; metabolism; metabolite profiling; aroma compounds

Citation: Pinu, F.R.; Stuart, L.; Topal, T.; Albright, A.; Martin, D.; Grose, C. The Effect of Yeast Inoculation Methods on the Metabolite Composition of Sauvignon Blanc Wines. *Fermentation* 2023, 9, 759. https://doi.org/10.3390/fermentation9080759

Academic Editors: Maria Dimopoulou, Spiros Paramithiotis, Yorgos Kotseridis and Jayanta Kumar Patra

Received: 27 June 2023
Revised: 8 August 2023
Accepted: 11 August 2023
Published: 14 August 2023

1. Introduction

Winemaking is a complex process that involves interactions between different biotic and abiotic factors, including temperature, pH, juice composition, and wine yeasts. Wine yeast plays a vital role in determining the quality and style of wine by directly affecting the production of aroma and other flavour active compounds during fermentation [1,2]. Wine yeast cells are also exposed to multiple stresses (e.g., ethanol toxicity, starvation, heat, and oxidative stress) during winemaking that affect their growth, viability, fermentation capability, and, as a result, end-product (e.g., aroma compounds) formation [3,4]. Moreover, uneven distribution of nutrients and conditions present within the fermentation tank also may affect the growth and metabolism of wine yeasts.

The inoculation of fermentation tanks with a starter culture of wine yeast is a relatively new practice that has only been employed since the early 1970s. Now, many wineries follow this practice mainly because it offers better process control by providing successful completion of fermentation and consistent end-product formation [5]. Although rehydration of commercially available active dry yeasts (ADY) is mostly used by wineries to inoculate the grape must, many researchers are still concerned about the viability and fermentation performances of those wine yeasts [6,7]. The cell membranes of wine yeasts can be damaged during rehydration of ADY, thus causing the loss of cytoplasmic contents [8]. Moreover, inappropriate storage temperatures and the condition of the ADY also affect the lipid composition of the cell membranes, especially by increasing the concentrations of saturated fatty acids, which in turn reduces the viability of the yeast cells [9].

It has been reported previously that ADY cultures are less viable than wet yeast cultures, thus leading to decreased fermentation performance [6]. Therefore, some studies have already been conducted to determine the most appropriate and effective ways of wine yeast rehydration, including addition of different nutrients during the rehydration, and optimization of rehydration conditions [6,10–15]. For instance, Kontkanen, Inglis, Pickering and Reynolds [10] studied the effects of inoculation rate, acclimatisation, and nutrient addition on ice wine fermentation, and they found that stepwise-acclimatized (in grape juice) yeast cells fermented more sugar and thus produced more ethanol than direct inoculum. They also noted that nutrient addition to the yeast inoculum increased the biomass production, decreased fermentation time, and reduced ethanol and acetic acid concentrations in the ice wines. In particular, nutrient addition (e.g., glutathione, vitamins, and unsaturated fatty acids) while rehydrating the ADY has been recommended to increase the vitality and to improve the aroma profiles of wines [7,16,17]. Other researchers have shown that the amount of yeast inoculum significantly affects wine fermentation by changing the production of aroma compounds [18,19]. Therefore, the yeast inoculum preparation step is important and plays a direct role in the fermentation kinetics and in the production of different fermentation end products.

Although ADY is commonly used in commercial wine fermentation in New Zealand, most of the laboratory-scale winemaking here is still performed by using a pre-inoculum of wine yeast grown on an enriched medium (e.g., yeast extract peptone dextrose) under highly aerated conditions [20–22]. It is noteworthy that good aeration of the pre-inoculum provides the molecular oxygen essential for the development of the components of yeast membranes, such as ergosterol [23]. As a result, cell viability increases significantly, and the yeast cells can survive different stresses (e.g., osmotic and ethanol stress) effectively; thus, they might show better fermentation efficiency than the ADY cells. Therefore, the formation of different aroma compounds and other end products (e.g., ethanol, glycerol, and acetic acid) also could be affected by the type and quantity of inoculum used for winemaking. However, more research is needed to confirm this.

To fill the current knowledge gap, our main goal for this project was to determine the effects of two different wine yeast inoculation protocols—rehydration of ADY and inoculum preparation using enriched medium—on fermentation time and other wine parameters. We determined the primary and secondary metabolites in the resulting wines to understand how the inoculum method and quantity impacts the formation of different fermentation end products and thus overall wine composition.

2. Materials and Methods

2.1. Collection and Characterization of Sauvignon Blanc Juice

A commercial, cold-settled Sauvignon blanc (SB) grape juice was collected in the 2017 harvesting year from the Pernod Ricard winery (Marlborough, New Zealand). The juice was brought to room temperature (20 °C) and chemically sterilised by adding 400 µL/L of food grade dimethyl dicarbonate (DMDC, Sigma-Aldrich, Germany), followed by vigorous mixing. In aqueous conditions, DMDC reacts rapidly with proteins to inactivate yeast cells, and any unreacted DMDC then breaks down into methanol and carbon dioxide [24]. DMDC-treated juices were kept at 20 °C overnight prior to starting the fermentation process [25].

The total soluble solids content (°Brix) of the starting fresh juice was determined on a Mettler Toledo RM40 refractometer (Mettler Toledo, Columbus, OH, USA), while a Mettler Toledo T70 autotitrator was used for determining the acidity (pH) and acid content (titratable acidity). Total acid content was measured using an equivalence point titration with aqueous sodium hydroxide (0.1 M) as the titrant and calculated in tartaric acid equivalents (g/L) [26].

Glucose and fructose contents of the fresh juice were quantified by enzymatic assay kit purchased from Megazyme (Wicklow, Ireland) based on the reduction of nicotinamide adenine dinucleotide phosphate (NADP). Samples were appropriately diluted and quantified in duplicate against an eight-point standard curve ($R^2 > 0.98$) [27].

Primary amino acids (PAA) were quantified in duplicate in isoleucine (N) equivalents using the orthophthaldialdehyde (NOPA) method adapted for the plate reader. Quantification was performed using a five-point calibration curve ($R^2 > 0.98$). Ammonium content of the fresh juice was quantified by enzymatic assay kit obtained from Megazyme (Bray, Ireland) by monitoring the deprotonation of NADPH at 340 nm. Juice was appropriately diluted (usually two-fold) and quantified in duplicate using a five-point standard curve ($R^2 > 0.98$). Yeast available nitrogen (YAN) was calculated as the sum of PAA plus ammonium expressed in mg N/L.

Optical density of the juice was determined in duplicate in a UV-transparent 96-well micro-plate at 280, 320, and 420 nm. Absorbance at 280 nm was used to quantify total phenolics against a gallic acid standard curve (five-point, $R^2 > 0.98$) [28].

All spectrophotometric assays were run on a Molecular Devices Spectramax 384 Plus (San Jose, CA, USA) with a 1 cm path length cuvette reference correction.

2.2. Different Inoculum Preparation Methods

Saccharomyces cerevisiae X5 (Laffort, Bordeaux, France) was used for the inoculation of the microferments. Two different inoculation protocols were used in this study as described below and shown in Figure 1.

Treatment 1: Rehydration of active dry yeast (ADY) cells using winery protocol

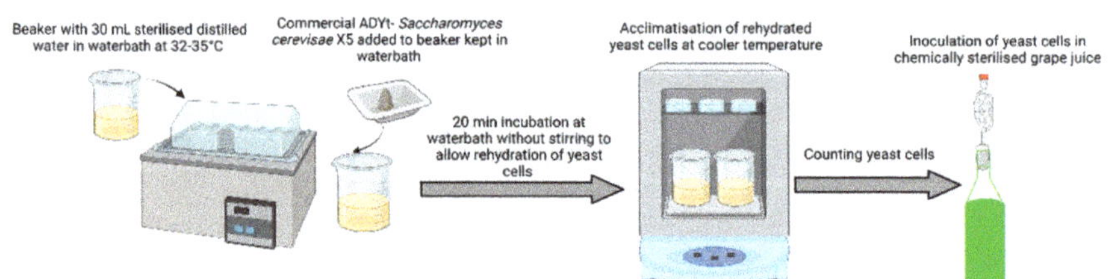

Treatment 2: Pre-inoculumn preparation in enriched medium using laboratory protocol

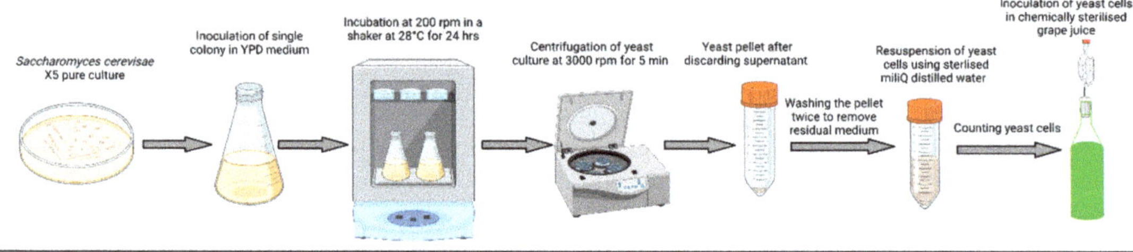

Winemaking experimental design

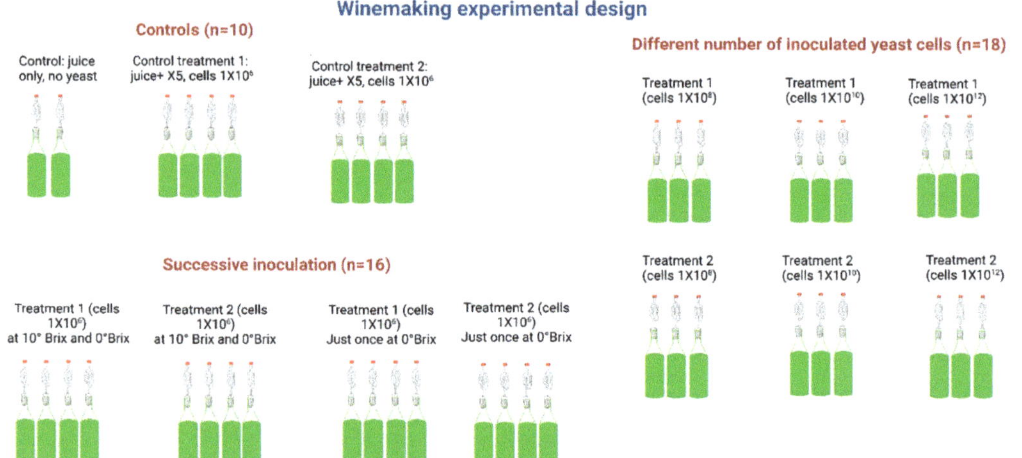

Figure 1. Two yeast inoculation preparation methods and winemaking experimental design used in this study. Yeast cells were counted as cells/mL.

2.2.1. Rehydration of Commercial ADY (RY)

Commercial ADY (*S. cerevisiae* X5) was rehydrated following the protocol used by the research winery of The New Zealand Institute for Plant and Food Research Limited [28]. Briefly, 30 mL amount of distilled water in a beaker was kept in a water bath at 32–35 °C, and 0.25 gm to 3 gm of ADY was added to the beaker depending on inoculated cell numbers and kept for 20 min without stirring. Then, the beaker was removed from the water bath and stirred and then sat on the bench for another 15 min. The beaker with rehydrated wine yeasts was transferred to a chiller room to acclimatise them to a cooler temperature (15 °C). Inoculation of grape juices was carried out within 45 min to avoid the loss of viability of the rehydrated wine yeasts.

2.2.2. Preparation of Pre-Inoculum in a Rich Growth Medium (PI)

S. cerevisiae X5 was grown in a rich culture medium, YPD broth (1% bacto-yeast extract, 2% bacto-yeast peptone and 2% D-glucose), as described in Pinu, Jouanneau, Nicolau, Gardner and Villas-Boas [25]. A single colony of the yeast grown in YPD agar was inoculated into 30 mL of YPD broth in a series of 150 mL conical flasks ($n = 50$) and was incubated at 28 °C, with shaking at 200 rpm for 24 h. Good aeration of the pre-inoculum provides molecular oxygen, which is essential for the development of components of yeast membranes, such as ergosterol, which helps yeast cells to tolerate high concentrations of ethanol under anaerobic conditions [29]. Pre-inoculums were centrifuged at 3000 rpm for 5 min, and the YPD supernatant was discarded, and the yeast pellets were then washed twice with sterilised milliQ distilled water to remove any YPD residues on the cells. The cell pellet was resuspended with sterilised milliQ distilled water as required prior to inoculating the grape juice [22]. Cell counts and viability testing were carried out prior to determining the required amount of inoculum for each treatment described in Section 2.3.

2.3. Counting of Yeast Cells and Viability Testing

Prior to conducting the final fermentation experiments, a number of trial experiments were performed to determine the cell numbers and viability for both of the inoculation methods. As available methods for cell counting have both merits and demerits, especially when counting the higher number of cells we used in this study, we compared two methods to determine the viability of cells: using YPD medium and using a hemocytometer as described below. Yeast cells were prepared as described in Sections 2.2.1 and 2.2.2. YPD medium was inoculated with yeast cell suspension (100 μL) from both protocols and incubated at 28 °C for 36 h. After that, visible colonies were counted as CFU/mL. While awaiting for yeast cells to grow in YPD medium, the same suspensions were used to determine the number of yeast cells from both methods using a Neubauer hemocytometer as described in Pinu [30]. Briefly, a 1/20 dilution of culture was made and transferred to hemocytometer for counting cell number per mL; the dilution was needed so that cell numbers of 10 squares ranged between 30 and 400. A cover slip was placed on the hemocytometer, and yeast cells were counted in five of the large squares in two separate counting chambers under a 40× objective. Budding yeast cells were counted as one cell unless they were the same size as the parent cell. Yeast cells on the top and left edges of the squares were counted, but cells on the bottom and right edges were not counted. From these cell count data, the volume of pre-culture required to inoculate 1×10^6 to 1×10^{12} yeast cells per mL of the final fermentation was determined (Figure 1). Methylene blue (0.1%) was used to assess the number of viable yeast cells. A 1:1 dilution of pre-culture was mixed with methylene blue and incubated for 5 min at room temperature and yeast cells were counted as described before. Viable cells appeared as colourless, as they reduced methylene blue, and dead cells were stained blue. The viability of cells from both counting methods was determined. Viability for RY culture was around 75% in cells grown in YPD medium, while it was 80% for PI. Approximately 75% of cells were viable for both inoculation protocols in cell counting by using a Neubauer hemocytometer. As the results were within similar ranges, and growing yeast cells in YPD medium needed 36 h of incubation time, we decided to use hemocytometer to count the cells during our final experiment. However, it is noteworthy that viability testing using YPD medium is more reliable than the microscopy-based methods.

2.4. Experimental Design and Wine Fermentation

Fermentation was carried out in 750 mL bottles using 700 mL of sterilised grape juice (Figure 1). The inoculum for each bottle was prepared separately to adjust to the right amount of cell numbers, and inoculation was also performed separately with 250 μL of inoculum to avoid dilution and extra nutrient inoculation in case of RY. The experiment was designed to determine the effects of two different yeast inoculation protocols on different primary and secondary metabolites produced by wine yeasts compared with control wines.

The control wines for both methods (RY and PI; $n = 4$) were inoculated with approximately 1×10^6 yeast cells/mL. A set of ferments was inoculated with increased numbers of approximate yeast cells, namely 1×10^8 ($n = 3$), 1×10^{10} ($n = 3$) and 1×10^{12} ($n = 3$) in the must, as shown in Figure 1. Successive inoculation was also carried out for both methods: treatment 1= 1×10^6, then 1×10^6 at 10 and 0 °Brix ($n = 4$) and treatment 2 = 1×10^6, then 1×10^6 at 0 °Brix ($n = 4$) (Figure 1). Yeast cells and their viability were determined within 3 h of inoculating the ferments using hemocytometer (see Section 2.3, data in Table S1).

All the microferments were kept in a chiller room at 15 °C without any agitation. However, the contents were slightly mixed every day by swirling the bottles slowly to avoid any loss of fermenting grape juice. Soluble solids content (measured as °Brix) and temperature were monitored daily during fermentation using an Anton-Paar DMA 35 portable density meter. Once °Brix values dropped below 0, residual sugars were monitored daily using Clinitest®. When residual sugars reached less than 2.0 g/L as determined by an enzymatic assay kit, the ferments were stopped with the addition of 50 ppm SO_2 (as potassium metabisulphite) and kept at the chiller (15 °C) overnight and centrifuged at 4600 rpm for 5 min using a Thermo Scientific Heraeus Multifuge 3SR+ centrifuge (Thermo Fisher Scientific, Waltham, MA, USA) prior to taking them to the laboratory for chemical analysis.

2.5. Analysis of the Resulting Wines

All the resulting wines were analysed to determine their acidity (pH and titratable acidity), sugar content (reducing sugars, glucose, and fructose), and total phenolics, as described in Section 2.2. Alcohol was determined using an Anton Parr Wine Alcolyzer (Graz, Austria).

2.5.1. Varietal Thiols

Thiols, 3-mercaptohexanol (3-MH), 3-mercaptohexyl acetate (3-MHA), and 4-mercapto-4-methylpentan-2-one (4-MMP) were extracted and analysed by Hill Laboratories Limited according to Green et al. [31]. In summary, varietal thiols in defrosted wines were analysed by headspace–solid-phase micro-extraction (HS-SPME) with gas chromatography–two-dimensional mass spectrometry (GC-MS/MS). Wine sub-samples were pipetted into 10 mL headspace vials with NaCl added to "salt-out" thiols into the headspace. Vials were capped, and then, deuterated internal standards for 3MHA and 3MH were robotically added to each vial through the cap septa. Each sample was buffered to pH 6.5–7.0 with phosphate buffer before analysis to avoid losses of thiols. Samples and calibration standards were extracted by SPME (polyacrylate, 85 lm coating supplied by Supelco, Bellefonte, PA, USA), using a robotic CTC CombiPal auto-sampler (Agilent Technologies, Santa Clara, CA, USA). For analysis, the SPME fibre was inserted into the hot (270 °C) GC inlet to desorb extracted thiols. Injections were in splitless mode, with a 0.75 mm i.d. glass liner (Restek, Bellefonte, PA, USA) at a temperature of 270 °C. Thiols were separated from co-extracted wine volatiles on a HP-5MS capillary column (30 m × 0.25 mm i.d. × 0.25 lm film thickness, Agilent Technologies, Santa Clara, CA, USA).

2.5.2. Aroma Compounds

Wine aroma compounds belonging to esters, higher alcohols, volatile fatty acids, c6 compounds, and terpenes were quantified following the method described by Herbst-Johnstone et al. [32] using SPME-GC-MS. Briefly, 10 mL of defrosted wine was mixed with 3.5 g of sodium chloride and transferred to an Agilent 20 mL amber screw cap vial. A mixture of deuterated internal standards in methanol was added; each sample was purged briefly with argon gas and sealed with a screw cap. Samples were placed for agitation and left for further automated analysis. The GC-MS system was an Agilent 7890A GC System coupled to 5975C mass selective detector. MassHunter software (v. B.05.00) (Agilent Technologies, Santa Clara, CA, USA) was used for data analysis, and aroma compounds were quantified using calibration curves. Odour activity value (OAV), which is the ratio of concentrations to perception thresholds, was determined to evaluate the impact of aroma compounds [22].

2.5.3. Organic Acids

A Shimadzu Prominence high-performance liquid chromatography (HPLC) (Shimadzu Corporation, Kyoto, Japan) system was used to quantify organic acids (tartaric, malic, ascorbic, shikimic, citric, and succinic acids) in wines using isocratic elution with a phosphate buffer (140 mM, pH 2.4) on an Allure Organic Acids Restek column (5 μm, 240 × 4.6 mm). Defrosted wine samples were diluted 10-fold in a solution containing internal standard (thiourea) and filtered through a 0.45 μm syringe filter before injection [33]. Samples were run in duplicate and quantified on a five-point standard curve. The correlation coefficient (R^2) of actual versus predicted concentration was >0.98.

2.5.4. Amino Acids

Quantification of amino acids in wines was performed on an Agilent 1200 series HPLC (Santa Clara, CA, USA) using a gradient elution programme of phosphate/borate buffer (10 mM each, pH 8.2) and organic solvent (MeOH: MeCN: H_2O, 45:45:10) on a Phenomenex Kinetix C18 column (5 μm, 240 × 4.6 mm) as described in Martin, Grose, Fedrizzi, Stuart, Albright, and McLachlan [28]. Briefly, online derivatization of primary amino acids was carried out with o-phthalaldehyde and 3-mercaptopropionic acid and detected by a diode array detector (DAD) at 340 nm excitation and 450 nm emission. Defrosted wine samples were treated with iodoacetic acid to encourage in the reduction of cysteine. Secondary amino acids were derivatized online with 9-fluorenylmethyl chloroformate and detected by DAD (260 nm excitation, 315 nm emission). A standard mix of 17 amino acids was purchased from Agilent (Santa Clara, CA, USA). All standards and samples contained internal standards sarcosine (100 mg/L) and α-aminobutyric acid (100 mg/L). Samples were diluted 4-fold in water and filtered through a 0.45 μm syringe filter before injection. Samples were run in duplicate and quantified on a four-point standard curve ($R^2 > 0.98$) [34].

2.6. Statistical Analysis

Independent Student's *t*-tests were performed to compare the changes in different parameters (e.g., alcohol content, varietal thiols concentrations, and other oenological parameters) of each treatment (e.g., two methods of inoculation, different numbers of cell inoculation, and successive inoculation) compared with their respective controls using an in-house R script. False-discovery rates (FDR) were calculated to account for multiple comparisons [27]. Principal component analysis (PCA), hierarchical cluster analysis (HCA), and two-way ANOVA were performed using a web interface, namely Metaboanalyst 5.0 (http://www.metaboanalyst.ca, accessed during 10–17 April 2023) [35]. Heatmaps were also produced to show the changes in different primary and secondary metabolites due to the inoculation method used in this study [35]. Pathway analysis was also performed based on the primary and secondary metabolites quantified in the resulting wines using *S. cerevisiae* as the pathway library to determine the pathways that were significantly affected due to inoculation methods. Global test and relative-betweenness centrality algorithms were selected for pathway enrichment and pathway topology analyses, respectively [27]. Debiased sparse partial correlation (DSPC) networks were created based on Basu et al. [36] using the Network analysis function in Metaboanalyst 5.0 with the quantified metabolites, for which KEGG ID is available.

3. Results and Discussion

Pre- and post-fermentation manipulations are often performed to produce a desired style of wine, often by changing the metabolism of wine yeasts [37,38]. Many other factors also can be modified and optimised (e.g., temperature, grape juice composition, and even the preparation of starting culture) to obtain different fermentation end products by wine yeasts. In this study, we determined how inoculum preparation protocols and inoculated yeast cell numbers affect overall fermentation behaviour of a commercial wine yeast strain and how these approaches change wine metabolite composition.

3.1. Fermentation Completion Time Depended on the Type of Inoculation Methods

Prior to fermentation, we determined the oenological properties of the starting commercial SB juice: pH 3.2, TA (total acidity) 8.1 g/L total soluble solids content 19.2 °Brix, and total reducing sugars 207.8 g/L (glucose 112.3 g/L and fructose 95.5 g/L). The YAN of the starting juice was 195 mg/L; therefore, no nitrogen adjustment prior to starting fermentation was performed, as YAN > 150 of the must is enough to avoid stuck fermentation [18]. Immediately after the yeast inoculation, we determined the cell numbers and their viability from each ferments. While approximately 75% cells were viable for inoculation protocols prior to inoculation, this number was around 70% in the ferments in the beginning of fermentation. This is not unexpected, as grape juice is a comparatively harsh media with low pH and high sugars. Moreover, typical New Zealand SB fermentation is carried out at 15 °C, which might have also affected the adaptation of the yeast cells into the ferments [37].

Our fermentation data clearly indicated that the ferments inoculated with rehydrated ADY (RY) were completed at least 4–5 days earlier than the ferments inoculated with pre-inoculum prepared in YPD medium (PI) (Table 1). Ferments inoculated with increased numbers of yeast cells also finished the fermentation one or two days earlier than the control ferments. Although an already-published study claimed that yeast cells were more viable and performed better during fermentation when a pre-inoculum was prepared in a highly aerated rich medium [6], our data from this experiment were not in agreement. However, ADY was also produced in highly aerated conditions, dried, and packed with nutrients and adjuvants, which might have a role in increased fermentation performance [7]. The ferments that underwent successive inoculation were deliberately kept for longer, although residual sugars were below 2 g/L. Therefore, we did not compare their fermentation completion time with other ferments. Overall, our data suggest that yeast inoculation protocol and inoculum quantity indeed affect the fermentation performance of wine yeasts.

Table 1. Fermentation completion time and basic oenological properties of all experimental wines.

Wine	Inoculated Yeast Cells	Completion Time	Alcohol (%v/v)	pH	Titratable Acidity (g/L)	Glucose (g/L)	Fructose (g/L)	Total Residual Sugar (g/L)	Phenolics (mg Gallic Acid/L)
				Rehydrated ADY					
Control RY	1×10^6	13	11.81 (0.01)	3.20 (0.01)	8.46 (0.13)	0.03 (0.01)	0.60 (0.28)	0.63 (0.27)	207.62 (6.50)
RY 1	1×10^8	12	11.76 (0.00)	3.21 (0.01)	8.41 (0.06)	0.05 (0.01)	1.09 (0.03) [a]	1.14 (0.03)	210.41 (3.38)
RY2	1×10^{10}	11	11.65 (0.01)	3.19 (0.01)	9.52 (0.17)	0.00	1.62 (0.32) [a]	1.62 (0.32)	201.19 (2.50)
RY3	1×10^{12}	11	11.64 (0.02)	3.19 (0.01)	9.51 (0.09)	0.00 [b]	0.87 (0.10) [b]	0.87 (0.10) [b]	204.95 (1.17)
SI RY 1	1×10^6, then 1×10^6 at 10 and 0 °Brix	14 *	11.86 (0.03)	3.28 (0.01)	8.48 (0.04)	0.00	0.04 (0.03) [a]	0.04 (0.03)	207.93 (4.47)
SI RY 2	1×10^6, then 1×10^6 at 0 °Brix	14 *	11.91 (0.00)	3.24 (0.01)	8.38 (0.11)	0.03 (0.04)	0.02 (0.02) [a]	0.05 (0.06)	211.37 (0.61)
				Pre-inoculum					
Control PI	1×10^6	17	11.91 (0.02)	3.24 (0.02)	9.01 (0.16)	0.05 (0.05)	0.87 (0.28)	0.92 (0.28)	209.82 (3.55)
PI 1	1×10^8	16	11.80 (0.08)	3.20 (0.02)	8.68 (0.10)	0.04 (0.01)	1.73 (0.74)	1.77 (0.74)	212.81 (2.83)
PI 2	1×10^{10}	17	11.83 (0.03)	3.24 (0.02)	8.56 (0.13)	0.01 (0.01)	1.05 (0.37)	1.06 (0.37)	209.70 (2.80)
PI 3	1×10^{12}	16	11.79 (0.01)	3.19 (0.04)	8.41 (0.06)	0.08 (0.02) [b]	1.54 (0.24) [b]	1.62 (0.21) [b]	212.64 (4.27)
SI PI 1	1×10^6, then 1×10^6 at 10 and 0 °Brix	17 *	11.90 (0.00)	3.28 (0.02)	8.29 (0.04)	0.02 (0.00)	0.06 (0.02) [a]	0.08 (0.02)	212.04 (2.73)
SI PI 2	1×10^6, then 1×10^6 at 0 °Brix	17 *	11.92 (0.01)	3.26 (0.02)	8.29 (0.10)	0.02 (0.00)	0.02 (0.01) [a]	0.04 (0.01)	206.93 (4.61)

Here, ADY, active dry yeast; RY, rehydrated yeast; SI, successive inoculation; PI, pre-inoculum. * denotes the ferments whose fermentation was not stopped although the residual sugar was below 2 g/L. [a] indicates the statistically significant differences in comparison to control ($p < 0.05$); [b] indicates the statistical differences between RY and PI when comparison was made with the same inoculated cell numbers.

3.2. No significant Change Was Observed on Different Oenological Properties of the Wines Based on Inoculation Methods

Table 1 also shows the oenological parameters of the wines produced in this study. We observed no significant difference in pH, TA, glucose, or total phenolic content of the wines compared with those of their respective control wines. As expected, glucose was almost completely consumed by the wine yeasts regardless of the inoculum preparation method or other treatments. However, fructose consumption varied among the ferments. For instance, less fructose was consumed when ferments were inoculated with increased numbers of yeast cells (for both RY and PI). This observation was expected, as it is well known that glucose is the most preferred and utilised carbon source for *S. cerevisiae* during fermentation [39]. On the other hand, yeast cells in ferments that were successively inoculated used more fructose than their controls. This is mainly because these ferments were kept for longer times, although their residual sugar concentrations were less than 2 g/L. Under fermentation condition, *S. cerevisiae* tends to utilise glucose at the beginning of fermentation, and once glucose is depleted, the cells start to consume fructose and other preferred carbon sources [39]. Our data suggest that fermentation stop time can influence the consumption of major sugars by wine yeasts during fermentation.

All the ferments had less than 12% *v/v* of alcohol, which is slightly lower than a typical "full-strength" New Zealand SB (12.5–13.5% *v/v*) [40]. As the TSS of the starting juice was 19.2 °Brix, the ethanol content reported here was within expectation. Although no statistically significant change was observed in the alcohol contents of the resulting wines because of the different treatments used in this study, we noted a trend of reduced alcohol when increased numbers of yeast cells were inoculated in the ferments (both RY and PI) (Table 1). Potential reasons behind this pattern could be the formation of other fermentation end products instead of ethanol or the metabolism of alcohol by higher number of yeast cells used during the experiment. However, as we found no statistical difference between the alcohol measurements, we did not explore this further here.

3.3. Inoculation Methods Altered the Overall Metabolite Composition of the Resulting Wines

To understand the effect of different wine yeast inoculation strategies, we determined absolute quantification of 24 primary metabolites (2 major sugars, and16 amino and 6 carboxylic acids) and 37 aroma compounds, including varietal thiols, higher alcohols, esters, terpenoids, and norisoprenoids. We analysed varietal thiol data separately given that they are key aroma compounds for SB wines (see Section 3.4). The concentrations of these compounds were within the ranges of previously published data of SB wines [22,40,41]. While comparing controls from two inoculation methods, we found that 21 primary metabolites were significantly different. The evaluation of secondary metabolites revealed that 15 aroma compounds, including 11 esters, 3 alcohols, and 1 aldehyde, were significantly different ($p < 0.05$, Figure 2).

We performed PCAs using all the metabolites quantified in the resulting wines (Figure 3), where the clustering pattern of wines produced from both inoculation strategies was more prominent for primary metabolites (PC1 and PC2 accounted for 80% of total variance). On the other hand, the PCA score plot of secondary metabolites showed a separation pattern with a slight overlap between the wine profiles produced from two inoculation strategies (PC and PC2 accounted 69% of total variance). This type of difference based on types of metabolites is not unusual, as production of primary metabolites is more regulated in a cell system compared with the secondary metabolites [42,43]. Many primary metabolites directly and indirectly contribute as precursors for many secondary metabolites, and that is particularly true for the production of volatile compounds [44]. Moreover, the primary metabolites analysed in this study are considered more stable during analysis than aroma compounds. Altogether, our data indicated that inoculation methods indeed influenced the metabolite composition of the wines (Figure 3).

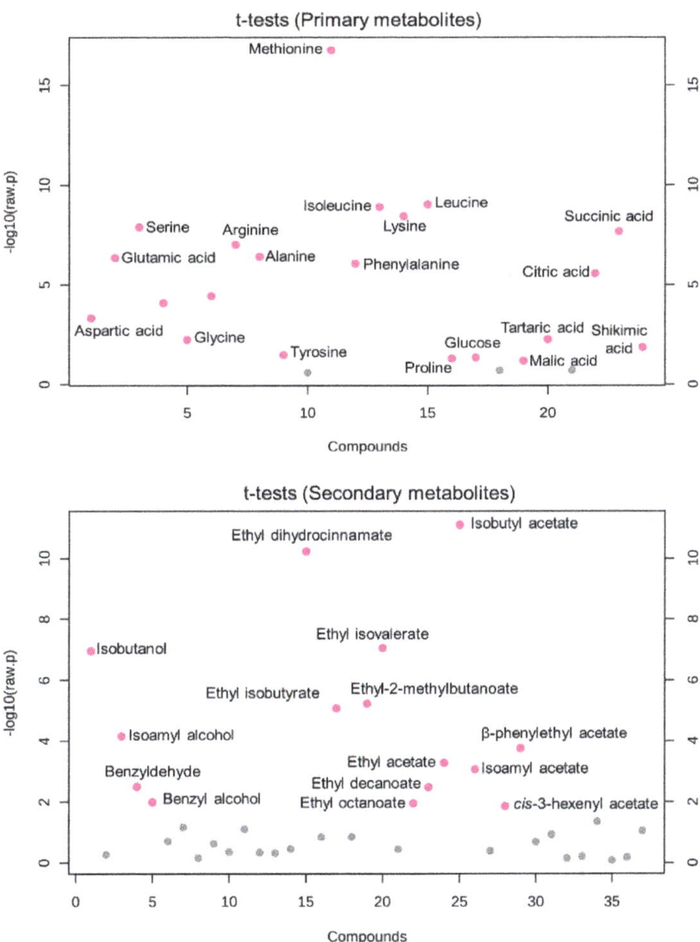

Figure 2. Primary and secondary metabolites that were significantly different between two inoculation methods used in the study. Pink dots with names indicate the metabolites that were statistically significant ($p < 0.05$) between two inoculation methods, while grey dots indicate insignificant metabolites ($p > 0.05$).

During commercial winemaking, little attention is usually paid towards the inoculated yeast cell numbers, while research winemaking is different, where similar numbers of yeast cells are inoculated. Published data that compare metabolites produced from commercial and research wines are limited. Therefore, we also investigated if inoculated yeast cell numbers had any impact on the metabolite composition of the resulting wines. As shown in Figure 4, the impact of inoculated cell numbers on metabolite composition varied between inoculation methods. For example, separation patterns were more visible for wines produced by different inoculum levels for RY, where PC1 and PC2 accounted for more than 60% of total variance. On the other hand, higher or lower levels of inoculum had little impact on wines produced from PI, which was evident by lack of separations based on cell numbers (Figure 4). Although this is an interesting observation, the reason behind this is not clear and will require further studies in future. Typically, it is recommended to use at least 10^5–10^6 CFU per mL [45]; in this experiment, we used 10^6 CFU per mL for the control wines, while other ferments were inoculated with higher cell numbers ranging from 10^8 to 10^{12} cells/mL. These are relatively high amounts of inoculation, and our data indicated that we should consider the inoculation method and the number of inoculated cells while conducting fermentation to achieve the expected outcomes.

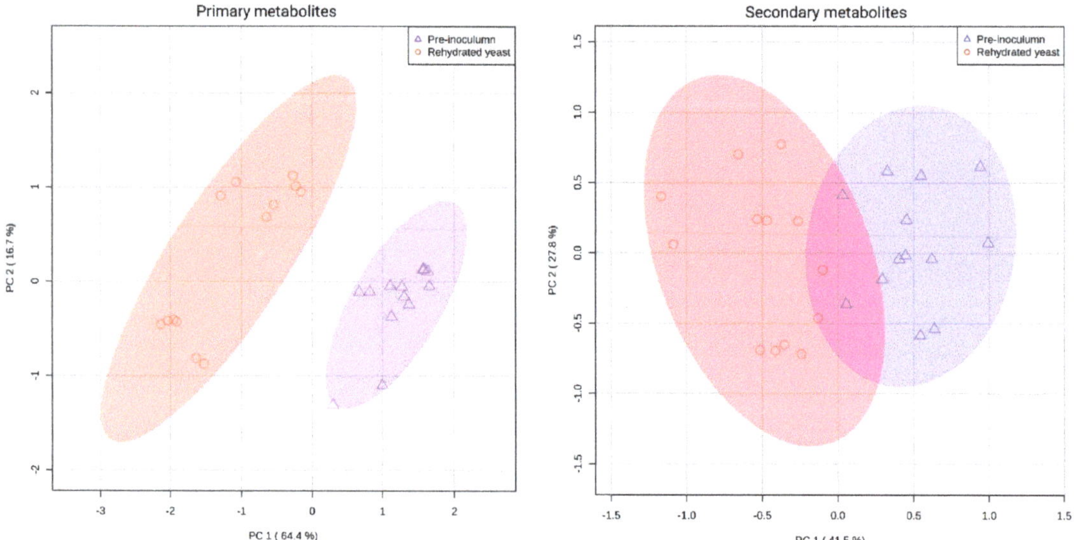

Figure 3. Two-dimensional projection of principal component analysis score plots based on primary and secondary metabolites showing the differences in the Sauvignon blanc wine composition produced by commercial *Saccharomyces cerevisiae* X5 strain using two different inoculation methods: pre-inoculum in rich medium and rehydrated active dry yeasts.

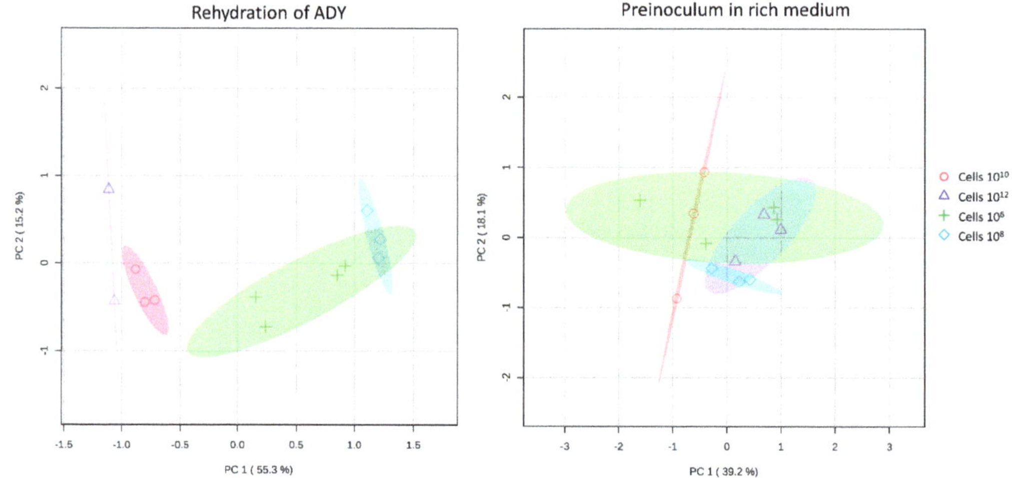

Figure 4. Two-dimensional projection of principal component analysis (PCA) score plots based on primary and secondary metabolites showing the effect of inoculated cell numbers during Sauvignon blanc fermentation by commercial *Saccharomyces cerevisiae* X5. ADY, active dry yeasts.

3.4. Varietal Thiols Are Affected by the Inoculation Methods

Varietal thiols are one of the most important aroma compounds in SB wines, and New Zealand SBs are known to have higher concentration of these tropical aroma compounds [40]. As their concentration in the wines is well within ng/L as compared to µg or mg/L for other aroma compounds analysed in this study, we analysed the varietal thiols data separately to determine the real impact of wine yeast preparation on this group of compounds. The perception threshold of these varietal thiols is very small: 0.8 ng/L for 4MMP, 4 ng/L for 3MHA, and 60 ng/L for 3MH [41], and OAV well-exceeded the ranges where the human olfactory system can detect them: OAV- 30–165 for 4MMP, 730–828 for 3MH, and 202–254 for 3MH. It is noteworthy that even if there was a difference of 20–30 ng/L of these thiols in wine, it would influence the overall aroma of the resulting wines.

Varietal thiol production, specifically 3MH and 4MMP, during fermentation was affected by the method of inoculum preparation (Table 2). The control wines produced using the PI contained almost 3000 ng/L more 3MH than the control wines inoculated with RY. The opposite was found for 4MMP, and its concentration was much higher in all the wines produced from RY. Moreover, we observed a positive correlation between the numbers of yeast cells inoculated and 3MH production regardless of the inoculation method, although this increase was not linear. In the case of 4MMP, this increase was linear for all the wines produced from RY when higher numbers of cells were inoculated. For 3MHA, the inoculation method and numbers of inoculated cells had little impact. However, successive inoculation seemed to have increased the concentration of 3MHA while reducing the concentration of 4MMP in wines for RY treatments. However, this was not the case for PI wines. All these observations suggest that the influence of method of inoculation and numbers of inoculated cells is dependent on the individual varietal thiol.

Table 2. Three major varietal thiols in Sauvignon blanc wines made after different yeast fermentations.

Wine	No of Inoculated Yeast Cells	3MH (ng/L)	3MHA (ng/L)	4MMP (ng/L)
Inoculation of rehydrated ADY				
Control RY	1×10^6	*11,498 (2717)*	3242 (843)	*62 (41)*
RY 1	1×10^8	*12,657 (1414)* [a]	3197 (250)	*69 (14)*
RY2	1×10^{10}	14,217 (829) [b]	3146 (175)	*105 (15)* [a]
RY3	1×10^{12}	14,066 (742) [b]	2943 (77)	*157 (59)* [b]
SI RY 1	1×10^6, then 1×10^6 at 10 and 0 °Brix	13,947 (1210) [a]	3588 (492)	29 (5) [b]
SI RY 2	1×10^6, then 1×10^6 at 0 °Brix	14,626 (636) [b]	*4037 (131)* [a]	27 (9) [b]
Inoculation of pre-inoculum				
Control PI	1×10^6	*14,382 (802)*	3147 (335)	*20 (8)*
PI 1	1×10^8	*15,573 (596)* [a]	3131 (216)	*25 (2)*
PI 2	1×10^{10}	14,502 (2010)	3095 (301)	*21 (8)*
PI 3	1×10^{12}	15,914 (692) [a]	3078 (117)	*38 (8)* [b]
SI PI 1	1×10^6, then 1×10^6 at 10 and 0 °Brix	13,879 (2846)	3091 (522)	17 (5)
SI PI 2	1×10^6, then 1×10^6 at 0 °Brix	15,679 (859) [a]	*3382 (305)*	24 (4)

p-values are shown as superscripts that were calculated by comparing with respective control wines; [a] < 0.05 and [b] < 0.01. ADY, active dry yeast; RY, rehydrated yeast; SI, successive inoculation; PI, pre-inoculum; 3MH, 3-mercaptohexanol; 3MHA, 3-mercaptohexylacetate; 4MMP, 4-methyl-4-mercaptopentan-2-one. Standard deviations of replicates in each treatment and control wines are shown within brackets. Numbers shown in italics indicate the statistical differences when comparison was made between RY and PI with same inoculated cell numbers (RY vs. PI).

3.5. Interactions between Inoculation Methods and Inoculated Cell Numbers on Metabolite Composition

As we observed differences in metabolite composition based on inoculation methods, and a method depended on the influence of inoculated cell numbers, we carefully looked at both primary and secondary metabolite data to determine which compounds are more affected by these factors. Figures 5 and 6 show the heatmaps produced using primary and secondary metabolites, respectively. When interaction between inoculation methods and inoculated cell numbers was considered using a two-way ANOVA, 9 out of 24 quantified primary metabolites, including lysine, arginine, leucine, isoleucine alanine, glutamic acid, glycine, glucose, and tartaric acid, were statistically different (p-value < 0.05) between treatments. However, the method of inoculation affected 19 of them, as discussed earlier (see Section 3.3), while cell numbers influenced 10 of the primary metabolites. This indicated that the method of inoculation (wine yeast preparation) was a more important factor than the number of inoculated cells that drives the differentiation in primary metabolite composition of wines. We observed a general trend of increased concentration of most of the amino acids in wines produced from cells grown in rich medium (shown as PI in Figure 5). Amino acid concentrations in these wines also increased when a higher number of cells were inoculated. This was expected and may result from the released amino acids from yeast autolysis at the end of fermentation [46]. The concentration of proline and valine was relatively higher in wines produced using RY. As proline is one of the most abundant amino acids in wines, total amino acid contents of resulting wines did not vary significantly between the inoculation methods and cell numbers used as inoculum.

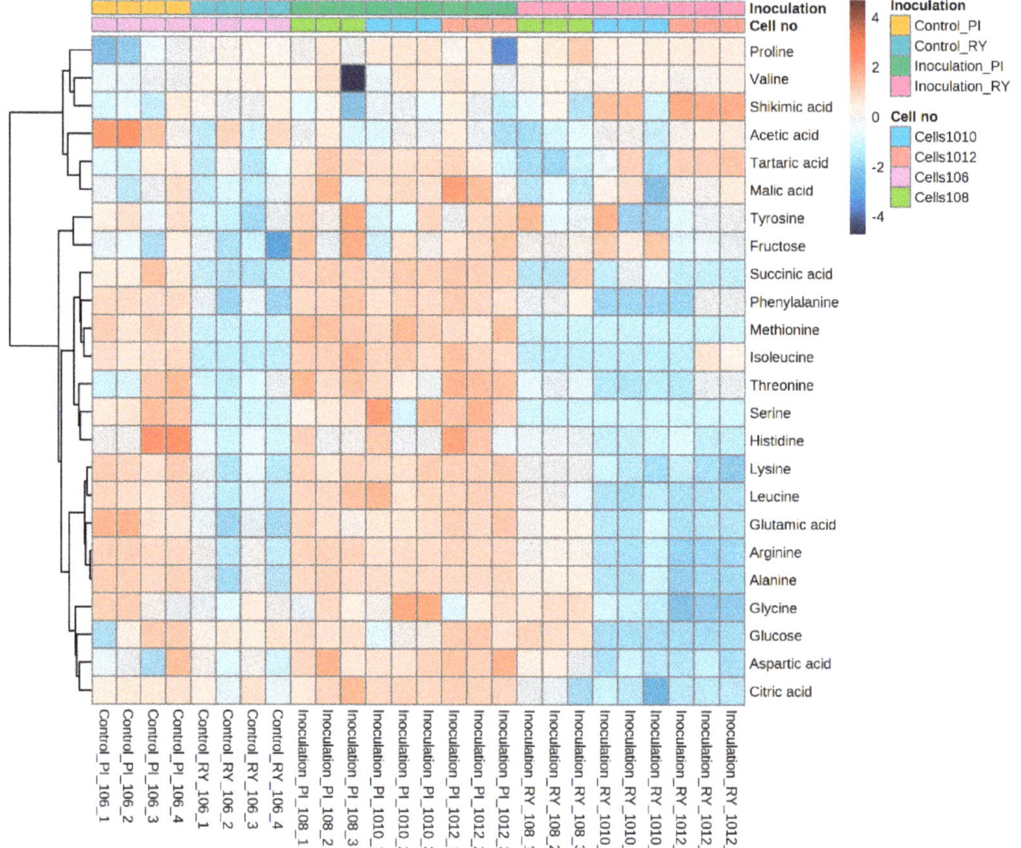

Figure 5. Heatmaps showing the concentration of primary metabolites in different Sauvignon blanc wines produced by using commercial *Saccharomyces cerevisiae* X5. Two different inoculation methods were used: pre-inoculum prepared in rich media (noted as PI) and rehydrated active dry yeasts (noted as RY). Inoculated cell numbers are also shown as: Cells106, Cells 10^6 (control wines); Cells108, Cells 10^8; Cells1010, Cells 10^{10}; Cells1012, Cells 10^{12}.

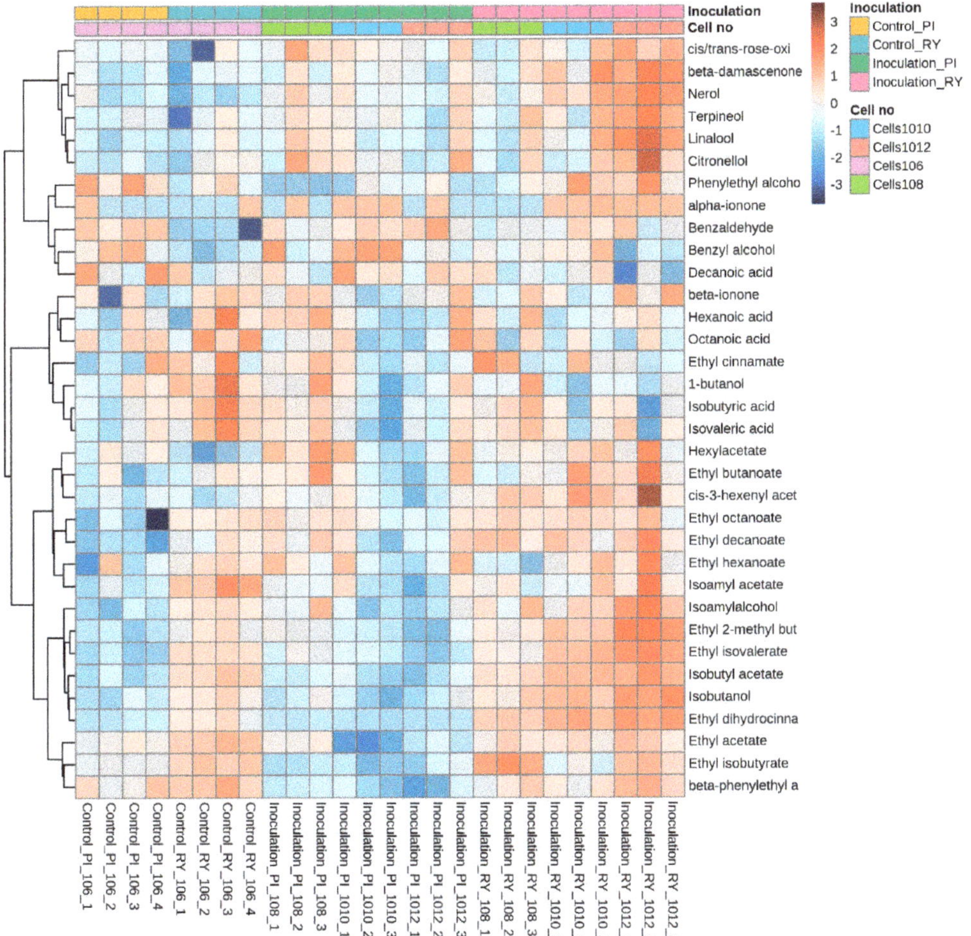

Figure 6. Heatmaps showing the concentration of secondary metabolites in different Sauvignon blanc wines produced by using commercial *Saccharomyces cerevisiae* X5. Two different inoculation methods were used: pre-inoculum prepared in rich media (noted as PI) and rehydrated active dry yeasts (noted as RY)). Inoculated cell numbers are also shown as: Cells106, Cells 10^6 (control wines); Cells108, Cells 10^8; Cells1010, Cells 10^{10}; Cells1012, Cells 10^{12}.

We determined the OAV of all the aroma compounds analysed in this study to evaluate if these compounds can be perceived by the human olfactory system [41]. Out of 34 aroma compounds belonging to esters, higher alcohols, volatile fatty acids, monoterpenoids, and norisoprenoids, 16 of them had OAV > 1 (Table S2), indicating their importance in the overall aroma perception of wine. As expected, OAVs for most of the monoterpenoid and norisoprenoid aroma compounds were below 1, while esters, higher alcohols, and a few volatile fatty acids were the compounds with higher levels of OAV (>1). The results we observed here in terms of relevance to aroma perception are in accordance with previously published data on SB wine aroma [41,47]. Results from the two-way ANOVA using secondary metabolites showed that 6 out of 34 aroma compounds were affected when the interaction between inoculation methods and inoculated cell numbers was considered. These compounds were ethyl dihydrocinnamate (spicy, fruity notes), ethyl isovalerate (apple note), ethyl isobutyrate (fruity note), ethyl acetate (nail polish remover), ethyl 2-

methyl butanoate (berries and apples), and nerol (sweet rose odour) [30]. Similar to what we observed for primary metabolites, 18 aroma compounds were influenced by the inoculation method used, while only 3 (ethyl dihydrocinnamate, ethyl isovalerate, and ethyl acetate) of the aroma compounds were affected by cell numbers. This again suggested that choice of suitable inoculation method was important to produce different aroma compounds, while cell numbers should also be considered, as this factor seemed to affect the production of a few important ethyl esters.

As shown in Figure 6, we observed another general trend that inoculation method using RY increased the production of most of the aroma compounds in the resulting wines, particularly the concentrations of esters that contributed towards the fruity and sweet note of the wines. In comparison, the concentration of most of these aroma compounds was generally lower in the PI wines. The reason behind these observations could be traced back to the primary metabolite data, where we found higher concentration of most of the amino acids in PI wines and lower concentration in RY wines. As mentioned earlier, the primary and secondary metabolisms are interlinked [42]. In central carbon metabolism, different types of amino acids (e.g., leucine, isoleucine, valine, and phenylalanine) contribute to the production of various intermediate molecules (e.g., acetyl-CoA) and aroma compounds, including higher alcohols and medium-chain volatile fatty acids [48]. These groups of compounds then result in the production of esters either by condensation reactions between acetyl-CoA and higher alcohols (acetate esters) or by the esterification of ethanol and acyl-CoA intermediates because of esterase and transferase enzyme activity (ethyl esters) [48]. One reason behind observing higher esters and lower amino acids in RY wines could be the metabolic activity of the wine yeasts during fermentation, where conditions in the RY ferments were more suitable for producing relevant aroma compounds [18,19,37,38]. In the case of PI ferments, the opposite trend was found, indicating the impact of tailoring different parameters, including fermentation conditions, media, and yeast preparation, on the development of targeted aroma compounds in the resulting wines.

3.6. Pathway Analysis Indicates the Changes in Metabolic Pathways Due to the Differences in Inoculation Methods

Our data suggest that inoculation methods indeed influence the overall metabolite composition of the resulting wines; therefore, metabolic pathways are most likely to be affected as well. As inoculation methods seemed to affect wine composition more than the inoculated cell numbers, we only performed pathway analysis to gain insights on the changes in yeast metabolism due to the two wine yeast preparation methods. As shown in Figure 7a, 11 metabolic pathways related mainly to nitrogen and sulphur metabolism were significantly influenced by the inoculation methods ($p < 0.05$).

Two metabolic pathways, aspartate and glutamate metabolism and cysteine methionine metabolism, particularly stood out compared with other pathways when their impact (shown in the x-axis in Figure 7a) and significance (shown in the y-axis in Figure 7a) were taken into account. Both pathways were upregulated for PI, which could be traced back to the presence of relatively higher concentrations of amino acids (nitrogen compounds) and varietal thiols (sulphur-containing compounds) in the wines produced using PI. On the other hand, the metabolic pathway related to aromatic amino acids (phenylalanine, tyrosine, and tryptophan biosynthesis) was slightly upregulated for RY, even though the concentration of most of these amino acids was comparatively lower in RY wines than PI. However, levels of esters, medium-chain fatty acids, and higher alcohols were higher in RY wines, indicating these amino acids were catabolized and aided in the production of these aroma compounds [49]. We also performed DSPC network analysis to determine the relationship between the metabolites analysed in the wines. There were 3 subnetworks: subnetwork 1 with 21 different notes (Figure 7b), subnetwork 2 with 5 nodes, and subnetwork 3 with 3 nodes. Subnetwork 1 provided the comprehensive picture that supported our observation from the pathway analysis, as we could see the positive correlation (red lines) among most of the primary metabolites and negative correlation of primary metabolites with the major aroma compounds (blue lines). This again confirmed that most of the primary metabolites were consumed by the wine yeasts during the fermentation that ultimately resulted in the production of the associated aroma compounds, including esters and higher alcohols (Figure 7b).

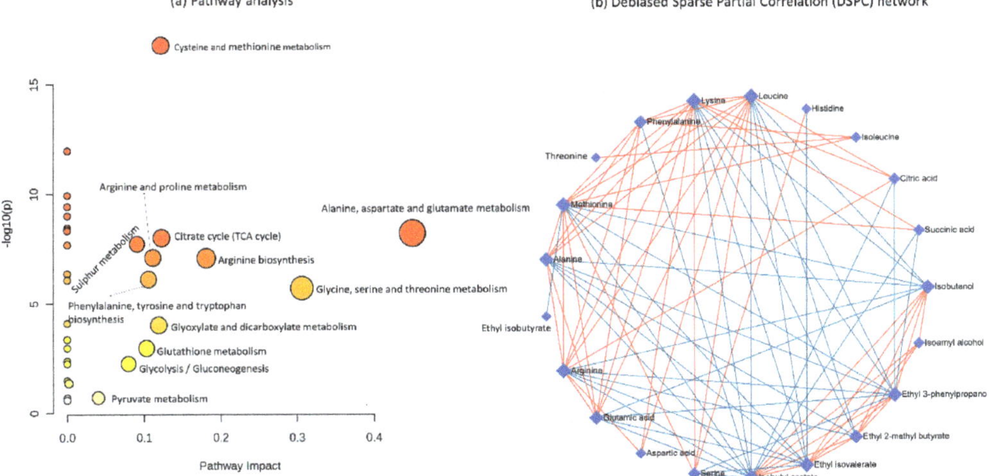

Figure 7. Impact of inoculation methods on different metabolic pathways and associated metabolites. Pathway analysis (**a**) was carried out using all the quantified metabolites with Kyoto Encyclopedia of Genes and Genomes (KEGG) ID using *Saccharomyces cerevisiae* as the pathway library from KEGG. Colour gradient is based on *p*-values: yellow (lower) and red (higher). Debiased sparse partial correlation (DSPC) network (**b**) showing the correlation between significant primary and secondary metabolites. The red line indicates a positive correlation, while the blue line shows the negative correlation.

4. Conclusions

Based on previously published studies, we know that wine yeast metabolism during fermentation changes depending on the pre-fermentative manipulations, fermentation medium, and conditions [18,27,38,50]. Our investigation here involved the use of two different inoculation strategies to generate understanding of how two wine yeast preparation methods either used commercially (RY) or in a research laboratory (PI) affect overall metabolism of wine yeasts. We indeed found differences in fermentation time, metabolite composition, and related metabolic pathways depending on inoculation methods. Although the impact of inoculum size during wine fermentation is a less-studied area, we provided some insights on how differences in inoculated cell numbers also affect the production of different classes of aroma compounds. Our findings agreed with Carrau, Medina, Farina, Boido, and Dellacassa [18], who also found an unpredictable pattern of impacts on wine composition based on inoculated cells in two different wine yeast strains. We found that the method of inoculation was a more impactful determinant of fermentation end-product formation than inoculum quantity. Interestingly, and in contrast to previously published work, inoculation with RY increased the fermentation performance of wine yeasts, as evident from the fermentation completion time [6]. While wines produced by RY contained a higher amount of esters, higher alcohols, and 4MMP, 3MH and amino acid concentrations were considerably higher in PI wines. Successive inoculation also resulted in the production of higher concentration of varietal thiols in the wines. Altogether, these data suggest that production of wines with desirable aroma compounds is achievable via adopting a suitable wine yeast inoculation protocol, while attention should also be paid towards inoculum size wherever possible.

Supplementary Materials: The following supporting information can be downloaded at: https://www.mdpi.com/article/10.3390/fermentation9080759/s1. Table S1: Viability testing carried out for each ferments and treatments using Neubauer hemocytometer and methylene blue (0.1%) dye. Table S2: Title: Odour activity value (OAV) of aroma compounds quantified in Sauvignon blanc wines. RY, rehydrated active dry yeast; PI, pre-inoculum prepared in rich medium.

Author Contributions: Conceptualisation, F.R.P.; methodology, F.R.P., C.G., L.S., A.A. and T.T.; formal analysis, F.R.P.; investigation, F.R.P., C.G. and L.S.; resources, F.R.P. and D.M.; data curation, F.R.P.; writing—original draft preparation, F.R.P.; writing—review and editing, F.R.P., L.S., A.A., C.G. and D.M.; visualisation, F.R.P.; project administration, F.R.P.; funding acquisition, F.R.P. and D.M. All authors have read and agreed to the published version of the manuscript.

Funding: This research was funded as part of New Zealand Lighter Wines Programme (P/471777/04) co-funded by New Zealand Winegrowers, and the Ministry for Primary Industries Primary Growth Partnership.

Institutional Review Board Statement: Not applicable.

Informed Consent Statement: Not applicable.

Data Availability Statement: Access to data can be requested by contacting the corresponding author.

Acknowledgments: Our sincere thanks to Andy Frost, Pernod Ricard Winery (Marlborough, New Zealand), for providing the Sauvignon blanc juice. We also thank the Plant and Food Research research winery members during the 2017 vintage, Franzi Grab, Victoria Raw, Judy Zhao, Linlin Yang, Giulia Decordi, and Alessandro Serughetti for their help with daily monitoring of the ferments. We are grateful to Gert-Jan Moggre, Kevin Sutton, and Plant and Food Research's Science Publishing Unit for providing feedback on the manuscript.

Conflicts of Interest: The authors declare no conflict of interest.

References

1. Antalick, G.; Perello, M.C.; de Revel, G. Esters in Wines: New Insight through the Establishment of a Database of French Wines. *Am. J. Enol. Vitic.* **2014**, *65*, 293–304. [CrossRef]
2. Hirst, M.B.; Richter, C.L. Review of Aroma Formation through Metabolic Pathways of Saccharomyces cerevisiae in Beverage Fermentations. *Am. J. Enol. Vitic.* **2016**, *67*, 361–370. [CrossRef]
3. Cardona, F.; Carrasco, P.; Perez-Ortin, J.E.; del Olmo, M.L.; Aranda, A. A novel approach for the improvement of stress resistance in wine yeasts. *Int. J. Food Microbiol.* **2007**, *114*, 83–91. [CrossRef] [PubMed]
4. Pizarro, F.; Vargas, F.A.; Agosin, E. A systems biology perspective of wine fermentations. *Yeast* **2007**, *24*, 977–991. [CrossRef] [PubMed]
5. Diaz-Hellin, P.; Ubeda, J.; Briones, A. Improving alcoholic fermentation by activation of Saccharomyces species during the rehydration stage. *LWT-Food Sci. Technol.* **2013**, *50*, 126–131. [CrossRef]
6. Jenkins, D.M.; Powell, C.D.; Fischborn, T.; Smart, K.A. Rehydration of Active Dry Brewing Yeast and its Effect on Cell Viability. *J. Inst. Brew.* **2011**, *117*, 377–382. [CrossRef]
7. Schmidt, S.A.; Henschke, P.A. Production, reactivation and nutrient requirements of active dried yeast in winemaking: Theory and practice. *Aust. J. Grape Wine Res.* **2015**, *21*, 651–662. [CrossRef]
8. Monk, P.R. Rehydration and propagation of active dry wine yeast. *Aust. Wine Ind. J.* **1986**, *1*, 35.
9. Redon, M.; Guillamon, J.M.; Mas, A.; Rozes, N. Effect of active dry wine yeast storage upon viability and lipid composition. *World J. Microbiol. Biotechnol.* **2008**, *24*, 2555–2563. [CrossRef]
10. Kontkanen, D.; Inglis, D.L.; Pickering, G.J.; Reynolds, A. Effect of yeast inoculation rate, acclimatization, and nutrient addition on icewine fermentation. *Am. J. Enol. Vitic.* **2004**, *55*, 363–370. [CrossRef]
11. Diaz-Hellin, P.; Ubeda, J.; Briones, A. Study of two wine strains rehydrated with different activators in sluggish fermentation conditions. *Eur. Food Res. Technol.* **2013**, *237*, 547–554. [CrossRef]
12. Andujar-Ortiz, I.; Chaya, C.; Martin-Alvarez, P.J.; Moreno-Arribas, M.V.; Pozo-Bayon, M.A. Impact of using new commercial glutathione enriched inactive dry yeast oenological preparations on the aroma and sensory properties of wines. *Int. J. Food Prop.* **2014**, *17*, 987–1001. [CrossRef]
13. Novo, M.; Beltran, G.; Rozes, N.; Guillamon, J.M.; Sokol, S.; Leberre, V.; Francois, J.; Mas, A. Early transcriptional response of wine yeast after rehydration: Osmotic shock and metabolic activation. *FEMS Yeast Res.* **2007**, *7*, 304–316. [CrossRef] [PubMed]
14. Winter, G.; Henschke, P.A.; Higgins, V.J.; Ugliano, M.; Curtin, C.D. Effects of rehydration nutrients on H2S metabolism and formation of volatile sulfur compounds by the wine yeast VL3. *AMB Express* **2011**, *1*, 11. [CrossRef] [PubMed]
15. Soubeyrand, V.; Julien, A.; Sablayrolles, J.M. Rehydration protocols for active dry wine yeasts and the search for early indicators of yeast activity. *Am. J. Enol. Vitic.* **2006**, *57*, 474–480. [CrossRef]

16. Vaudano, E.; Noti, O.; Costantini, A.; Garcia-Moruno, E. Effect of additives on the rehydration of Saccharomyces cerevisiae wine strains in active dry form: Influence on viability and performance in the early fermentation phase. *J. Inst. Brew.* **2014**, *120*, 71–76. [CrossRef]
17. Rodriguez-Porrata, B.; Novo, M.; Guillamon, J.; Rozes, N.; Mas, A.; Otero, R.C. Vitality enhancement of the rehydrated active dry wine yeast. *Int. J. Food Microbiol.* **2008**, *126*, 116–122. [CrossRef]
18. Carrau, F.; Medina, K.; Farina, L.; Boido, E.; Dellacassa, E. Effect of Saccharomyces cerevisiae inoculum size on wine fermentation aroma compounds and its relation with assimilable nitrogen content. *Int. J. Food Microbiol.* **2010**, *143*, 81–85. [CrossRef]
19. Erten, H.; Tanguler, H.; Cabaroglu, T.; Canbas, A. The influence of inoculum level on fermentation and flavour compounds of white wines made from cv. Emir. *J. Inst. Brew.* **2006**, *112*, 232–236. [CrossRef]
20. Deed, N.K.; van Vuuren, H.J.J.; Gardner, R.C. Effects of nitrogen catabolite repression and di-ammonium phosphate addition during wine fermentation by a commercial strain of S. cerevisiae. *Appl. Microbiol. Biotechnol.* **2011**, *89*, 1537–1549. [CrossRef]
21. Harsch, M.J.; Benkwitz, F.; Frost, A.; Colonna-Ceccaldi, B.; Gardner, R.C.; Salmon, J.M. New Precursor of 3-Mercaptohexan-1-ol in Grape Juice: Thiol-Forming Potential and Kinetics during Early Stages of Must Fermentation. *J. Agric. Food Chem.* **2013**, *61*, 3703–3713. [CrossRef] [PubMed]
22. Pinu, F.R.; Edwards, P.J.B.; Jouanneau, S.; Kilmartin, P.A.; Gardner, R.C.; Villas-Boas, S.G. Sauvignon blanc metabolomics: Grape juice metabolites affecting the development of varietal thiols and other aroma compounds in wines. *Metabolomics* **2014**, *10*, 556–573. [CrossRef]
23. Shekhawat, K.; Bauer, F.F.; Setati, M.E. Impact of oxygenation on the performance of three non-Saccharomyces yeasts in co-fermentation with Saccharomyces cerevisiae. *Appl. Microbiol. Biotechnol.* **2017**, *101*, 2479–2491. [CrossRef] [PubMed]
24. Ancin-Azpilicueta, C.; Jimenez-Moreno, N.; Moler, J.A.; Nieto-Rojo, R.; Urmeneta, H. Effects of reduced levels of sulfite in wine production using mixtures with lysozyme and dimethyl dicarbonate on levels of volatile and biogenic amines. *Food Addit. Contam. Part A-Chem.* **2016**, *33*, 1518–1526. [CrossRef] [PubMed]
25. Pinu, F.R.; Jouanneau, S.; Nicolau, L.; Gardner, R.C.; Villas-Boas, S.G. Concentrations of the Volatile Thiol 3-Mercaptohexanol in Sauvignon blanc Wines: No Correlation with Juice Precursors. *Am. J. Enol. Vitic.* **2012**, *63*, 407–412. [CrossRef]
26. Iland, P.; Bruer, N.; Wilkes, E. *Chemical Analysis of Grapes and Wine: Techniques and Concepts*; Patrick Iland Wine Promotions: Adelaide, Australia, 2004.
27. Pinu, F.R.; Villas-Boas, S.G.; Martin, D. Pre-fermentative supplementation of fatty acids alters the metabolic activity of wine yeasts. *Food Res. Int.* **2019**, *121*, 835–844. [CrossRef]
28. Martin, D.; Grose, C.; Fedrizzi, B.; Stuart, L.; Albright, A.; McLachlan, A. Grape cluster microclimate influences the aroma composition of Sauvignon blanc wine. *Food Chem.* **2016**, *210*, 640–647. [CrossRef]
29. Hansen, E.H.; Nissen, P.; Sommer, P.; Nielsen, J.C.; Arneborg, N. The effect of oxygen on the survival of non-Saccharomyces yeasts during mixed culture fermentations of grape juice with Saccharomyces cerevisiae. *J. Appl. Microbiol.* **2001**, *91*, 541–547. [CrossRef]
30. Pinu, F.R. Sauvignon Blanc Metabolomics: Metabolite Profile Analysis before and after Fermentation. Ph.D. Thesis, University of Auckland, Auckland, New Zealand, 2013.
31. Green, J.A.; Parr, W.V.; Breitmeyer, J.; Valentin, D.; Sherlock, R. Sensory and chemical characterisation of Sauvignon blanc wine: Influence of source of origin. *Food Res. Int.* **2011**, *44*, 2788–2797. [CrossRef]
32. Herbst-Johnstone, M.; Araujo, L.D.; Allen, T.A.; Logan, G.; Nicolau, L.; Kilmartin, P.A. *Effects of Mechanical Harvesting on 'Sauvignon Blanc' Aroma*; International Society for Horticultural Science (ISHS): Leuven, Belgium, 2013; pp. 179–186.
33. Shi, S.; Condron, L.; Larsen, S.; Richardson, A.E.; Jones, E.; Jiao, J.; O'Callaghan, M.; Stewart, A. In situ sampling of low molecular weight organic anions from rhizosphere of radiata pine (*Pinus radiata*) grown in a rhizotron system. *Environ. Exp. Bot.* **2011**, *70*, 131–142. [CrossRef]
34. Henderson, J.W.J.; Brooks, A. *Improved Amino Acid Methods using Agilent ZORBAX Eclipse Plus C18 Columns for a Variety of Agilent LC Instrumentation and Separation Goals*; Agilent Application Note 5990–4547EN; Agilent Technologies, Inc.: Wilmington, NC, USA, 2010.
35. Pang, Z.; Chong, J.; Zhou, G.; de Lima Morais, D.A.; Chang, L.; Barrette, M.; Gauthier, C.; Jacques, P.-É.; Li, S.; Xia, J. MetaboAnalyst 5.0: Narrowing the gap between raw spectra and functional insights. *Nucleic Acids Res.* **2021**, *49*, W388–W396. [CrossRef] [PubMed]
36. Basu, S.; Duren, W.; Evans, C.R.; Burant, C.F.; Michailidis, G.; Karnovsky, A. Sparse network modeling and metscape-based visualization methods for the analysis of large-scale metabolomics data. *Bioinformatics* **2017**, *33*, 1545–1553. [CrossRef] [PubMed]
37. Pinu, F.R.; Stuart, L.; Grose, C. *Effects of Different Yeast Preparation, Inoculation Timing and Ferment Oxygenation Methodologies on Ethanol Yield and Varietal Aroma Compound Formation*; A Plant & Food Research Report; New Zealand Winegrowers: Auckland, New Zealand, 2019.
38. Rollero, S.; Bloem, A.; Camarasa, C.; Sanchez, I.; Ortiz-Julien, A.; Sablayrolles, J.M.; Dequin, S.; Mouret, J.R. Combined effects of nutrients and temperature on the production of fermentative aromas by Saccharomyces cerevisiae during wine fermentation. *Appl. Microbiol. Biotechnol.* **2015**, *99*, 2291–2304. [CrossRef]
39. Pinu, F.R.; Edwards, P.J.B.; Gardner, R.C.; Villas-Boas, S.G. Nitrogen and carbon assimilation by Saccharomyces cerevisiae during Sauvignon blanc juice fermentation. *FEMS Yeast Res.* **2014**, *14*, 1206–1222. [CrossRef]

40. Pinu, F.R.; Tumanov, S.; Grose, C.; Raw, V.; Albright, A.; Stuart, L.; Villas-Boas, S.G.; Martin, D.; Harker, R.; Greven, M. Juice Index: An integrated Sauvignon blanc grape and wine metabolomics database shows mainly seasonal differences. *Metabolomics* **2019**, *15*, 3. [CrossRef]

41. Benkwitz, F.; Takatoshi, T.; Kilmartin, P.A.; Lund, C.; Wohlers, M.; Nicolau, L. Identifying the Chemical Composition Related to the Distinct Aroma Characteristics of New Zealand Sauvignon blanc Wines. *Am. J. Enol. Vitic.* **2012**, *63*, 62. [CrossRef]

42. Pinu, F.R.; Villas-Boas, S.G.; Aggio, R. Analysis of Intracellular Metabolites from Microorganisms: Quenching and Extraction Protocols. *Metabolites* **2017**, *7*, 53. [CrossRef]

43. Drew, S.W.; Demain, A.L. Effect of Primary Metabolites on Secondary Metabolism. *Annu. Rev. Microbiol.* **1977**, *31*, 343–356. [CrossRef] [PubMed]

44. Nielsen, J. Cell factory engineering for improved production of natural products. *Nat. Prod. Rep.* **2019**, *36*, 1233–1236. [CrossRef] [PubMed]

45. Pinu, F.R.; Villas-Boas, S.G. Extracellular Microbial Metabolomics: The State of the Art. *Metabolites* **2017**, *7*, 43. [CrossRef] [PubMed]

46. Lombardi, S.J.; De Leonardis, A.; Lustrato, G.; Testa, B.; Iorizzo, M. Yeast Autolysis in Sparkling Wine Aging: Use of Killer and Sensitive Saccharomyces cerevisiae Strains in Co-Culture. *Recent Pat. Biotechnol.* **2015**, *9*, 223–230. [CrossRef] [PubMed]

47. Tumanov, S.; Pinu, F.R.; Greenwood, D.R.; Villas-Bôas, S.G. Effect of free fatty acids and lipolysis on Sauvignon Blanc fermentation. *Aust. J. Grape Wine Res.* **2018**, *24*, 398–405. [CrossRef]

48. Prusova, B.; Humaj, J.; Sochor, J.; Baron, M. Formation, Losses, Preservation and Recovery of Aroma Compounds in the Winemaking Process. *Fermentation* **2022**, *8*, 93. [CrossRef]

49. Cordente, A.G.; Schmidt, S.; Beltran, G.; Torija, M.J.; Curtin, C.D. Harnessing yeast metabolism of aromatic amino acids for fermented beverage bioflavouring and bioproduction. *Appl. Microbiol. Biotechnol.* **2019**, *103*, 4325–4336. [CrossRef] [PubMed]

50. García-Ríos, E.; Gutiérrez, A.; Salvadó, Z.; Arroyo-López, F.N.; Guillamon, J.M. The Fitness Advantage of Commercial Wine Yeasts in Relation to the Nitrogen Concentration, Temperature, and Ethanol Content under Microvinification Conditions. *Appl. Environ. Microbiol.* **2014**, *80*, 704. [CrossRef] [PubMed]

 *fermentation*

Article

Development of Fermented Kombucha Tea Beverage Enriched with Inulin and B Vitamins

Yuliya Frolova [1],*, Valentina Vorobyeva [1], Irina Vorobyeva [1], Varuzhan Sarkisyan [1], Alexey Malinkin [2], Vasily Isakov [3] and Alla Kochetkova [1]

[1] Laboratory of Food Biotechnology and Foods for Special Dietary Uses, Federal State Budgetary Scientific Institution Federal Research Center of Nutrition, Biotechnology and Food Safety, 109240 Moscow, Russia
[2] Laboratory of Food Chemistry, Federal State Budgetary Scientific Institution Federal Research Center of Nutrition, Biotechnology and Food Safety, 109240 Moscow, Russia
[3] Department of Gastroenterology & Hepatology, Federal State Budgetary Scientific Institution Federal Research Center of Nutrition, Biotechnology and Food Safety, 109240 Moscow, Russia
* Correspondence: y.operarius@yandex.ru

Abstract: Kombucha is a sweet and sour beverage made by fermenting a liquid base with a symbiotic culture of bacteria and yeast. Different tea substrates, carbohydrate sources, and additional ingredients are used to create beverages with different physical and chemical characteristics. The purpose of this work was to create a recipe and technology to study the properties of the beverage based on kombucha with a given chemical composition. The content of added functional ingredients (vitamins and inulin) in quantities comparable with reference daily intake was the specified parameter characterizing the distinctive features of the enriched beverages. For fermentation using symbiotic cultures of bacteria and yeast, a black tea infusion sweetened with sucrose was used as a substrate. The changes in the physicochemical characteristics of the fermented tea beverage base were evaluated. The dynamics of changes in pH, acidity, the content of mono- and disaccharides, ethanol, organic acids, polyphenolic compounds, and volatile organic substances were shown. The fermentation conditions were selected (pH up to 3.3 ± 0.3, at T = 25 ± 1 °C, process duration of 14 days) to obtain the beverage base. Strawberry and lime leaves were used as flavor and aroma ingredients, and vitamins with inulin were used as functional ingredients. Since the use of additional ingredients changed the finished beverage's organoleptic profile and increased its content of organic acids, the final product's physical–chemical properties, antioxidant activity, and organoleptic indicators were assessed. The content of B vitamins in the beverages ranges from 29 to 44% of RDI, and 100% of RDI for inulin, which allows it to be attributed to the category of enriched products. The DPPH inhibitory activity of the beverages was 82.0 ± 7%, and the ethanol content did not exceed 0.43%. The beverages contained a variety of organic acids: lactic (43.80 ± 4.82 mg/100 mL), acetic (205.00 ± 16.40 mg/100 mL), tartaric (2.00 ± 0.14 mg/100 mL), citric (65.10 ± 5.86 mg/100 mL), and malic (45.50 ± 6.37 mg/100 mL). The technology was developed using pilot equipment to produce fermented kombucha tea enriched with inulin and B vitamins.

Keywords: fermented beverages; kombucha; black tea; physical and chemical characteristics; vitamins; inulin; volatile organic substances

 check for updates

Citation: Frolova, Y.; Vorobyeva, V.; Vorobyeva, I.; Sarkisyan, V.; Malinkin, A.; Isakov, V.; Kochetkova, A. Development of Fermented Kombucha Tea Beverage Enriched with Inulin and B Vitamins. *Fermentation* **2023**, *9*, 552. https://doi.org/10.3390/fermentation9060552

Academic Editor: Mutamed Ayyash

Received: 9 May 2023
Revised: 30 May 2023
Accepted: 6 June 2023
Published: 8 June 2023

1. Introduction

Kombucha is a beverage obtained through the fermentation of sweetened black or green tea using a symbiotic culture of bacteria and yeast (SCOBY) [1–3]. It is believed that the first references to the use of such a beverage date back to 220 BC in Manchuria [1,2,4]. The beneficial health properties attributed to kombucha have led to its present distribution in Europe, the United States, and other countries [4–6]. Kombucha is a popular beverage in many countries, as evidenced by the growth in its production, the constant expansion of its range, and numerous studies [7,8]. According to the results of a multidimensional scaling of the results of articles searched using the "kombucha" keyword (Figure S1) obtained using the KH Coder software [9], several research areas can be identified: composition of symbiotic cultures of bacteria and yeast; factors influencing development and metabolism; the influence of technological parameters on quality and safety; antioxidant properties, including changes in polyphenolic composition during fermentation; and factors influencing the formation of bacterial cellulose, its application in various industries, and in vivo beverage studies.

Various studies of the kombucha production process and kombucha-based beverages are aimed at increasing the functional properties of the end product by selecting optimal production parameters (temperature [10,11], oxygen access [11], fermentation duration [10–12], the ratio of SCOBY to tea substrate, and size and shape of the fermentation tank [11]); basic ingredient composition (nitrogen sources [10,13–15], carbohydrate sources [13,16], and microbiological composition of SCOBY [17]); and additional flavor and functional ingredients [7,8,18].

An algorithm for the development of food products and the development of beverages includes the stage of the choice of ingredients based on the study of their chemical composition, properties, and physiological functions that provide health benefits. Apples [19,20], lemons [21], snake fruit [22], goji berry [15], cherry [23], dragon fruit [24], kiwi [24], red raspberry [25], blackthorn fruits [25], black carrot [25], strawberry tree [26], and others in fresh form, or as juices, purees, and extracts, are used as ingredients that form the taste and aroma of kombucha. [2,8]. Fresh and dried herbs and spices such as ginger [27], cherry leaf [28], sage [29,30], mint [29,31], linden [29], cinnamon [32], cardamom [32], thyme [30,32], echinacea [33], and functional ingredients [34] can be added at the stage of tea base preparation before its fermentation by SCOBY [31,32] or in the fermented base [7] for secondary fermentation and formation of the sensory profile of the beverage. These ingredients can be added at the stage of base preparation before fermentation by SCOBY [31,32] or in the fermented base [7] for secondary fermentation and formation of the sensory profile of the beverage. The use of combinations of ingredients allows not only to obtain a better taste of the beverages but also to regulate the composition of biologically active substances with beneficial health effects [35,36].

Along with the positive effects, kombucha also has negative effects associated with high acidity, in particular, the development of halitosis and potential negative effects on tooth enamel due to enamel degradation [37]. In this regard, it is reasonable to enrich kombucha in order to minimize its negative effects. Direct reduction of its acidity is not acceptable because a sour taste is expected and desired by the consumer. Therefore, searching for indirect ways to correct the negative effects of kombucha is an urgent task.

In order to provide additional health benefits, it is advisable to enrich kombucha-based beverages with physiologically important ingredients, which include soluble dietary fiber, prebiotics of polysaccharide nature, and vitamins.

Vitamins are essential micronutrients, which are low-molecular-weight organic compounds necessary for the enzymatic catalysis mechanisms, the normal course of metabolism, the maintenance of homeostasis, and the biochemical support of all vital functions of the body [38]. The need for vitamins is significantly influenced by a person's age, health condition, nature, and intensity of work. Inadequate consumption of vitamins is a risk factor for many nutrition-dependent diseases [39]. In this regard, it is advisable to include vitamins in the composition of beverages in sufficient amounts to correct their insufficient consumption.

Among the water-soluble ingredients are vitamin C and the complex of B vitamins, whose antioxidant effect neutralizes the reactive oxygen species (ROS) causative of many oral pathologies [40]. Vitamin C is the most common vitamin in kombucha, and its content in a single serving of kombucha varies within the reference daily intake (RDI). Supplementation with this vitamin is, therefore, not necessary. In turn, the content of B vitamins is an order of magnitude less than the RDI, so enrichment with vitamins of this group can increase the positive effects of kombucha consumption and is advisable [3].

In addition, for healthy bones and teeth, adequate dietary calcium intake is essential [41,42]. The direct introduction of soluble calcium salts will lead to an increase in pH and consequently to a deterioration of the consumption properties of kombucha; therefore, the solution may be to use substances that increase the adsorption of calcium from the normal diet. This substance is inulin, a soluble dietary fiber (DF). The role of DF as a nutritional factor that actively influences metabolic processes in the human body has been repeatedly proven in a number of biological and clinical studies [43–45]. Among the proven physiological effects manifested by DF, the most pronounced include the normalization of the motor and evacuatory functions of the large intestine, prebiotic action, and influence on the state of lipid and carbohydrate metabolism. Soluble DF (alginates, pectin, β-glucans, gum arabic, some types of hemicelluloses, and modified cellulose) has a positive effect on lipid and carbohydrate metabolism [43,46], as does inulin. Inulin is contained in the tubers and roots of dahlias, artichokes, chicory, dandelion leaves, and asparagus; it is a mixture of oligomers and polymers of fructose and belongs to the group of fructosans [47,48]. As a prebiotic, inulin has a favorable effect on the human body in that it selectively stimulates the growth and activity of probiotic bacteria in the large intestine, thereby increasing the adaptive capabilities of the organism [47,49]. Besides stimulating the growth and activity of bifido- and lactobacilli, inulin increases calcium absorption in the large intestine, which has a positive effect on bone mineralization [47]. Inulin has a low glycemic index (from 4 to 7) and a low caloric value (1 kcal/g), as well as a distinctly sweet taste, neutral color, and odor [50]. This would replace some of the sucrose in the kombucha formulation, which would also have a positive effect on caries prevention with prolonged consumption of sugary beverages [41]. Inulin can also be used in the diet of patients with type 2 diabetes, as well as patients with impaired glucose tolerance. Similar to dietary fiber, inulin is resistant to the effects of digestive enzymes in the stomach and small intestine. This is due to their structure since, in the human body, there are no enzymes specific to the cleavage of bonds present in the molecules of such polysaccharides [50,51].

The introduction of these components can significantly change the properties of kombucha; therefore, the purpose of this work was to develop a kombucha containing vitamin B complex and inulin and evaluate its properties.

2. Materials and Methods

2.1. Materials

To obtain the fermented beverages, we used black Indian big leaf tea (Moscow, Russia), sugar (Moscow, Russia), a symbiotic culture of bacteria and yeast—SCOBY (Moscow, Russia), frozen strawberry fruit and Kaffir lime leaves (provided by a local supermarket), inulin (90% of the main substance) (Belgium), and vitamin premix RUS 28,174 by DSM Nutritional Products Europe Ltd. (Basel, Switzerland) (the composition is shown in Table 1). For analytical studies, we used NaOH, Folin-Ciocalteu reagent (Sigma-Aldrich), gallic acid (98%) (Diaem), Na_2CO_3, and 2,2-diphenyl-1-picrylhydrazyl (DPPH) (ABCR GmbH and Co. KG, Karlsruhe, Germany). Analytical grade reagents (Sigma-Aldrich, St. Louis, MO, USA) were used for chromatography.

Table 1. Composition of vitamin premix.

Compound	Concentration, mg/kg
Thiamine (Vitamin B_1)	48,750
Riboflavin	44,000
Pyridoxine (Vitamin B_6)	57,000
Niacin	479,998
Folic acid	7500

2.2. Fermented Beverage Base Production Technology

Black tea (0.5 wt.%) was added to boiled water at $94 \pm 2\ °C$ and brewed for 10 min with occasional stirring. Then, sugar was added in an amount of 5.0 wt.% and stirred for

5 min until complete dissolution. The obtained solution was filtered to remove tea leaves and cooled to 23 ± 2 °C. Then, the SCOBY culture was added (10.0 wt.%), and the container was covered with a permeable cloth. Fermentation was carried out in 25 L fermenters (Brew Bucket Brewmaster, Pittsburg, CA, USA). The duration of the fermentation process was 14 days at 25 ± 1 °C. After fermentation, the resulting base was filtered through a filter (70 μm) and pasteurized at 72 ± 2 °C for 40 min. The resulting base was then cooled to a temperature of 25–30 °C and used to prepare a beverage or stored in hermetically sealed containers at 5 °C until use. During fermentation with a periodicity of 0 days, 3 days, 7 days, and 14 days, acidity, pH, total content of polyphenolic compounds, content of mono- and disaccharides, organic acids, ethanol, and the profile of volatile substances were controlled.

2.3. Methods

2.3.1. Titratable Acidity Determination

The acidity value was determined using the titrimetric method [52]. The method is based on titration with NaOH solution (0.1 N) of all acidic substances. Before the study, the samples were freed from carbon dioxide formed during fermentation by boiling. Acidity was expressed in cubic centimeters of sodium hydroxide solution with a concentration of 1 N, which was used to titrate 100 mL of the beverages.

2.3.2. pH Determination

The pH changes were monitored using an electronic pH meter S20_K Mettler Toledo (Greifensee, Switzerland) [31].

2.3.3. Determination of Carbohydrates Profile

The composition of carbohydrates (mono- and disaccharides) was determined using standard high-performance liquid chromatography on an Agilent Technologies 1260 chromatograph (Agilent Technologies, Santa Clara, CA, USA) with a refractometric detector Agilent 1260 RID, G1362A (Agilent Technologies, Santa Clara, CA, USA) according to [53]. Before analysis, the samples were centrifuged at 990× g for 15 min (if necessary, diluted with water at a ratio of 1:5 (by volume)). In order to determine the mass concentration of sucrose, glucose, and fructose, the samples were diluted with distilled water in a 1:20 ratio. Then, 1–2 mL of the sample was taken and filtered through a filter with a 0.45 μm pore diameter into a vial.

2.3.4. Determination of the Dry Matter Content

Total soluble substances were determined using an Atago refractometer (Tokyo, Japan) and expressed in degrees Brix (°Brix), according to [54].

2.3.5. Determination of Ethanol Content

The determination of ethanol content in the samples was carried out on an Agilent Technologies 7890A gas chromatograph with a flame ionization detector (Agilent Technologies, Santa Clara, CA, USA) and an analytical column Supelcowax 10 (60 m × 0.53 mm × 1 μm) from Supelco (Bellefonte, PA, USA), according to [55] with modifications. Temperature program: 78 °C for 2 min, heating to 110 °C at 3 °C/min, rising to 220 °C at 30 °C/min, with a 10 min delay. The carrier gas was helium, and the flow rate was 2.8 mL/min. A 20 mL vial was placed on the scale, 5 mL of water was added, 2 mL of sample was added (weight of the added sample was recorded with an accuracy of 1 mg), then 1 mL of internal standard solution (0.8% 1-butanol solution, the weight of internal standard solution was recorded with an accuracy of 1 mg) was added, and another 10 mL of water was added. After being stirred, 2 mL was taken into a 2 mL centrifuge tube and centrifuged at 18,400× g. Then, 1 mL was taken into a 2 mL glass vial, and 2 mL was injected into the chromatograph. The chromatograph was calibrated in the concentration range of 3 to 300 μg/g.

2.3.6. Determination of Organic Acids

Organic acids were determined by standard reversed-phase high-performance liquid chromatography on an Agilent Technologies 1100 chromatograph (Agilent Technologies, Santa Clara, CA, USA), according to [56]. Organic acids were separated in a chromatographic column filled with octadecyl silica gel. Concentrations were determined using a spectrophotometric detector at 210 nm using the external standard method.

2.3.7. Determination of Total Polyphenol Content

The total content of polyphenolic compounds during fermentation was determined by the Folin-Ciocalteu spectrophotometric method using a SpectroQuest 2800 spectrophotometer (UNICO, Suite E, Dayton, NJ, USA), according to [57] with modifications. Before analysis, the beverage sample was diluted with distilled water in a ratio of 1:9. To 1 mL of the sample, 5 mL of 10% Folin-Ciocalteu solution was added and incubated for 5 min. After this time, 4 mL of 7.5% Na_2CO_3 solution was added, thoroughly mixed, and incubated in a dark place for 1 h. The optical density of the samples was measured at 765 nm. Gallic acid was used as a standard.

2.3.8. Determination of Volatile Substances

The analysis was performed on an Agilent Technologies 7890A gas chromatograph with an Agilent Technologies 5975C mass detector (Agilent Technologies, Santa Clara, CA, USA) and a Supelcowax 10 60 m × 0.53 mm × 1 μm chromatographic column (Supelco, Bellefonte, PA, USA), according to [58] with modifications. Divinylbenzene/carboxene/polydimethylsiloxane (50/30 μm) fiber with manual holder (Supelco, Bellefonte, PA, USA) was used for the extraction of aroma components. The fiber was preconditioned before the analyses, according to the instructions of the manufacturer. A total of 10 mL of sample was placed in a 20 mL headspace vial, sealed with a septum and an aluminum cap. The fiber was placed in the space above the sample and incubated for 30 min on a tile heated to 110 °C. Then, it was placed into the gas chromatograph injector, and the analysis was performed. Temperature program: 35 °C for 5 min, heating to 220 °C at a rate of 4 °C/min, isotherm 40 min. Helium was used as a carrier gas. Injector was operated in splitless mode at a temperature of 225 °C. Operation parameters of the mass detector: scanning range 35–400 m/z, ionization source temperature 230 °C, quadrupole temperature 150 °C, and electron impact ionization with energy 70 eV. The results were processed using the program "MSD ChemStation E02.02.1431" (Agilent Technologies, Santa Clara, CA, USA), the program "The NIST Mass Spectral Search Program for the NIST/EPA/NIH Mass Spectral Library Ver. 2.0 gm" (National Institute of Standards and Technology, Gaithersburg, MD, USA), and a set of commercially available mass spectral libraries. Components with a spectral coefficient of agreement with the library of more than 700 were considered to be identified. Subsequent processing of the obtained data was performed using the "Microsoft Office Excel 2007 SP3 MSO" software package.

2.3.9. Modeling of Reaction Kinetics

An analysis of reaction kinetics was performed according to the principle outlined in [59]. In order to determine the reaction rate constant (k), the conformity of changes in the concentration of volatile substances with the models of zero (Equation (1)), first (Equation (2)), and second (Equation (3)) order reactions were evaluated:

$$C - C_0 = kt \tag{1}$$

$$\ln\left(\frac{C}{C_0}\right) = kt \tag{2}$$

$$\frac{1}{C} - \frac{1}{C_0} = kt \tag{3}$$

where C_0 is the initial concentration (%), C is the concentration (%) at time t (days), and k is the reaction constant.

The value of the reaction constant was calculated using the linear equation of the relationship between concentration and reaction time. In addition to the reaction constant, we also determined the half-life ($t_{1/2}$) of the studied substances according to Equations (4) and (5) for first- and second-order reactions, respectively.

$$t_{1/2} = \frac{\ln 2}{k} \tag{4}$$

$$t_{1/2} = \frac{1}{k} \cdot \frac{1}{C_0} \tag{5}$$

2.4. The Technology of Fermented Beverage Production

Before use, the prepared fermented base (technology of obtaining described above) was brought to a temperature of 25–30 °C. Preliminary organoleptic studies (data not shown) allowed us to identify preferences and determine the number of ingredients that form the taste and aroma of fermented beverages: frozen strawberries and lime leaves. Strawberries and lime leaves were crushed, added to the prepared fermented base, and left for 24 h at 23 ± 2 °C for infusion and secondary fermentation. Further filtration was carried out in stages from 70 μm to 5 μm. For enrichment, pre-dissolved inulin and vitamin premix, which included water-soluble B vitamins B_1, B_2, B_6, PP, and folic acid, were added to the beverages after filtration. Then, the beverages were placed into a cylindroconical tank "SS Brewtech Unitank" (USA) equipped with a circulating cooler "Termex" and a bottling nozzle (Tomsk, Russia) for carbonization of the beverages (10 °C, 1.1 MPa) with subsequent bottling into glass bottles. To stop further fermentation and ensure microbiological purity, the bottles with the beverages were pasteurized (72 ± 2 °C, 40 min) in a Binder BD53 (Tuttlingen, Germany). The line with the pilot equipment is shown in Figure 1.

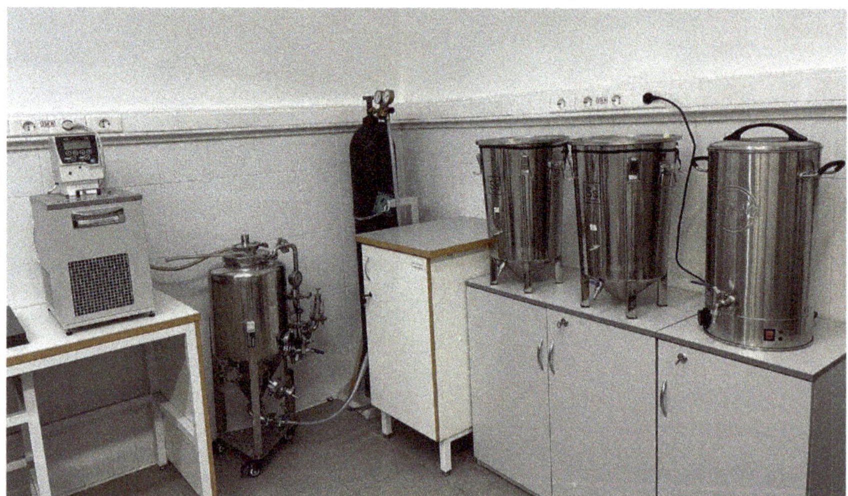

Figure 1. Pilot equipment line for the production of fermented beverages.

The finished beverages were characterized according to the indicators: acidity, pH, total soluble substances, the content of mono-, disaccharides, ethanol, organic acids (the methods for determining the indicators are presented above), inulin, vitamins, antioxidant activity, and organoleptic evaluation.

2.5. Methods for Investigation of Fermented Beverages

2.5.1. Determination of Inulin Content

Inulin content in the finished beverages was determined with a standard method using high-performance liquid chromatography on an Agilent Technologies 1260 (Agilent Technologies, Santa Clara, CA, USA) chromatograph with refractometric detector Agilent 1260 RID, G1362A (Agilent Technologies) [60].

2.5.2. Determination of Vitamin Content

The content of vitamins B_1, B_2, B_6, PP, and folic acid was determined according to the method [61].

2.5.3. Assessment of Antioxidant Activity

The study of the antioxidant activity of the finished beverages was carried out using free radical DPPH [62]. Before the study, the beverages were filtered and diluted twice. The filtered beverages in an amount of 100 µL were mixed with 2 mL of DPPH solution in ethanol (250 µM) and incubated in a dark place for 1 h at 24 ± 1 °C. The obtained solutions were studied on an electron paramagnetic resonance spectrometer SPINSCAN X (Republic of Belarus) operating at a frequency of X-band. Conditions for obtaining electron paramagnetic resonance (EPR) spectra: central field 336 mT, sweep amplitude 15 mT, modulation amplitude 200 µT, and power 20 dB. The obtained spectra were processed using the program e-Spinoza (Republic of Belarus). The percent inhibition of the EPR spectrum was calculated according to the following equation: % inhibition = $[(I_0 - I)/I0] \times 100\%$, where I_0 is the area of the EPR spectrum of DPPH (control sample), and I is the area of the EPR spectrum of DPPH with sample.

2.5.4. Organoleptic Evaluation

The sensory assessors were selected, trained, and monitored according to ISO 8586 standards [63,64]. The sensory panel consisted of 8 trained assessors (5 females and 3 males; mean age: 37.2 ± 1.3). They were required to be healthy and not smoke, drink coffee, or eat spicy food for 3 h before analysis.

The conditions of the sensory evaluation environment, including the area of sample preparation, sensory evaluation, and concentrative discussion, met the requirements of the ISO 8589 standard [65]. The sensory evaluation area was kept in an adequately air-conditioned environment, and the temperature in the booths was controlled at about 25 °C. The conventional sensory profile of the kombucha samples was established according to the ISO 13299 standard [66]. Organoleptic evaluation of the finished beverages was carried out on a 5-point scale by indicators: appearance, smell, consistency, taste, and color.

2.6. Statistical Processing

Obtained data are presented in the form of average values. Statistical data processing and graphing were carried out using the software package R studio (version 3.5.3) [67], using the module ggplot2 [68]. The Kruskal–Wallis test with the Mann–Whitney test as post hoc was used to compare changes of measured parameters during the fermentation (the first day was used as control) [69]. Results are shown as H and p-values. All measurements were conducted in triplicate. p values < 0.05 were considered significant.

3. Results and Discussion

Studies on kombucha and beverages based on it are mainly conducted on a laboratory scale, with volumes ranging from 200 mL to 2 L [11]. However, a number of researchers studied the fermentation process in larger volumes [70,71], and it was shown that regardless of the size and volume of the vessel, with a constant value of the interfacial surface, it is possible to obtain a beverage with similar properties [11]. In order to optimize the industrial production of enriched kombucha, the main fermentation process was carried out in 25 L fermenters.

3.1. Preparation and Study of Fermented Beverage Bases

The processes of obtaining fermented tea beverages have their own technological features, which depend on cultural traditions, the raw materials used, the concentration of tea and sugar, and the duration of fermentation [72]. However, the basic technological steps are common and consist of brewing the tea, adding sucrose or other carbohydrate sources, filtering and cooling to room temperature, adding SCOBY and/or starter fluid (fermented kombucha from a previous batch), and fermentation [72,73]. The fermentation process of sweetened tea begins after the addition of SCOBY and/or starter liquids. Under the action of enzymes formed in yeast cells, hydrolysis of sucrose to glucose and fructose occurs at room temperature, which are subsequently converted to ethanol and carbon dioxide as a result of alcoholic fermentation [74]. The change in the profile of carbohydrates during fermentation is shown in Figure 2.

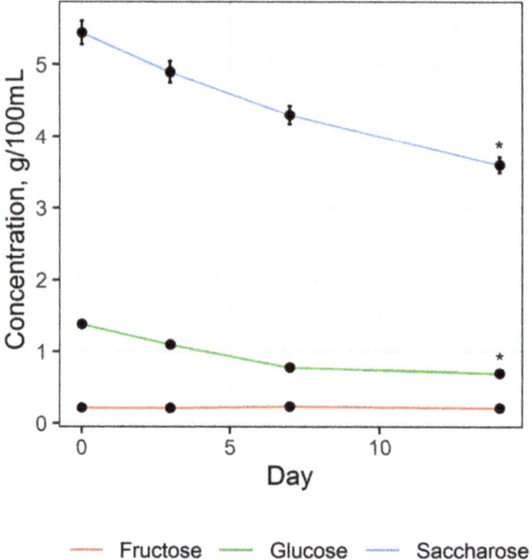

Figure 2. Changes in carbohydrate profile during fermentation (significant ($p < 0.05$) differences from the starting point are shown by "*" symbol).

Figure 2 shows that during the fermentation process for 14 days, there is a decrease in saccharose content by an average of 45% (H = 8.897, p = 0.045). The concentration of glucose also significantly decreased (H = 9.051, p = 0.029), but the change in the fructose concentration had only an insignificant tendency (H = 0.723, p = 0.868), indicating the different metabolism of these carbohydrates by SCOBY microorganisms. The change in the concentration of glucose and fructose during fermentation depends on the composition of the SCOBY [5]. Thus, acetic acid bacteria assimilate fructose and glucose to produce various organic acids and form a cellulose film [74], while the assimilation of glucose and fructose is not the same [75]. Yeast mainly assimilates glucose with the formation of carbon dioxide and ethanol; however, yeast of the genus *Saccharomyces* prefers glucose, whereas certain yeast of the genus *Zygosaccharomyces* prefer fructose. In our study, the fermentation process was continued for 14 days, after which the fermented base contained residual sugars. However, there are alternative methods. For example, prolonged fermentation for more than 30 days allows the use of a fermented base with 0% sugar content, due to which its caloric value is lower than that of a base containing sugar. For the preparation of the beverages, the base is necessarily diluted with juices, herbal infusions, etc., while for sweetening, stevia extract or other sweeteners can be used. In our study, the criterion for the completion of the primary fermentation process was the pH value. Residual sugar in the studied base gave a sour-sweet taste and reduced the perception of the sour taste of the beverages. During fermentation, due to the accumulation of organic acids, mainly acetic acid, the pH value decreases, and the acidity increases [74]. Control of these indicators allows for the preparation of a beverage with the desired organoleptic characteristics. The results of changes in pH and acidity during the fermentation of the beverage base are shown in Figure 3.

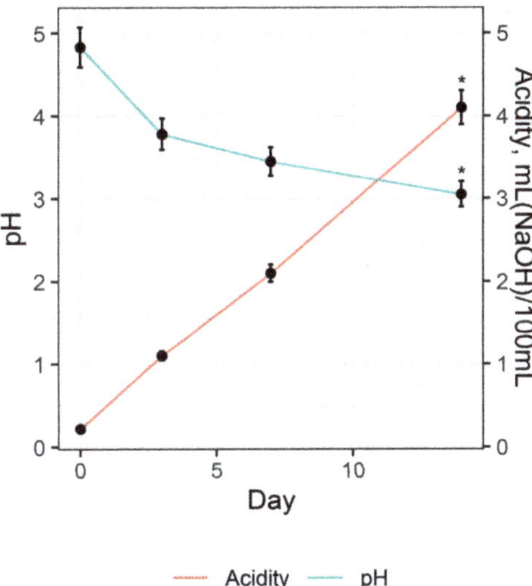

Figure 3. Changes in pH and acidity of the beverage base during fermentation (significant ($p < 0.05$) differences from the starting point are shown by "*" symbol).

As shown in Figure 3, the rate of change in the pH and acidity of the beverage base during fermentation is non-linear. At the beginning of the fermentation process (up to 3 days), there is an intense decrease in pH values with a subsequent decrease in rate (H = 13.5, $p = 0.0091$). At the same time, the change in acidity is reversed: during the first 7 days, acidity increases slightly, then increases sharply, which is associated with the accumulation of organic acids in the substrate, in particular acetic acid (Figure 4). It should be noted that the sharp increase in acidity (H = 13.5, $p = 0.0091$) does not have a strong correlation (r = -0.864, $p < 0.001$) correlate with the change in pH value, which may be due to the manifestation of buffering properties of the fermented beverages [76]. Fermentation was continued until reaching pH values of 3.3 ± 0.3.

Due to alcoholic fermentation, ethanol was formed during the first stages of the fermentation process [74]. It was found that the ethanol content on day 7 of fermentation was 0.12%, and on day 14, it did not exceed 0.22%. When selecting a tea substrate for kombucha, the dynamics of ethanol accumulation during fermentation should be taken into account [77,78].

The fermentation process begins with the hydrolysis of sucrose into fructose and glucose by enzymes produced in the yeast cells. Part of the glucose and fructose is then used by the yeast to produce ethanol and carbon dioxide [11,79]. Another part of the fructose and glucose and the resulting ethanol is used by acetic acid bacteria to produce cellulose and organic acids, of which acetic acid is the main one [11,79]. In addition to acetic acid, other organic acids, such as gluconic acid, lactic acid, malic acid, citric acid, tartaric acid, and others, are accumulated in much smaller amounts during fermentation [80]. Regulation of the concentration and type of organic acids formed is achieved by varying the component composition of the tea base and carbohydrate sources [81,82]. In this work, to obtain a fermented beverage base, traditional raw materials were used: black tea and sugar; therefore, the main organic acid formed in the process of fermentation was acetic acid (Figure 4).

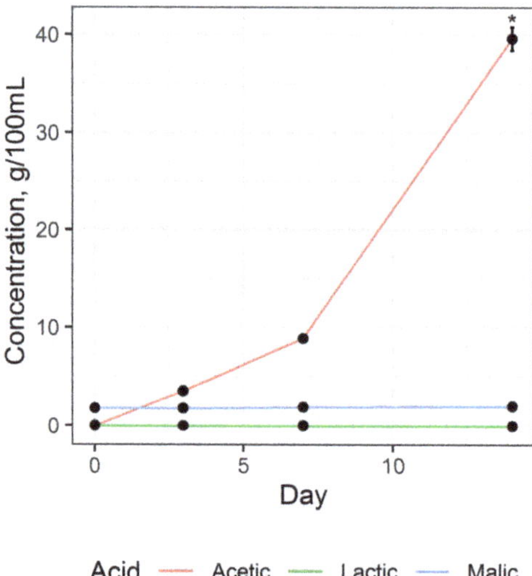

Figure 4. Changes in the content of the main organic acids during fermentation (significant ($p < 0.05$) differences from the starting point are shown by "*" symbol).

It was found that the accumulation of acetic acid in the first three days of fermentation occurs slowly, and subsequently, the rate of accumulation increases (H = 10.531, $p = 0.015$). The character of the curve of acetic acid accumulation correlates with the change in the acidity of the base during fermentation (Figure 3). In the period from the 10th to the 14th day, there is an accumulation of acetic acid, which negatively affects the organoleptic properties of the base due to the appearance of a sharp smell and sour taste. Accumulation of other major organic acids was much lower and was statistically significant for lactic acid (H = 9.581, $p = 0.023$) and insignificant for malic acid (H = 1.049, $p = 0.789$).

Consumption of beverages with a high organic acid content is unsafe and poses potential risks to consumers. To eliminate the negative influence of the temperature factor on the technological process and the formation of undesirable fermentation products in fermenters, they were maintained at a temperature of 25 ± 1 °C.

The popularity of fermented tea beverages is due not only to their original taste and odor but also to their health-promoting properties, in particular, their antioxidant properties. The biological activity of SCOBY is primarily related to the chemical composition of the tea itself used in the preparation of the beverage [6]. Flavonoids such as flavanols (flavan-3-ols), flavonols, flavones, flavanones, and anthocyanidins are the main components of tea leaves, which together account for up to 30% of the dry weight of tea leaves [83]. The composition of polyphenolic compounds and their content depend on the type of tea, growing conditions, processing and storage technology, etc. [84]. The results of the evaluation of the total content of polyphenolic compounds in the beverage base and their change during fermentation are shown in Figure 5.

During the fermentation process under the influence of SCOBY during the first 7 days, there is an increase in the total content of polyphenolic compounds (H = 7.308, $p = 0.042$). As the process continues, their content decreases. Data from other researchers support our conclusions [85]. The transformation of phenolic compounds from their conjugated forms into their free forms, as well as the oxidation and dimerization of tea catechins with the increased catalytic ability of polyphenol oxidase, are all linked to changes in this indicator during fermentation [85]. The conversion of flavonoids during fermentation, e.g., thearubigin to theaflavin, leads to a change in the color of the fermented base from dark to light [86].

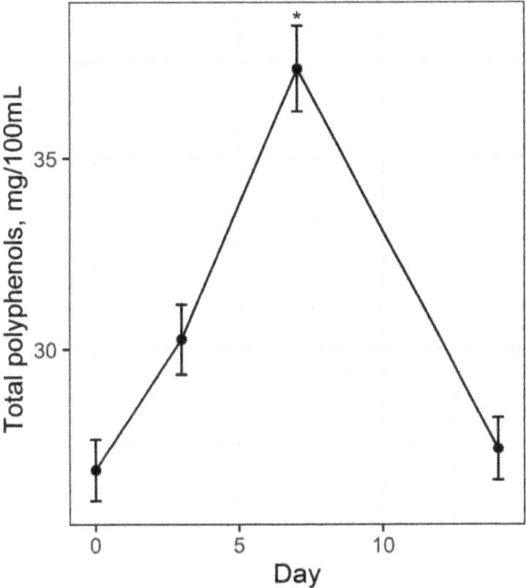

Figure 5. Changes in the content of total polyphenolic compounds during fermentation in terms of gallic acid (significant ($p < 0.05$) differences from the starting point are shown by "*" symbol).

The type of tea plant, growing conditions, and a variety of other factors affect how the flavor profile of the finished product develops [87]. In the process of fermentation, there is the formation of volatile compounds that form a specific category of beverages' flavor profile. As a result of this research, 70 volatile compounds were identified in the initial infusion of black tea before the beginning of the fermentation process. On the 14th day of fermentation (the completion of the fermentation process), 55 volatile compounds were identified. From the total list of identified volatile organic compounds, substances with a high content in the aromatic profile were selected. These substances were subdivided into four main groups: aldehydes, acids, alcohols, and terpenes (Figure 6).

During the fermentation, the profile of volatile substances changes: some identified in the initial tea infusion disappeared while the concentration of others increased. As can be seen from Figure 6, at the starting point, aldehydes and terpenes have the highest concentrations in the tea base, with hexanal (12.1%) and linalool (8.2%) predominating in these groups, respectively. The presence of hexanal and linalool imparts an herbaceous and floral aroma to the beverages, which disappears during fermentation. On the third day of fermentation, due to the alcoholic fermentation, the content of ethanol in the vapor phase increases rapidly (59%). On the 7th day, the content of organic acids increases significantly, mainly due to acetic (13.5%), pentanoic (4.3%), and octanoic (3.1%) acids. On the 7th day, the concentration of alcohols (phenylethyl, ethanol, dimethylphenol, and 3-methyl-1-butanol) decreases naturally. The maximum content was observed for ethanol (40%). On the 14th day of the study, the alcohol content decreased to below 3%, except for ethanol, where the concentration in the vapor phase above the fermented beverage sample was 35.2%. However, the ethanol content in the fermented base was 0.22%. A low alcohol content of up to 1% may not be felt by the consumer because the threshold of taste for ethanol ranges from 1 to 2% [88], while a concentration of more than 1% may result in a bitter taste. The content of acetic acid on the 14th day increased to 23.4%, and the content of 3-methylbutanoic acid also increased significantly. The results obtained are generally characteristic of the process of tea fermentation by SCOBY [89] and indicate that aldehydes and terpenes are the main volatile organic compounds that are metabolized in the fermentation process. This may be an indication of the loss of the traditional black tea aroma in the final product.

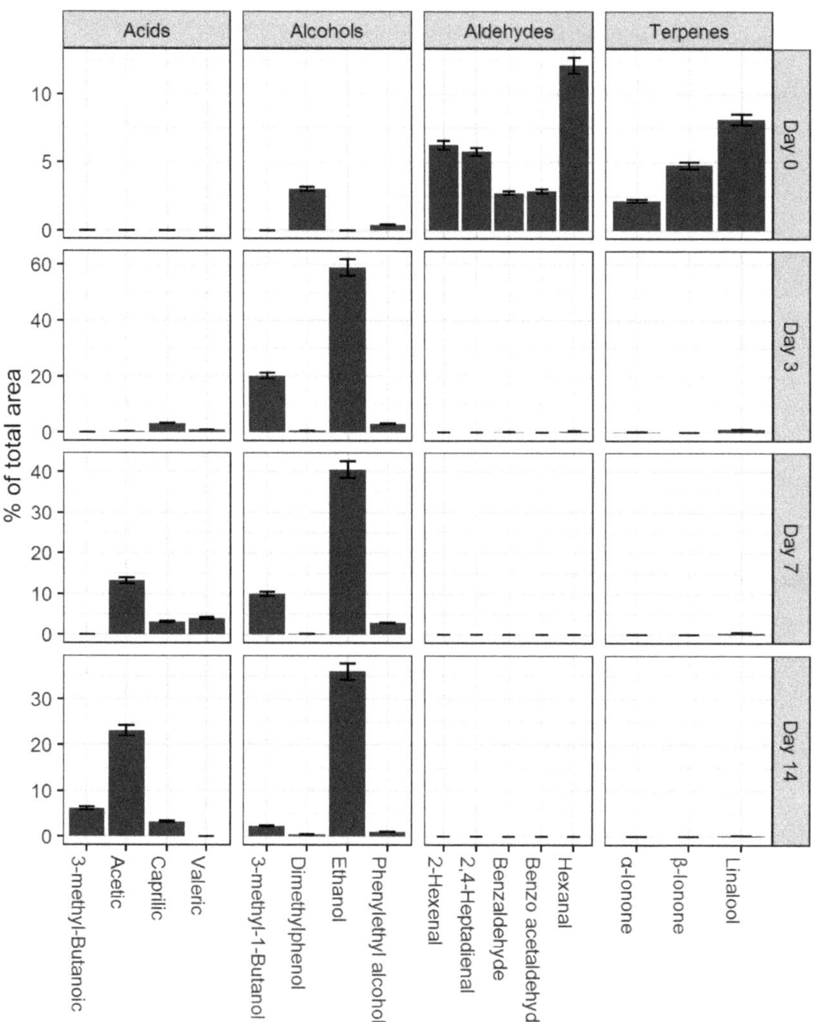

Figure 6. Diagram of the content of groups of volatile organic substances in the studied samples from the first to the fourteenth day of fermentation.

The kinetics of the accumulation and degradation of volatile substances in kombucha were analyzed. For substances whose content was measured throughout the experiment and varied at selected measurement intervals, information on the order and constants of reactions as well as on the time of semi-transformation of substances was obtained (Table 2). The table shows information on the models with the highest values of linear correlation coefficients (r^2).

Only the accumulation of acetic acid fitted well ($r^2 = 0.960$) with the zero-order reaction kinetic model. This indicates that the process of vinegar synthesis from ethanol under the action of acetic acid bacteria does not depend on ethanol concentration. Thus, we can say that up to the 14th day of fermentation, ethanol was in excess for the synthesis of acetic acid. This may be due, in particular, to the low rate of metabolization of ethanol, which has a half-life time of 18.05 days. The decrease in ethanol (after the third day) proceeds as a second-order reaction with the lowest reaction constant among those studied. In contrast to ethanol, the change in the content of 3-methyl-1-butanol proceeds faster as a first-order reaction with a half-life of 1.15 days.

Table 2. Kinetic parameters of degradation or synthesis of volatile compounds in kombucha.

Compound	Reaction Order	r^2	K	$t_{1/2}$, Days
Acetic acid	Zero	0.960 ± 0.012	1.22 ± 0.03 (M·d^{-1})	-
Ethanol	Second	0.854 ± 0.010	$9.39 \pm 0.11 \times 10^{-4}$ (M^{-1}·d^{-1})	18.05 ± 1.28
3-methyl-1-Butanol	First	0.999 ± 0.014	$5.99 \pm 0.02 \times 10^{-1}$ (d^{-1})	1.15 ± 0.16
Linalool	Second	0.996 ± 0.013	$2.95 \pm 0.07 \times 10^{-1}$ (M^{-1}·d^{-1})	0.42 ± 0.09
Hexanal	Second	0.984 ± 0.021	1.4 ± 0.06 (M^{-1}·d^{-1})	0.06 ± 0.01

"-"—not applicable.

The decrease in linalool and hexanal, which were initially present in the tea, also corresponds to the kinetics of the second-order reaction. They are characterized by a high rate of degradation: 0.42 days for linalool and 0.06 days for hexanal. The rate constants of these reactions are 3–4 orders of magnitude higher than the rate constant of the ethanol metabolization reaction. Nevertheless, the order of these reactions indicates that the decrease in the concentration of linalool and hexanal is not a simple process but is caused by a number of factors, such as the action of aldehyde dehydrogenase in acetic acid bacteria [90].

3.2. Preparation and Study of the Fermented Beverage

A kombucha base was used as a matrix for the beverages. The use of fresh berries, fruits, and vegetables in the composition of beverages is due to their content of vitamins, minerals, polyphenolic compounds, phenolic acids, tannins, etc. [91,92]. Lime leaves give a special flavor to the product due to the essential oils. Crushed strawberries and lime leaves were introduced into the prepared fermented base and incubated at 23 ± 2 °C for 24 h for infusion and secondary fermentation, followed by filtration. The composition of the fermented strawberry and lime beverages is shown in Table 3.

Table 3. Composition of the fermented strawberry and lime beverages.

Ingredient	Amount, g
Kombucha base	1038.0
Strawberry	173.0
Kaffir lime leaf	3.7
Total	1214.7
Yield after filtration	1000

The addition of flavoring ingredients leads to secondary fermentation and the extraction of biologically active substances. The effect of the addition of flavoring ingredients (strawberry fruit and lime leaves) and secondary fermentation on the physicochemical parameters of the fermented base is presented in Table 4.

Based on the data obtained (Table 4), it was found that the addition of ingredients leads to secondary fermentation, as evidenced by the decrease in sucrose content and increase in alcohol and organic acids. Changing the profile of organic acids is associated not only with the process of secondary fermentation but also with the extraction of acids from strawberry fruits. According to [93,94], strawberry fruits mainly contain citric and malic acids and, in small amounts, tartaric, oxalic, and fumaric acids. The increase in the content of citric, malic, and tartaric acids in the fermented beverages after the second fermentation relative to the fermented base (Table 4) indicates their extraction from strawberry fruits. The increase in alcohol content (from 0.22 to 0.43%) and decrease in sugar content (from 3.6 g/100 mL to 0.18 g/100 mL) indicate the progress of alcoholic fermentation. There is also a dramatic increase in acetic acid content after secondary fermentation. At the same time, the pH value increased insignificantly. As noted earlier, the change in pH is not always associated with a change in the content of organic acids because the fermented base may have buffering properties [76].

Table 4. Comparison of physicochemical parameters of fermented base and base after secondary fermentation.

Parameter	Value Kombucha Base	Value Fermented Beverages
Acidity, mL 1 mole/L NaOH/100 mL	4.10 ± 0.26	4.72 ± 0.38
pH	3.31 ± 0.08	3.57 ± 0.07
Solids content, °Brix	4.60 ± 0.25	4.95 ± 0.25
Ethanol content, %	0.22 ± 0.01	0.43 ± 0.01
Content of organic acids, mg/100 mL		
- lactic	0.025 ± 0.001	43.80 ± 4.82
- acetic	39.80 ± 1.82	205.00 ± 16.40
- tartaric	-	2.00 ± 0.14
- citric	-	65.10 ± 5.86
- malic	2.04 ± 0.11	45.50 ± 6.37
Carbohydrates contents (mono-, disaccharides), g/100 mL		
- glucose	0.70 ± 0.11	1.45 ± 0.13
- fructose	0.22 ± 0.12	1.47 ± 0.12
- saccharose	3.60 ± 0.04	0.18 ± 0.02

"-"—not detected.

In order to increase the nutritional value of the beverages, they were enriched with B vitamins and inulin, nutrients with clinically proven health benefits. Inulin was pre-dissolved in boiled water at 60–70 °C, and then the solution was cooled to room temperature. We added vitamin premix and stirred until it completely dissolved. When calculating the number of enriching ingredients, we proceeded from the recommendation to ensure the content in a beverage portion (220 g) of inulin was in the amount corresponding to the RDI for vitamins, from 15 to 50% of the RDI. The composition of the enriched fermented beverages is shown in Table 5.

Table 5. Composition of the enriched fermented beverage.

Ingredient	Amount, g
Filtered fermented beverage	903.34
Water	84.0
Inulin	12.6
Vitamin premix	0.06
Total	1000

The content of B vitamins in the enriched fermented beverages was as follows: vitamin B_1—0.27 ± 0.02 mg/100 g, vitamin B_2—0.25 ± 0.02 mg/100 g, vitamin B_6—0.34 ± 0.03 mg/100 g, folic acid—0.04 ± 0.01 mg/100 g, and vitamin PP—2.76 ± 0.22 mg/100 g. Inulin content was 1.21 ± 0.12%. In the manufacturing process of enriched food products, vitamins are subjected to physical and chemical treatment, which may adversely affect their safety. The results of analytical studies have stated that the technological processes of beverage production, in particular carbonization and pasteurization, have a negative impact on the stability of folic acid (12% loss).

The antioxidant properties of tea and tea beverages are influenced by their high polyphenolic compound content [6]. The DPPH inhibitory activity of the kombucha base was 89.9%. According to the proposed technology (Section 2.4), crushed strawberry fruits and lime leaves were added to the base, which can lead to the extraction of substances that have antioxidant properties. At the same time, according to the recipe (Table 3), dissolved inulin was added to the kombucha base, which could lead to dilution of the base and a slight decrease in antioxidant activity. Moreover, the technology of obtaining enriched fermented beverages involves a pasteurization stage (72 ± 2 °C, 40 min), in which the destruction of biologically active substances that have antioxidant properties may occur. Analysis of the enriched fermented beverages showed that the DPPH inhibitory activity was 82.0%.

Consumer properties of beverages are decisive for successful market promotion (or sale). In the finished product, the ethanol content did not exceed 0.43%, which classifies the developed beverages as non-alcoholic [95]. Organoleptic properties of the developed beverages were formed as a result of fermentation processes as well as the extraction of flavor and aromatic substances from introduced strawberries and lime leaves. Characteristics of the organoleptic and the characteristics of the developed enriched beverages are presented in Table 6.

Table 6. Organoleptic characteristics of kombucha base and fermented beverages.

Parameter	Description	Value	
		Kombucha Base	Fermented Beverages
Appearance	Non-transparent liquid. Sludge due to the characteristics of the raw material used is allowed, without foreign inclusions.	5.00 ± 0.00	5.00 ± 0.00
Color	Red with a brownish hue due to the color of the raw material used.	5.00 ± 0.00	5.00 ± 0.00
Odor	Inherent in the ingredients used with the aroma of a fermented beverage.	4.25 ± 0.71	4.71 ± 0.48
Taste	Sweet and sour with strawberry flavor and pronounced kaffir lime leaf flavor.	4.25 ± 0.71	4.71 ± 0.48
Consistency	Liquid with low viscosity and rich body.	4.38 ± 0.50	5.00 ± 0.00

The results of the study of organoleptic properties, shown in Table 6, show a high estimate of the developed beverages by the indicators of "odor", "consistency", and "taste".

At the stage of secondary fermentation, there is an extraction from the introduced ingredients of flavors, aromatics, sugars, vitamins, minerals, polyphenols, and other biologically active substances. This technological stage is very important for the formation of the organoleptic properties of the beverages. The resulting enriched fermented beverages have a pleasant color and exquisite aroma, not characteristic of traditional beverages. However, the dominant flavor and aroma of lime were noted by some tasters as a drawback.

The content of vitamins in a portion (220 g) of a developed beverage refers to enriched products. Inulin content does not exceed the upper allowable level of consumption (Figure 7).

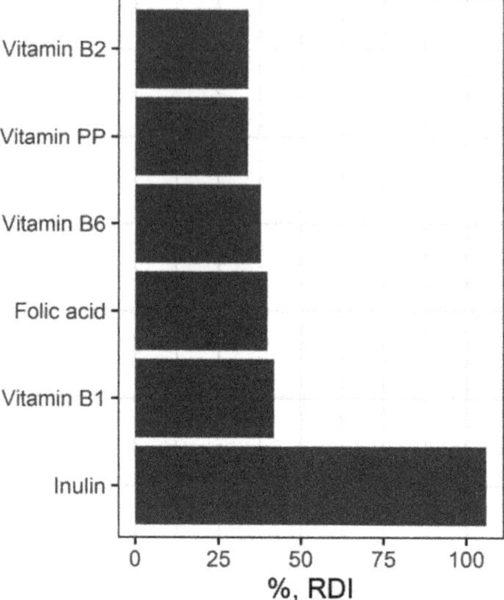

Figure 7. Meeting the average daily requirement of an adult's inulin and vitamins.

The technological scheme for producing kombucha enriched with inulin and B vitamins was developed (Figure 8).

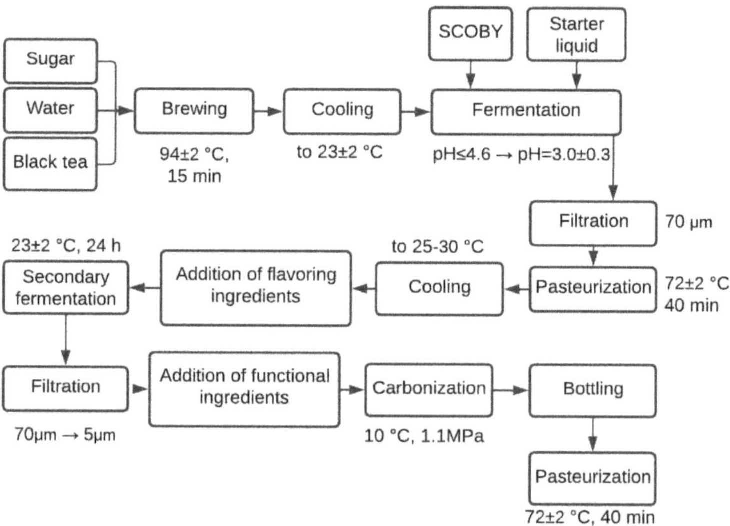

Figure 8. Technological scheme of enriched fermented beverage production.

Basic and auxiliary processes in fermented beverage technology can be distinguished [73]. The developed technology of obtaining fermented beverage kombucha includes the main stages: brewing tea, cooling, fermentation, and filtration, i.e., stages that allow obtaining classic unpasteurized fermented beverage without the additional introduction of other ingredients. The auxiliary stages of the process (pasteurization, addition of flavors and functional ingredients, secondary fermentation, and carbonation) allow for a broader range of beverages and additional health benefits. A clinical study of the developed beverage enriched with inulin and vitamins presented in [96] showed a decrease in the intensity of the complaints significant for irritable bowel syndrome with constipation due to the normalization of stool frequency and consistency.

4. Conclusions

Changes in physical and chemical indicators of the tea base sweetened with sucrose under the influence of a symbiotic culture of bacteria and yeast have been investigated. The conditions for obtaining fermented tea bases with stable physicochemical characteristics on the pilot equipment were described. The change in the profile of volatile flavor-forming substances in the beverages during fermentation has been studied, and the kinetics of their accumulation have been described. The formulation and technology of fermented tea beverage kombucha with strawberry and lime enriched with vitamins and inulin were developed. As practical recommendations for the preservation of vitamins during long storage, it is recommended to bottle the developed fermented beverage kombucha, enriched with vitamins and inulin, into dark glass bottles. The results of these studies testify that kombucha can serve as a basis for new specialized beverages with tailored consumer properties, and the addition of B vitamins and inulin leads to an acceptable organoleptic profile.

Supplementary Materials: The supporting information can be downloaded at: https://www.mdpi.com/article/10.3390/fermentation9060552/s1, Figure S1: Results of multidimensional scaling of the search results for articles on the keyword "kombucha".

Author Contributions: Conceptualization, A.K. and V.I.; methodology, Y.F., V.V., I.V. and V.S.; validation, V.V., I.V. and V.S.; formal analysis, Y.F. and A.M.; writing—original draft preparation, Y.F., V.S., V.V. and I.V.; writing—review and editing, A.K., V.S. and Y.F.; visualization, V.S. and Y.F.; supervision, A.K.; project administration, V.I. All authors have read and agreed to the published version of the manuscript.

Funding: This research was funded by the Russian Science Foundation, grant number 19-76-30014.

Institutional Review Board Statement: Not applicable.

Informed Consent Statement: Not applicable.

Data Availability Statement: Not applicable.

Conflicts of Interest: The authors declare no conflict of interest.

References

1. Nyhan, L.M.; Lynch, K.M.; Sahin, A.W.; Arendt, E.K. Advances in kombucha tea fermentation: A review. *Appl. Microbiol.* **2022**, *2*, 73–103. [CrossRef]
2. Bortolomedi, B.M.; Paglarin, C.S.; Brod, F.C.A. Bioactive compounds in kombucha: A review of substrate effect and fermentation conditions. *Food Chem.* **2022**, *385*, 132719. [CrossRef] [PubMed]
3. de Miranda, J.F.; Ruiz, L.F.; Silva, C.B.; Uekane, T.M.; Silva, K.A.; Gonzalez, A.G.M.; Fernandes, F.F.; Lima, A.R. Kombucha: A review of substrates, regulations, composition, and biological properties. *J. Food Sci.* **2022**, *87*, 503–527. [CrossRef] [PubMed]
4. Soares, M.G.; de Lima, M.; Schmidt, V.C.R. Technological aspects of kombucha, its applications and the symbiotic culture (SCOBY), and extraction of compounds of interest: A literature review. *Trends Food Sci. Technol.* **2021**, *110*, 539–550. [CrossRef]
5. Antolak, H.; Piechota, D.; Kucharska, A. Kombucha tea—A double power of bioactive compounds from tea and symbiotic culture of bacteria and yeasts (SCOBY). *Antioxidants* **2021**, *10*, 1541. [CrossRef] [PubMed]
6. Vargas, B.K.; Fabricio, M.F.; Ayub, M.A.Z. Health effects and probiotic and prebiotic potential of Kombucha: A bibliometric and systematic review. *Food Biosci.* **2021**, *44*, 101332. [CrossRef]
7. Kim, J.; Adhikari, K. Current trends in kombucha: Marketing perspectives and the need for improved sensory research. *Beverages* **2020**, *6*, 15. [CrossRef]
8. Frolova, Y.V. Russian market of fermented kombucha beverages. *Vopr. Pitan.* **2022**, *91*, 115–118. [CrossRef]
9. Higuchi, K. A Two-Step Approach to Quantitative Content Analysis: KH Coder Tutorial using Anne of Green Gables (Part II). *Ritsumeikan Soc. Sci. Rev.* **2017**, *53*, 137–147.
10. Li, S.; Zhang, Y.; Gao, J.; Li, T.; Li, H.; Mastroyannis, A.; He, S.; Rahaman, A.; Chang, K. Effect of Fermentation Time on Physiochemical Properties of Kombucha Produced from Different Teas and Fruits: Comparative Study. *J. Food Qual.* **2022**, *2022*, 2342954. [CrossRef]

11. Villarreal-Soto, S.A.; Beaufort, S.; Bouajila, J.; Souchard, J.P.; Taillandier, P. Understanding kombucha tea fermentation: A review. *J. Food Sci.* **2018**, *83*, 580–588. [CrossRef] [PubMed]

12. La Torre, C.; Fazio, A.; Caputo, P.; Plastina, P.; Caroleo, M.C.; Cannataro, R.; Cione, E. Effects of long-term storage on radical scavenging properties and phenolic content of Kombucha from black tea. *Molecules* **2021**, *26*, 5474. [CrossRef] [PubMed]

13. Kluz, M.I.; Pietrzyk, K.; Pastuszczak, M.; Kacaniova, M.; Kita, A.; Kapusta, I.; Zaguła, G.; Zagrobelna, E.; Struś, K.; Marciniak-Lukasiak, K.; et al. Microbiological and physicochemical composition of various types of homemade kombucha beverages using alternative kinds of sugars. *Foods* **2022**, *11*, 1523. [CrossRef] [PubMed]

14. Vitas, J.S.; Cvetanović, A.D.; Mašković, P.Z.; Švarc-Gajić, J.V.; Malbaša, R.V. Chemical composition and biological activity of novel types of kombucha beverages with yarrow. *J. Funct. Foods* **2018**, *44*, 95–102. [CrossRef]

15. Abuduaibifu, A.; Tamer, C.E. Evaluation of physicochemical and bioaccessibility properties of goji berry kombucha. *J. Food Process. Preserv.* **2019**, *43*, e14077. [CrossRef]

16. Francisco, Á.R.; Igor, H. Development of a no added sugar kombucha beverage based on germinated corn. *Int. J. Gastron. Food Sci.* **2021**, *24*, 100355. [CrossRef]

17. Li, R.; Xu, Y.; Chen, J.; Wang, F.; Zou, C.; Yin, J. Enhancing the proportion of gluconic acid with a microbial community reconstruction method to improve the taste quality of Kombucha. *LWT* **2022**, *155*, 112937. [CrossRef]

18. da Silva Júnior, J.C.; Magnani, M.; da Costa, W.K.A.; Madruga, M.S.; Olegário, L.S.; Borges, G.D.S.C.; Dantas, A.M.; Lima, M.S.; de Lima, L.C.; de Limo Brinto, I.; et al. Traditional and flavored kombuchas with pitanga and umbu-caja pulps: Chemical properties, antioxidants, and bioactive compounds. *Food Biosci.* **2021**, *44*, 101380. [CrossRef]

19. Zubaidah, E.; Yurista, S.; Rahmadani, N.R. Characteristic of physical, chemical, and microbiological kombucha from various varieties of apples. *IOP Conf. Ser. Earth Environ. Sci.* **2018**, *131*, 012040. [CrossRef]

20. Liamkaew, R.; Chattrawanit, J.; Danvirutai, P. Kombucha production by combinations of black tea and apple juice. *Prog. Appl. Sci. Technol.* **2016**, *6*, 139–146.

21. Zazueta, A.C.L.; Zavala, P.G.; Arguilez, C.G.Z. una Estandarización química de una bebida fermentada de Kombucha a base de té verde, té de limón ye infusión hojas de guayaba. *Rev. De Investig. Académica Sin Front. Div. Cienc. Económicas Soc.* **2022**, *38*. [CrossRef]

22. Zubaidah, E.; Dewantari, F.J.; Novitasari, F.R.; Srianta, I.; Blanc, P.J. Potential of snake fruit (*Salacca zalacca* (Gaerth.) Voss) for the development of a beverage through fermentation with the Kombucha consortium. *Biocatal. Agric. Biotechnol.* **2018**, *13*, 198–203. [CrossRef]

23. Yavari, N.; Assadi, M.M.; Larijani, K.; Moghadam, M.B. Response surface methodology for optimization of glucuronic acid production using kombucha layer on sour cherry juice. *Aust. J. Basic Appl. Sci.* **2010**, *4*, 3250–3256.

24. Tang, Z.; Zhao, Z.; Chen, S.; Lin, W.; Wang, Q.; Shen, N.; Qin, Y.; Xiao, Y.; Chen, H.; Bu, T.; et al. Dragon fruit-kiwi fermented beverage: In vitro digestion, untargeted metabolome analysis and anti-aging activity in *Caenorhabditis elegans*. *Front. Nutr.* **2022**, *9*, 1052818. [CrossRef] [PubMed]

25. Ulusoy, A.; Tamer, C.E. Determination of suitability of black carrot (*Daucus carota* L. spp. *sativus* var. *atrorubens* Alef.) juice concentrate, cherry laurel (*Prunus laurocerasus*), blackthorn (*Prunus spinosa*) and red raspberry (*Rubus ideaus*) for kombucha beverage production. *J. Food Meas. Charact.* **2019**, *13*, 1524–1536.

26. Tejedor-Calvo, E.; Morales, D. Chemical and Aromatic Changes during Fermentation of Kombucha Beverages Produced Using Strawberry Tree (*Arbutus unedo*) Fruits. *Fermentation* **2023**, *9*, 326. [CrossRef]

27. Budiari, S.; Mulyani, H.; Maryati, Y.; Filailla, E.; Devi, A.F.; Melanie, H. Chemical properties and antioxidant activity of sweetened red ginger extract fermented with kombucha culture. *Agrointek: J. Teknol. Ind. Pertan.* **2023**, *17*, 60–69. [CrossRef]

28. Nintiasari, J.; Ramadhani, M.A. Uji Kuantitatif flavonoid dan Aktivitas Antioksidan Teh Kombucha Daun Kersen (*Muntingia calabura*). *Indones. J. Pharm. Nat. Prod.* **2022**, *5*, 174–183.

29. Kayisoglu, S.; Coskun, F. Determination of physical and chemical properties of kombucha teas prepared with different herbal teas. *Food Sci. Technol.* **2020**, *41*, 393–397. [CrossRef]

30. Velićanski, A.; Cvetković, D.; Markov, S. Characteristics of Kombucha fermentation on medicinal herbs from Lamiaceae family. *Rom. Biotechnol. Lett.* **2013**, *18*, 8034–8042.

31. Tanticharakunsiri, W.; Mangmool, S.; Wongsariya, K.; Ochaikul, D. Characteristics and upregulation of antioxidant enzymes of kitchen mint and oolong tea kombucha beverages. *J. Food Biochem.* **2021**, *45*, e13574. [CrossRef] [PubMed]

32. Shahbazi, H.; Hashemi Gahruie, H.; Golmakani, M.T.; Eskandari, M.H.; Movahedi, M. Effect of medicinal plant type and concentration on physicochemical, antioxidant, antimicrobial, and sensorial properties of kombucha. *Food Sci. Nutr.* **2018**, *6*, 2568–2577. [CrossRef]

33. Özyurt, H. Changes in the content of total polyphenols and the antioxidant activity of different beverages obtained by Kombucha 'tea fungus'. *Int. J. Agric. Environ. Food Sci.* **2020**, *4*, 255–261. [CrossRef]

34. Tu, C.; Tang, S.; Azi, F.; Hu, W.; Dong, M. Use of kombucha consortium to transform soy whey into a novel functional beverage. *J. Funct. Foods* **2019**, *52*, 81–89. [CrossRef]

35. Martínez Leal, J.; Valenzuela Suárez, L.; Jayabalan, R.; Huerta Oros, J.; Escalante-Aburto, A. A review on health benefits of kombucha nutritional compounds and metabolites. *CyTA-J. Food* **2018**, *16*, 390–399. [CrossRef]

36. Kapp, J.M.; Sumner, W. Kombucha: A systematic review of the empirical evidence of human health benefit. *Ann. Epidemiol.* **2019**, *30*, 66–70. [CrossRef] [PubMed]

37. Bishop, P.; Pitts, E.R.; Budner, D.; Thompson-Witrick, K.A. Chemical composition of kombucha. *Beverages* **2022**, *8*, 45. [CrossRef]
38. Godswill, A.G.; Somtochukwu, I.V.; Ikechukwu, A.O.; Kate, E.C. Health benefits of micronutrients (vitamins and minerals) and their associated deficiency diseases: A systematic review. *Int. J. Food Sci.* **2020**, *3*, 1–32. [CrossRef]
39. Kodentsova, V.M.; Vrzhesinskaya, O.A.; Nikityuk, D.V.; Tutelyan, V.A. Vitamin status of adult population of the Russian Federation: 1987-2017. *Vopr. Pitan.* **2018**, *87*, 62–68.
40. Pietrangelo, L.; Magnifico, I.; Petronio Petronio, G.; Cutuli, M.A.; Venditti, N.; Nicolosi, D.; Perna, A.; Guerra, G.; Di Marco, R. A Potential "Vitaminic Strategy" against Caries and Halitosis. *Appl. Sci.* **2022**, *12*, 2457. [CrossRef]
41. Gondivkar, S.M.; Gadbail, A.R.; Gondivkar, R.S.; Sarode, S.C.; Sarode, G.S.; Patil, S.; Awan, K.H. Nutrition and oral health. *Disease-a-Month* **2019**, *65*, 147–154. [CrossRef] [PubMed]
42. Sewón, L.A.; Karjalainen, S.M.; Söderling, E.; Lapinleimu, H.; Simell, O. Associations between salivary calcium and oral health. *J. Clin. Periodontol.* **1998**, *25*, 915–919. [CrossRef] [PubMed]
43. Lattimer, J.M.; Haub, M.D. Effects of dietary fiber and its components on metabolic health. *Nutrients* **2010**, *2*, 1266–1289. [CrossRef] [PubMed]
44. Myhrstad, M.C.; Tunsjø, H.; Charnock, C.; Telle-Hansen, V.H. Dietary fiber, gut microbiota, and metabolic regulation—Current status in human randomized trials. *Nutrients* **2020**, *12*, 859. [CrossRef]
45. Murga-Garrido, S.M.; Hong, Q.; Cross, T.W.L.; Hutchison, E.R.; Han, J.; Thomas, S.P.; Vivas, E.I.; Denu, J.; Ceschin, D.G.; Tang, Z.Z.; et al. Gut microbiome variation modulates the effects of dietary fiber on host metabolism. *Microbiome* **2021**, *9*, 117. [CrossRef]
46. Pyryeva, E.A.; Safronova, A.I. The role of dietary fibers in the nutrition of the population. *Vopr. Pitan.* **2019**, *88*, 5–11.
47. Wan, X.; Guo, H.; Liang, Y.; Zhou, C.; Liu, Z.; Li, K.; Niu, F.; Zhai, X.; Wang, L. The physiological functions and pharmaceutical applications of inulin: A review. *Carbohydr. Polym.* **2020**, *246*, 116589. [CrossRef]
48. Redondo-Cuenca, A.; Herrera-Vázquez, S.E.; Condezo-Hoyos, L.; Gómez-Ordóñez, E.; Rupérez, P. Inulin extraction from common inulin-containing plant sources. *Ind. Crops Prod.* **2021**, *170*, 113726. [CrossRef]
49. Illippangama, A.U.; Jayasena, D.D.; Jo, C.; Mudannayake, D.C. Inulin as a functional ingredient and their applications in meat products. *Carbohydr. Polym.* **2022**, *275*, 118706. [CrossRef]
50. Gupta, N.; Jangid, A.K.; Pooja, D.; Kulhari, H. Inulin: A novel and stretchy polysaccharide tool for biomedical and nutritional applications. *Int. J. Biol. Macromol.* **2019**, *132*, 852–863. [CrossRef]
51. Ni, D.; Xu, W.; Zhu, Y.; Zhang, W.; Zhang, T.; Guang, C.; Mu, W. Inulin and its enzymatic production by inulosucrase: Characteristics, structural features, molecular modifications and applications. *Biotechnol. Adv.* **2019**, *37*, 306–318. [CrossRef] [PubMed]
52. Lončar, E.S.; Kanurić, K.G.; Malbaša, R.V.; Đurić, M.S.; Milanović, S.D. Kinetics of saccharose fermentation by Kombucha. *Chem. Ind. Chem. Eng. Q.* **2014**, *20*, 345–352. [CrossRef]
53. Bokov, D.O.; Sergunova, E.V.; Marakhova, A.I.; Morokhina, S.L.; Plakhotnaia, O.N.; Krasnyuk, I.I.; Bessonov, V.V. Determination of Sugar Profile in Viburnum Fruits and its Dosage Forms by HPLC-RID. *Pharmacogn. J.* **2020**, *12*, 103–108. [CrossRef]
54. Gamboa-Gómez, C.I.; Simental-Mendía, L.E.; González-Laredo, R.F.; Alcantar-Orozco, E.J.; Monserrat-Juarez, V.H.; Ramírez-España, J.C.; Gallegos-Infante, J.A.; Moreno-Jiménez, M.R.; Rocha-Guzmán, N.E. In vitro and in vivo assessment of anti-hyperglycemic and antioxidant effects of Oak leaves (*Quercus convallata* and *Quercus arizonica*) infusions and fermented beverages. *Food Res. Int.* **2017**, *102*, 690–699. [CrossRef] [PubMed]
55. Borai, A.; Sebaih, S.; Alshargi, A.; Albarzan, F.; Al-Ghamdi, A.; Al-Ghamdi, S.; Boraie, S.; Bahijri, S.; Al-Shareef, A.S.; Al-Armi, A.; et al. Ethanol content of a traditional Saudi beverage Sobia. *Int. J. Food Prop.* **2021**, *24*, 1790–1798. [CrossRef]
56. Tytelyan, V.A. *Guidance on Methods of Quality Control and Safety of Biologically Active Food Supplements*; Federal Center for State Sanitary and Epidemiological Surveillance of the Ministry of Health of Russia: Moscow, Russia, 2004; 240p.
57. Chu, S.C.; Chen, C. Effects of origins and fermentation time on the antioxidant activities of kombucha. *Food Chem.* **2006**, *98*, 502–507. [CrossRef]
58. Zou, C.; Li, R.Y.; Chen, J.X.; Wang, F.; Gao, Y.; Fu, Y.Q.; Xu, Y.Q.; Yin, J.F. Zijuan tea-based kombucha: Physicochemical, sensorial, and antioxidant profile. *Food Chem.* **2021**, *363*, 130322. [CrossRef]
59. Espenson, J.H. *Chemical Kinetics and Reaction Mechanisms*; McGraw-Hill: New York, NY, USA, 1995; p. 296.
60. Bokov, D.O.; Khromchenkova, E.P.; Sokurenko, M.S.; Vasilev, A.V.; Bessonov, V.V. Development of a technique for the determination of inulin in natural instant chicory after enzymatic hydrolysis by high-performance liquid chromatography. *Vopr. Pitan.* **2017**, *86*, 50–55.
61. Bendryshev, A.A.; Pashkova, E.B.; Pirogov, A.V.; Shpigun, O.A. Determination of water-soluble vitamins in vitamin premixes, bioacitve dietary supplements, and pharmaceutical preparations using high-efficiency liquid chromatography with gradient elution. *Mosc. Univ. Chem. Bull.* **2010**, *65*, 260–268. [CrossRef]
62. Polak, J.; Bartoszek, M.; Chorążewski, M. Antioxidant capacity: Experimental determination by EPR spectroscopy and mathematical modeling. *J. Agric. Food Chem.* **2015**, *63*, 6319–6324. [CrossRef]
63. *ISO 8586-1:1996*; Sensory Analysis-General Guidance for the Selection, Training and Monitoring of Assessors-Part 1: Selected Assessors. ISO: Geneva, Switzerland, 1996.
64. *ISO 8586-2:1996*; Sensory Analysis-General Guidance for the Selection, Training and Monitoring of Assessors-Part 2: Expert Sensory Assessors. ISO: Geneva, Switzerland, 1996.

65. *ISO 8589:2007*; General Guidance for the Design of Test Rooms. International Organization for Standardization. ISO: Geneva, Switzerland, 2007.

66. *ISO 13299:2016*; Sensory Analysis. Methodology. General Guidance for Establishing a Sensory Profile. International Organization for Standardization (ISO): Geneva, Switzerland, 2016.

67. R Core Team. *R: A Language and Environment for Statistical Computing*; R Foundation for Statistical Computing: Vienna, Austria, 2020; Available online: https://www.R-project.org/ (accessed on 5 June 2023).

68. Wickham, H. *ggplot2: Elegant Graphics for Data Analysis*; Springer: New York, NY, USA, 2016.

69. Granato, D.; de Araújo Calado, V.M.; Jarvis, B. Observations on the use of statistical methods in food science and technology. *Food Res. Int.* **2014**, *55*, 137–149. [CrossRef]

70. Cvetković, D.; Markov, S.; Djurić, M.; Savić, D.; Velićanski, A. Specific interfacial area as a key variable in scaling-up Kombucha fermentation. *J. Food Eng.* **2008**, *85*, 387–392. [CrossRef]

71. Malbaša, R.; Lončar, E.; Djurić, M.; Klašnja, M.; Kolarov, L.J.; Markov, S. Scale-up of black tea batch fermentation by kombucha. *Food Bioprod. Process.* **2006**, *84*, 193–199. [CrossRef]

72. Diez-Ozaeta, I.; Astiazaran, O.J. Recent advances in Kombucha tea: Microbial consortium, chemical parameters, health implications and biocellulose production. *Int. J. Food Microbiol.* **2022**, *377*, 109783. [CrossRef]

73. Vorobyeva, V.M.; Vorobyeva, I.S.; Sarkisyan, V.A.; Frolova, Y.V.; Kochetkova, A.A. Technological features of fermented beverages production using kombucha. *Vopr. Pitan.* **2022**, *91*, 115–120. [CrossRef]

74. Tran, T.; Grandvalet, C.; Verdier, F.; Martin, A.; Alexandre, H.; Tourdot-Maréchal, R. Microbiological and technological parameters impacting the chemical composition and sensory quality of kombucha. *Compr. Rev. Food Sci. Food Saf.* **2020**, *19*, 2050–2070. [CrossRef]

75. Chen, C.; Liu, B.Y. Changes in major components of tea fungus metabolites during prolonged fermentation. *J. Appl. Microbiol.* **2000**, *89*, 834–839. [CrossRef]

76. Jayabalan, R.; Subathradevi, P.; Marimuthu, S.; Sathishkumar, M.; Swaminathan, K. Changes in free-radical scavenging ability of kombucha tea during fermentation. *Food Chem.* **2008**, *109*, 227–234. [CrossRef]

77. Jakubczyk, K.; Kałduńska, J.; Kochman, J.; Janda, K. Chemical profile and antioxidant activity of the kombucha beverage derived from white, green, black and red tea. *Antioxidants* **2020**, *9*, 447. [CrossRef]

78. Gaggìa, F.; Baffoni, L.; Galiano, M.; Nielsen, D.S.; Jakobsen, R.R.; Castro-Mejía, J.L.; Bosi, S.; Truzzi, F.; Musumeci, F.; Dinelli, G.; et al. Kombucha beverage from green, black and rooibos teas: A comparative study looking at microbiology, chemistry and antioxidant activity. *Nutrients* **2018**, *11*, 1. [CrossRef]

79. Laureys, D.; Britton, S.J.; De Clippeleer, J. Kombucha tea fermentation: A review. *J. Am. Soc. Brew. Chem.* **2020**, *78*, 165–174. [CrossRef]

80. Jayabalan, R.; Malbaša, R.V.; Sathishkumar, M. Kombucha tea: Metabolites. In *Fungal Metabolites*; Springer: Cham, Switzerland, 2017; pp. 965–978.

81. Reiss, J. Influence of different sugars on the metabolism of the tea fungus. *Z. Für Lebensm.-Unters. Und-Forsch.* **1994**, *198*, 258–261. [CrossRef]

82. Kaewkod, T.; Bovonsombut, S.; Tragoolpua, Y. Efficacy of kombucha obtained from green, oolong, and black teas on inhibition of pathogenic bacteria, antioxidation, and toxicity on colorectal cancer cell line. *Microorganisms* **2019**, *7*, 700. [CrossRef]

83. Liu, Z.; Bruins, M.E.; de Bruijn, W.J.; Vincken, J.P. A comparison of the phenolic composition of old and young tea leaves reveals a decrease in flavanols and phenolic acids and an increase in flavonols upon tea leaf maturation. *J. Food Compos. Anal.* **2020**, *86*, 103385. [CrossRef]

84. Meng, X.H.; Li, N.; Zhu, H.T.; Wang, D.; Yang, C.R.; Zhang, Y.J. Plant resources, chemical constituents, and bioactivities of tea plants from the genus *Camellia* section *Thea*. *J. Agric. Food Chem.* **2018**, *67*, 5318–5349. [CrossRef] [PubMed]

85. Wang, S.; Zhang, L.; Qi, L.; Liang, H.; Lin, X.; Li, S.; Yu, C.; Ji, C. Effect of synthetic microbial community on nutraceutical and sensory qualities of kombucha. *Int. J. Food Sci. Technol.* **2020**, *55*, 3327–3333. [CrossRef]

86. Chakravorty, S.; Bhattacharya, S.; Chatzinotas, A.; Chakraborty, W.; Bhattacharya, D.; Gachhui, R. Kombucha tea fermentation: Microbial and biochemical dynamics. *Int. J. Food Microbiol.* **2016**, *220*, 63–72. [CrossRef] [PubMed]

87. Feng, Z.; Li, Y.; Li, M.; Wang, Y.; Zhang, L.; Wan, X.; Yang, X. Tea aroma formation from six model manufacturing processes. *Food Chem.* **2019**, *285*, 347–354. [CrossRef] [PubMed]

88. Bishop, P.; Pitts, E.R.; Budner, D.; Thompson-Witrick, K.A. Kombucha: Biochemical and microbiological impacts on the chemical and flavor profile. *Food Chem. Adv.* **2022**, *1*, 100025. [CrossRef]

89. Jayabalan, R.; Marimuthu, S.; Swaminathan, K. Changes in content of organic acids and tea polyphenols during kombucha tea fermentation. *Food Chem.* **2007**, *102*, 392–398. [CrossRef]

90. Matsushita, K.; Adachi, O. Quinoprotein aldehyde dehydrogenases in microorganisms. In *Principles and Applications of Quinoproteins*; CRC Press: Boca Raton, FL, USA, 2020; pp. 65–72.

91. Akimov, M.Y.; Bessonov, V.V.; Kodentsova, V.M.; Eller, K.I.; Vrzhesinskaya, O.A.; Beketova, N.A.; Kosheleva, O.V.; Bogachuk, M.N.; Malinkin, A.D.; Makarenko, M.A.; et al. Biological value of fruits and berries of Russian production. *Vopr. Pitan.* **2020**, *89*, 220–232. [PubMed]

92. Tutel'ian, V.A.; Lashneva, N.V. Biologically active substances of plant origin. Flavonols and flavones: Prevalence, dietary sourses and consumption. *Vopr. Pitan.* **2013**, *02*, 4–22. [PubMed]

93. Ikegaya, A.; Toyoizumi, T.; Ohba, S.; Nakajima, T.; Kawata, T.; Ito, S.; Arai, E. Effects of distribution of sugars and organic acids on the taste of strawberries. *Food Sci. Nutr.* **2019**, *7*, 2419–2426. [CrossRef]

94. Koyuncu, M.A.; Dilmaçünal, T. Determination of vitamin C and organic acid changes in strawberry by HPLC during cold storage. *Not. Bot. Horti Agrobot. Cluj-Napoca* **2010**, *38*, 95–98.

95. Mukherjee, A.; Gómez-Sala, B.; O'connor, E.M.; Kenny, J.G.; Cotter, P.D. Global regulatory frameworks for fermented foods: A review. *Front. Nutr.* **2022**, *9*, 902642. [CrossRef] [PubMed]

96. Pilipenko, V.I.; Isakov, V.A.; Morozov, S.V.; Vlasova, A.V.; Kochetkova, A.A. Efficacy of newly developed kombucha-based specialized food product for treatment of constipation-predominant irritable bowel syndrome. *Vopr. Pitan.* **2022**, *91*, 95–104.

fermentation

Communication

Effect of *Botrytis cinerea* Activity on Glycol Composition and Concentration in Wines

Eszter Antal [1], Miklós Kállay [2], Zsuzsanna Varga [2,*] and Diána Nyitrai-Sárdy [2]

[1] Diagnosticum Ltd., Attila Str. 126., 1047 Budapest, Hungary; eszter@diagnosticum.hu
[2] Institute of Viticulture and Enology, Hungarian University of Agriculture and Life Sciences (MATE), Ménesi Str. 45., 1118 Budapest, Hungary; kallay.miklos@uni-mate.hu (M.K.); nyitraine.sardy.diana.agnes@uni-mate.hu (D.N.-S.)
* Correspondence: varga.zsuzsanna@uni-mate.hu

Abstract: The content of 2,3-butanediol ((R,R) and meso isomers) and 1,2-propanediol in grape berries and "liquid samples" (all non-berry extracts) from the Tokaj wine region of Hungary was investigated. Our aim was to find out how the activity of *Botrytis cinerea* influences the concentrations of these compounds compared with healthy grapes. Based on the measured concentrations, we can make a distinction between healthy berries and noble, rotted, so-called aszú berries. We also investigated if there is a difference between finished aszú wines and liquids intended for aszú production. We wanted to investigate the amount and distribution of the stereoisomers of 2,3-butanediol and their proportions. The results of the HS-SPME-GC-FID analysis of the samples showed significant differences in the 2,3-butanediol content between healthy and botrytised, aszú berries and between liquid samples for aszú production and aszú wines. In the berry samples, meso-2,3-butanediol could not be detected, whereas in the liquid samples, we found good amounts of this isomer. This may be due to the fact that the appearance of the meso form of 2,3-butanediol is a consequence of alcoholic fermentation. Significant differences were found between wines from healthy grapes and wines from botrytised grapes in terms of the levo-2,3-butanediol content, so that from an analytical point of view, a difference can be made between wines from healthy and botrytised grapes. No significant differences were found between berry and liquid samples in terms of 1,2-propanediol concentrations during our tests.

Keywords: *Botrytis cinerea*; levo-2,3-butanediol; meso-2,3-butanediol; 1,2-propanediol

Citation: Antal, E.; Kállay, M.; Varga, Z.; Nyitrai-Sárdy, D. Effect of *Botrytis cinerea* Activity on Glycol Composition and Concentration in Wines. *Fermentation* **2023**, *9*, 493. https://doi.org/10.3390/fermentation9050493

Academic Editor: Maria Dimopoulou

Received: 16 April 2023
Revised: 18 May 2023
Accepted: 19 May 2023
Published: 22 May 2023

1. Introduction

Alcoholic fermentation produces a number of secondary products, and 2,3-butanediol is the second most abundant as a normal constituent of wine, making it an important potential source of aroma. Because of its very high threshold value (about 150 mg/L), 2,3-butanediol does not usually have a noticeable effect on the organoleptic properties of alcoholic beverages. In contrast, its amount in wine can considerably vary, with concentrations ranging from 0.2 to 3 g/L, averaging around 0.57 g/L. This high concentration can affect the wine's aroma due to its slightly bitter taste and the wine's body due to its viscosity [1]. During fermentation processes, measurable amounts of 1,2-propanediol (propylene glycol (1,2-PG)) are also formed. These compounds are virtually absent in unfermented must but are present in wines within certain limits [2].

1,2-Propanediol is well-known as a legal solvent in the food and beverage, pharmaceutical and cosmetic industries and is, therefore, found in many products of our daily lives. As a common solvent, it is used in the flavouring industry and can also be found in flavoured drinks and in flavoured drinks containing wine [3]. However, since wine cannot be flavoured per se [4], knowledge of its formation and presence in wine is also important from a legal point of view, especially with regard to the authenticity of wine. Previous results [5] have given naturally occurring concentrations between 10 and 30 mg/L in white and red wines, but concentrations between 40 and 50 mg/L have also been described [5]. Some wines made with higher concentrations of crushed or enriched berry material showed elevated concentrations up to 140 mg/L. An important consideration in authenticity studies is the possible influence of wine fermentation conditions and yeast strains on the final concentration or enantiomeric ratio of 1,2-PG [5] and on malolactic fermentation (MLF). Taking into account the possibility of the natural production of 1,2-PG, some yeast species, considered as wild yeasts, such as *Hansenula*, *Candida*, *Pichia* or *Rhodoturola*, have been described to metabolise 1,2-PG under aerobic conditions in synthetic media. Su-zuki and Onishi described the ability of several different yeast genera and species to convert L-ramnose to 1,2-PG under aerobic conditions [6].

2,3-Butanediol is a well-known by-product of fermentation. A 2,3-butanediol content of between 0.4 and 1.0 g/L is an indicator of fermentation and shows whether ethyl alcohol and glycerol are the result of fermentation. It nominally occurs in four stereoisomeric forms (Figure 1), but two are identical mesoisomers, which form an easily crystallisable hydrate. The mesoisomer is optically inactive [7].

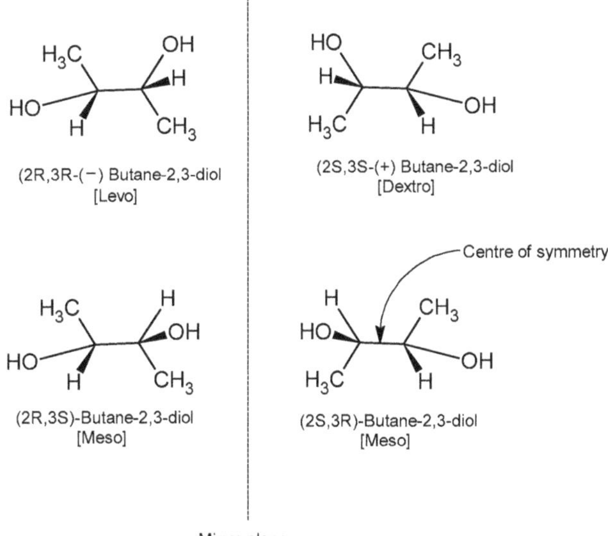

Figure 1. Stereoisomers of 2,3-butanediol.

Jung and co-workers [7] concluded that only the (R,R)- and meso isomers of 2,3-butanediol occur in wine, and that (S,S)-2,3-butanediol is not the true compound of wine. In addition, (R,R)-2,3-butanediol is the dominant isomer of 2,3-butanediol in wine. Its average value is 3.1 ± 0.7–1 g/L. The (R,R)/meso-butanediol ratio ranges from 1.6 to 5.2 g/L. It is considered that, in the future, the (R,R)/meso-2,3-butanediol ratio can be used to judge the authenticity of wines.

This seems to be contradicted by the study of Dankó and colleagues [8], who identified illiquid components in berries kept under conditions that induce nematode or bunch rot. In total, 30 volatile components were quantified. The Furmint berry samples affected by bunch rot also contained volatile organic compounds enriched in volatile matter [S,S]-2,3-butanediol. It was found that the progression of grey mould disease of furmint grapes was associated with increased volatile compound emissions, including [S,S]-2,3-butanediol.

The epidermal tissue, enzymatically loosened or destroyed by *Botrytis*, becomes permeable to must or surface saprobiont mycobacteria, so that other moulds, yeasts and acetic acid bacteria are also involved in the rotting process. Dankó et al. [8] observed that the presence of acetic acid in must increases the formation of acetoin and its derivative 2,3-butanediol, possibly as a result of yeast activity.

The appearance of 2,3-butylene glycol in wine is, in fact, derived from acetoin, much of it being produced by yeasts during the fermentation of carbohydrates by the enzyme reduction of acetoin. Acetoin is produced by *Saccharomyces cerevisiae* in the early stages of fermentation, reaching a maximum about halfway through the fermentation process, and then rapidly decreases in the final stages of fol-yamat as it is reduced to 2,3-butylene glycol [9].

Several studies [8,10–12] have addressed the relationship between the mould *Botrytis cinerea* and 2,3-butanediol. *B. cinerea* is one of the most studied pen-fungi in viticulture because of its negative and positive effects. It is widespread and responsible for serious vineyard diseases that can damage yields and wine quality. Several studies have addressed the impact of *B. cinerea* on wine aroma, either in its harmful form or in its positive effects in causing noble rot [10].

The development of noble rot is considered as a beneficial process, resulting in the formation of bursting, chocolate-brown berries—aszú grains. These raisin-like berries are the main sources of flavour and aroma in botrytised sweet wines such as Hungarian aszú, French Sauternes and German Trockenbeerenauslese.

Hungary's most famous wine variety is Tokaji aszú. Its unique, delicate aroma and flavour are the result of a special wine-making technique involving the extraction of botrytised (aszú) grapes from noble rot, the extraction of the aszú grapes with new dry wine (sometimes with fermenting must) and a few years of ageing in small oak barrels.

Infection of grapes with *B. cinerea* leads to a change in the chemical composition of the grapes and associated wine, thus affecting the quality of the wine. The metabolic effect of *Botrytis* infection in Champagne base wine was investigated by Hong and colleagues [12]. They identified a number of components in grape must and wine, including 2,3-butanediol, and found that these components contributed to the metabolic differences between healthy and botrytised wines, both from an analytical and sensory point of view [12].

Although noble rot caused by *B. cinerea* is desirable for the production of good-quality sweet white wines, grey rot leads to undesirable or negative wine quality effects mainly through the degradation of aroma compounds and the production of off-flavours such as "mouldy" and "earthy" aromas.

Lower levels of 2,3-butanediol were found in the tested botrytised base wines than in healthy base wines [12].

Leskó and colleagues (2015) [13] investigated the amount of 2,3-butanediol and 1,2-propanediol produced after the fermentation of high sugar musts. They observed that higher initial sugar concentration promoted the formation of these fermentation by-products up to a certain point, but above the 250 g/L sugar content, the formation of these compounds was inhibited. Using a gas chromatography technique, they were also able to determine the stereoisomers of the resulting 2,3-butanediol [13].

Guymon and Crowell (1967) [14] investigated the levo and meso stereoisomers of 2,3-butanediol formed during the fermentation of grape juice (must) by gas chromatography. The effect of various parameters was investigated. It was found that the formation of 2,3-butanediol significantly increased with increasing fermentation temperature and with increasing initial sugar content of must [14].

The aim of this study is to investigate how the 2,3-butanediol content and isomer (levo/meso ratio) distribution of grape berries and liquid samples from the Tokaj-Hegyalja region of Hungary and the 1,2-propanediol content of grape berries and liquid samples accordingly evolve, whether it is a healthy berry or a berry neat rotten by *B. cinerea*, and whether it is a liquid of healthy grapes used for making aszú wine or aszú wine. Our question was if there is a significant difference in the amount of components tested according to whether the sample is botrytised or healthy.

2. Materials and Methods

2.1. Materials

All chemicals were analytical reagents. NaCl was purchased from Merck (Merck Life Science Kft., an affiliate of Merck KGaA, Darmstadt, Germany). Alcohol standard solutions of levo-2,3-butanediol, meso-2,3-butanediol, 1,2-propanediol and 1,3-propanediol (Internal standard) were from Merck. A stock solution (10 g/L of each compound) was prepared from the relative standard solutions and stored at 4 °C. Solutions at different concentrations were obtained by diluting stock standard solutions at the desired concentration.

2.2. Wine Sampling

Grape berry and "liquid samples" (everything that is not berry extract: wine, must) collected from the Tokaj wine region (Hungary) were analysed. The 62 different grape berries were healthy (33 berries) and botrytised (29 berries). Our liquid samples, 26 in total, were liquids for aszú production (10 types) and fermented aszú wines (16 types). The liquids for aszú production were must from healthy grape berries, partially fermented grape must, new wine in the process of fermentation and wine of the same vintage as the aszú berries; all of them are accepted in practice according to the "Product description of the Tokaj protected designation of origin", Version 9 [15]. The samples were provided by the producers, and, therefore, especially in the case of the berries, we accepted and did not overrule their judgement.

2.3. Preliminary Chemical Analysis on Wine Samples

The pH, total acidity (g/L as tartaric acid) and reducing sugar (g/L) were analysed according to the official methods established by the European Commission (EC) [16].

2.4. Sample Preparation

The berry samples had to be extracted for analytical tests. The extraction was performed with a 100 g berry sample. It was mixed with 100 mL of 10% v/v ethanol [11]. After standing for half an hour, it was centrifuged and then filtered. The filtrate was collected in a 200 mL volumetric flask and then filled with 10% v/v ethanol. Ethanol was chosen as the extraction agent because ethanol is also produced during alcoholic fermentation, and the concentration of ethanol determines the solubility of glycols in the final product. In summary, we have tried to model the extraction process that occurs during the winemaking process.

2.5. HS-SPME-GC-FID Analysis

The concentrations of 2,3-butanediol (levo- and meso-) and 1,2-propanediol were determined by gas chromatography (GC) using a flame ionisation detector (FID) and an automatic headspace (HS) solid-phase micro-extraction (SPME) procedure. For sample preparation, automated headspace–solid-phase micro-extraction (HS-SPME) analysis was used. About 10 mL of wine samples was pipetted and placed into a 20 mL headspace vial with 2 g of NaCl. Each sample was spiked with 500 μL of a solution of 1,3-propanediol (10 g/L in distilled water containing 15% v/v of ethanol). The HS-SPME extraction procedure was carried out with a Combi PAL autosampler (from CTC Analytics AG, Zwingen, Switzerland) using an 85 μm CAR/PDMS fibre coating of 1 cm in length. Sample conditioning, extraction and headspace sampling were conducted using the agitator of the Combi PAL autosampler. The samples were incubated for 15 min at 85 °C followed by an extraction time of 3 min at 85 °C at a stirring rate of 400 rpm. Desorption was performed in the split/splitless injector of the gas chromatograph (GC) at a temperature of 200 °C. Desorption time was 5 min. Cleaning and conditioning of the fibre was conducted at 280 °C for 5 min under a nitrogen flow of 6 mL/min in the conditioning station of the autosampler before and after each analysis. Gas chromatographic analyses were conducted with a Shimadzu-2030 with FID detection. The separation column used was an HP 20M capillary column with 25.0 m × 0.20 mm i.d. and 0.20 μm film thickness. The GC was used in splitless mode (splitless time in 2 min). The carrier gas was helium at a flow rate of 0.80 mL/min. The detector temperature was 250 °C. The GC oven was programmed as follows: the initial temperature was 50 °C (held for 2.40 min), then ramped at 13.3 °C/min to 200 °C and held for 12.05 min. The total run time was 25.73 min.

2.6. Statistical Analysis

Statistical evaluation of the measurement results was performed using the ANOVA program of Microsoft Excel (version 18.2106.12410.0, license: Microsoft Corporation, Redmond, Washington). The analytical results were evaluated by one-factor analysis of variance. The amounts of levo-2,3-butanediol (R,R-isomer), meso-2,3-butanediol and 1,2-propanediol were measured in berry samples and liquid samples of different qualities, and the evolution of concentrations is illustrated by boxplot diagrams. We compared whether there was a significant difference at the 95% ($p = 0.05$) probability level between the measured concentrations depending on whether the berries were healthy or infected with *B. cinerea* and whether the sample was a basic wine made from healthy grapes or an aszú.

3. Results

We tested mature, healthy and botrytised clusters from several vineyards in the Tokaj wine region, producing high-quality berries with noble rot. We also tested samples of grape juice from the same area at different stages of aszú wine production (liquid samples for soaking aszú berries). The mean values of the basic analysis and the ±standard deviation are given in Table 1.

Table 1. Basic analysis of the samples, mean and ±standard deviation of the measured values. Concentrations are given per unit berry weight (g/kg) for berry samples and per unit volume (g/L) for liquids.

		Reducing Sugar (g/kg), (g/L)	pH	Total Acidity (g/kg), (g/L)
Berry samples	Healthy	208.8 ± 21.7	3.4 ± 0.2	5.8 ± 0.3
	Noble rotten	305.5 ± 63.2	3.3 ± 0.2	8.2 ± 0.9
Liquid samples	Liquid for aszú Production	15.7 ± 37.4	3.0 ± 0.2	7.0 ± 0.9
	Aszú wine	142.5 ± 34.6	3.1 ± 0.2	8.9 ± 1.6

In case of reducing the sugar content, the large variation between the results is due to the fact that there was a very large difference in the sugar content of the samples tested, both in terms of the sugar content of the berries at harvest and also among the liquid samples: there were some that were fermented completely dry and some that still had a low sugar content. In the course of the work, we did not require that the sugar contents were similar.

In the analysis of the 2,3-butanediol content of the berries, the meso-form could not be detected. Figure 2 shows the evolution of the levo-2,3-butanediol (R,R-isomer; hereafter referred to as R,R-isomer) content of berry samples at different stages. When tested by one-factor analysis of variance, it was found that there was a significant difference between the healthy grape berries and aszú berries at the 95% significance level ($p = 0.05$). Thus, the amount of levo-2,3-butanediol distinguishes these berry varieties from each other. The highest amount of le-vo-2,3-butanediol (1219.8 mg/kg) was measured in the aszú berries.

When analysing "liquid samples" (everything that is not berry extract: wine, must), we were able to measure both the levo and meso isomers of 2,3-butanediol. This leads to the primary conclusion that the appearance of the meso form of 2,3-butanediol is a consequence of alcoholic fermentation. The aszú wines contained both levo-2,3-butanediol (717.3 mg/L) and meso-2,3-butanediol (411.7 mg/L) in the highest concentrations.

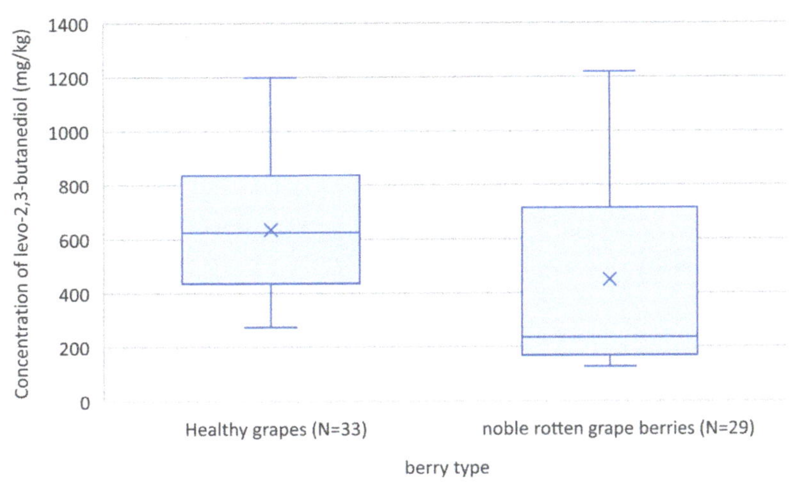

Figure 2. Boxplot of levo-2,3-butanediol concentrations of different types of berries.

The average levo- and meso-2,3-butanediol concentrations of the liquid and berry samples are shown in Figure 3. The amount of meso-2,3-butanediol was lower than that of levo-2,3-butanediol in both liquid types. In case of berry samples, however, the levo-2,3-butanediol content was higher in the healthy berries.

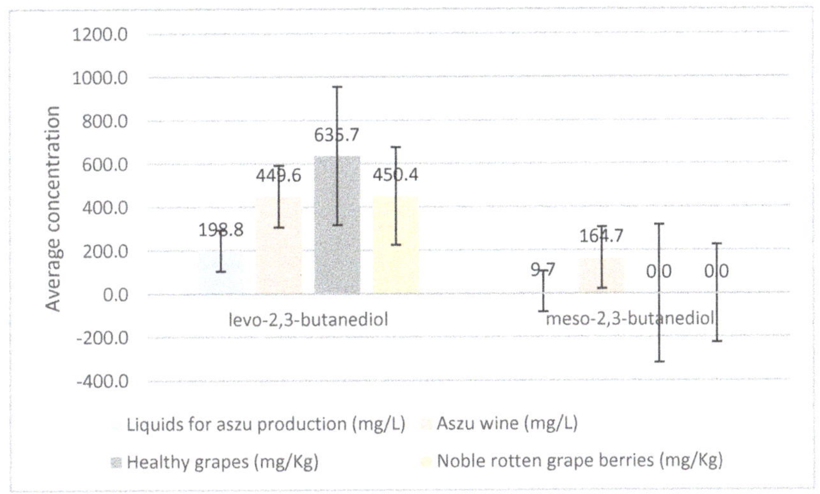

Figure 3. Comparison of average amounts of levo- and meso-2,3-butanediol measured in liquid and berry samples.

The results were analysed by one-factor analysis of variance and showed that at the 95% ($p = 0.05$) significance level, there was a difference between the aszú wines and the liquors for aszú production, both in terms of the amount of levo-2,3-butanediol and the amount of meso-2,3-butanediol.

We examined how the average amounts of levo-2,3-butanediol and meso-2,3-butanediol in the liquid samples relate to each other (Figure 4).

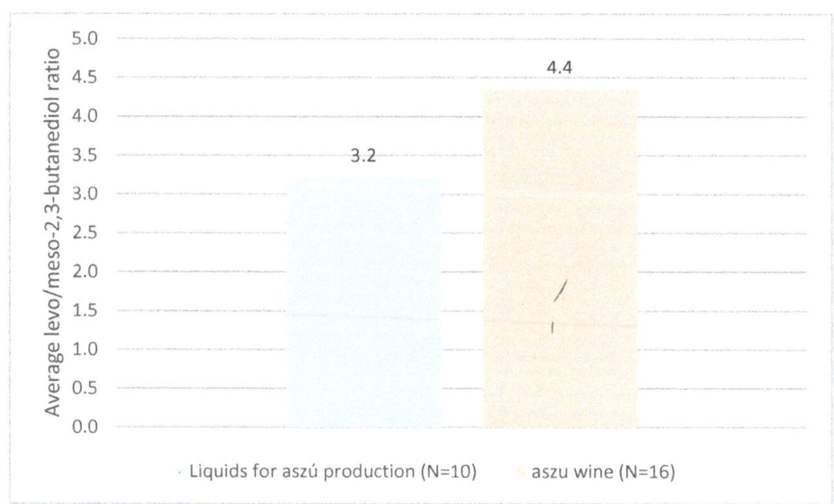

Figure 4. Ratio of average concentrations of levo- and meso-2,3-butanediol measured in liquid samples.

The trend of the levo/meso ratio shows that the ratio is higher for aszú wines (Figure 4).

Figures 5 and 6 show the evolution of 1,2-propanediol concentrations in the berry and liquid samples. The 1,2-propanediol content in berry samples varies within a wide range (healthy berries: 36.6–248.2 mg/kg, fortified berries: 21.6–510.8 mg/kg) (Figure 5), and the highest concentration was measured in the case of the aszú berries (510.8 mg/kg). By comparing the healthy berries and dried berries by one-factor analysis of variance, we found that there was no significant difference between them in terms of their 1,2-propanediol concentration at the 95% (p = 0.05) probability level.

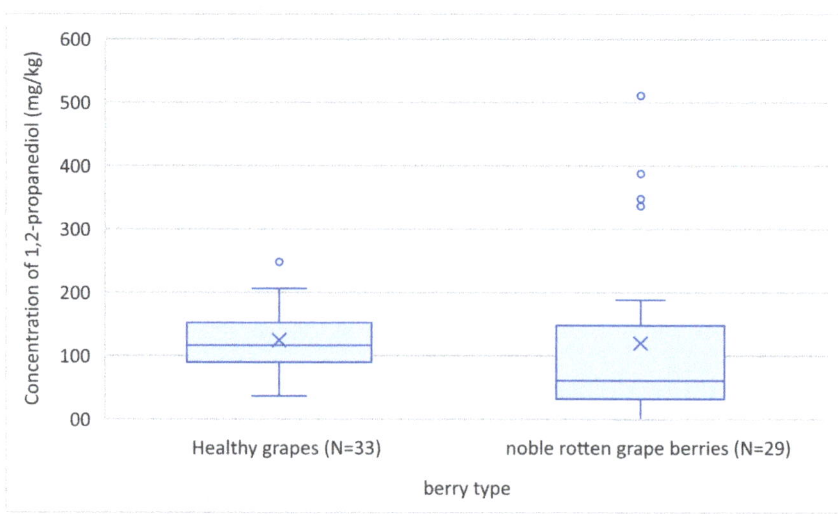

Figure 5. Boxplot of 1,2-propanediol concentrations of different types of berries (°outliers).

When looking at the concentrations of 1,2-propanediol in the liquids (Figure 6), it is observed that, apart from a few outliers (693.5; 394.4; 224.1; 103.1 mg/L), the concentrations measured in the samples remain below 100 mg/L. In case of liquids for aszú production, we measured concentrations between 43.1 and 103.1 mg/L and between 30.9 and 693.5 mg/L in aszú wines, which is a very wide range.

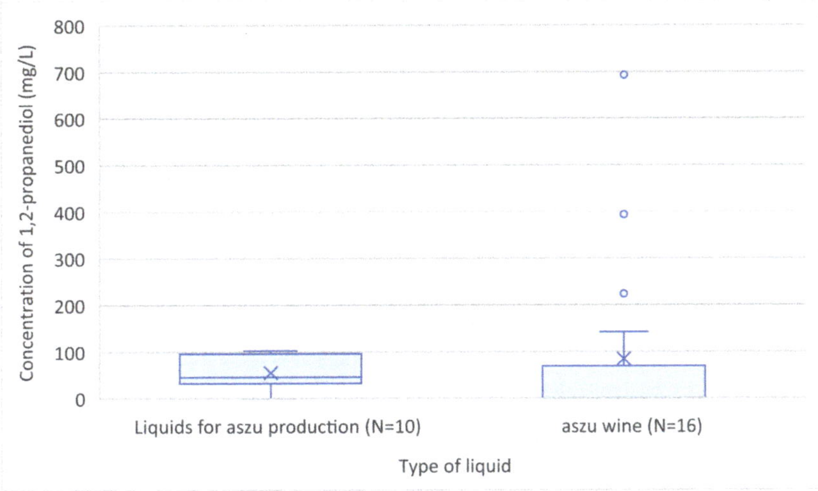

Figure 6. Boxplot of 1,2-propanediol concentrations of different types of liquid (°outliers).

There is no significant difference at the 95% ($p = 0.05$) probability level between the 1,2-propanediol content of liquids intended for aszú production and aszú wines.

4. Discussion

The tests were carried out on grapes subjected to noble rot (botrytisation) and wines from these grapes. The concentrations of 2,3-butanediol isomers and 1,2-propanediol were compared with healthy grapes and wines from healthy grapes. The aim of the work was to find out how the amount of levo-2,3-butanediol (R,R-isomer) and meso-2,3-butanediol and 1,2-propanediol in the samples tested varied depending on whether the sample was botrytised or not.

First, we analysed the results of the berry samples and found that meso-2,3-butanediol could not be detected, whereas for the liquid samples, we found good measurable amounts of this isomer. The primary conclusion drawn was that the appearance of the meso form of 2,3-butanediol is a consequence of alcoholic fermentation.

The mean levo-2,3-butanediol content of the healthy berry samples was 635.7 mg/kg ($N = 33$), and the mean levo-2,3-butanediol content of the aszú berry samples was 397.7 mg/kg ($N = 29$). The average levo-2,3-butanediol concentration in the liquids for aszú was 198.8 mg/L ($N = 10$), and for aszú wines, it was 449.6 mg/L ($N = 16$) (Figure 3) [17]. The average levo-2,3-butanediol content of healthy berries was higher than that measured in the aszú berries, yet the aszú wines had the higher levo-2,3-butanediol content compared with the samples used for aszú production. This can be explained by the higher sugar content of the aszú berries compared with the sugar content of the healthy berries. Son and colleagues (2009) [17] found that the 2,3-butanediol level in wines may be related to the sugar level in grapes, with a positive correlation. In addition to this, infection of grapes with *B. cinerea* is always associated with the development of acetic acid bacteria on the berries [18,19]. These bacteria produce acetic acid, which during the fermentation process is converted into acetoin and then into 2,3-butanediol. This results in a higher 2,3-butanediol content in botrytised wines compared with healthy wines.

For both berry samples and liquid samples, results of the statistical analysis showed that there were significant differences for both isomers tested (levo- and meso-2,3-butanediol); thus, these isomers contribute to the differences between healthy and botrytised grapes and their wines (healthy and botrytised wines) from an analytical point of view. Thus, the amounts of levo- and meso-2,3-butanediol distinguish the berry varieties and liquid samples tested. This result is consistent with the observations of Hong and colleagues (2011) [12].

We found that botrytised wines had higher concentrations of both isomers (levo- and meso-) than liquids intended for distillation. These liquids for distillation are musts from healthy grapes, partially fermented grape musts, new wines still in fermentation and wines. It was also observed that levo-isomer is produced in higher quantities than meso-isomer. This contradicts the finding of Hong and colleagues (2011) [12]. Hong and co-workers used 1H NMR techniques to investigate the amount of 2,3-butanediol in *Botrytis*-infected sparkling base wine and healthy base wine. They found lower 2,3-butanediol levels in botrytised base wines than in healthy base wines.

When looking at the levo/meso ratio in the liquids, the trend showed that this ratio is higher in aszú wines than in liquids intended for aszú production. In order to obtain a more accurate picture, it is necessary to investigate the levo- and meso-2,3-butanediol content of more wines from healthy grapes and aszú wines.

No significant differences were found between berry samples and liquid samples in terms of 1,2-propanediol concentrations in the samples tested.

Author Contributions: Conceptualisation, E.A. and M.K.; methodology, D.N.-S.; resources, E.A.; writing—original draft preparation, E.A. and M.K.; writing—review and editing, Z.V. and D.N.-S.; supervision, M.K. All authors have read and agreed to the published version of the manuscript.

Funding: This research received no external funding.

Institutional Review Board Statement: Not applicable.

Informed Consent Statement: Not applicable.

Data Availability Statement: The data presented in this study are available on request from the corresponding author.

Conflicts of Interest: The authors declare no conflict of interest.

References

1. Romano, P.; Brandolini, V.; Ansaloni, C.; Menziani, E. The Production of 2,3-Butanediol as a Differentiating Character in Wine Yeasts. *World J. Microbiol. Biotechnol.* **1998**, *14*, 649–653. [CrossRef]
2. International Organisation of Vine and Wine (OIV). *Compendium of International Methods of Wine and Must Analysis*, 2022nd ed.; International Organisation of Vine and Wine (OIV): Paris, France, 2022; Volume 2.
3. Langen, J.; Fischer, U.; Cavalar, M.; Coetzee, C.; Wegmann-Herr, P.; Schmarr, H.-G. Enantiodifferentiation of 1,2-Propanediol in Various Wines as Phenylboronate Ester with Multidimensional Gas Chromatography-Mass Spectrometry. *Anal. Bioanal. Chem.* **2016**, *408*, 2425–2439. [CrossRef] [PubMed]
4. Commission Regulation (EC) No 606/2009 of 10 July 2009 Laying down Certain Detailed Rules for Implementing Council Regulation (EC) No 479/2008 as Regards the Categories of Grapevine Products, Oenological Practices and the Applicable Restrictions. *Off. J. Eur. Union L* **2009**, *193*, 1–59.
5. Girard, B.; Yuksel, D.; Cliff, M.A.; Delaquis, P.; Reynolds, A.G. Vinification Effects on the Sensory, Colour and GC Profiles of Pinot Noir Wines from British Columbia. *Food Res. Int.* **2001**, *34*, 483–499. [CrossRef]
6. Suzuki, T.; Onishi, H. Aerobic Dissimilation of L-Rhamnose and the Production of L-Rhamnonic Acid and 1,2-Propanediol by Yeasts. *Agric. Biol. Chem.* **1968**, *32*, 888–893. [CrossRef]
7. Jung, J.; Jaufmann, T.; Hener, U.; Münch, A.; Kreck, M.; Dietrich, H.; Mosandl, A. Progress in Wine Authentication: GC–C/P–IRMS Measurements of Glycerol and GC Analysis of 2,3-Butanediol Stereoisomers. *Eur. Food Res. Technol.* **2006**, *223*, 811–820. [CrossRef]
8. Dankó, T.; Szelényi, M.; Janda, T.; Molnár, B.P.; Pogány, M. Distinct Volatile Signatures of Bunch Rot and Noble Rot. *Physiol. Mol. Plant Pathol.* **2021**, *114*, 101626. [CrossRef]
9. Romano, P.; Suzzi, G. Acetoin Production in Saccharomyces Cerevisiae Wine Yeasts. *FEMS Microbiol. Lett.* **1993**, *108*, 23–26. [CrossRef] [PubMed]
10. Lopez Pinar, A.; Rauhut, D.; Ruehl, E.; Buettner, A. Effects of Botrytis Cinerea and Erysiphe Necator Fungi on the Aroma Character of Grape Must: A Comparative Approach. *Food Chem.* **2016**, *207*, 251–260. [CrossRef] [PubMed]
11. Miklósy, É.; Kerényi, Z. Comparison of the Volatile Aroma Components in Noble Rotted Grape Berries from Two Different Locations of the Tokaj Wine District in Hungary. *Anal. Chim. Acta* **2004**, *513*, 177–181. [CrossRef]

12. Hong, Y.-S.; Cilindre, C.; Liger-Belair, G.; Jeandet, P.; Hertkorn, N.; Schmitt-Kopplin, P. Metabolic Influence of Botrytis Cinerea Infection in Champagne Base Wine. *J. Agric. Food Chem.* **2011**, *59*, 7237–7245. [CrossRef] [PubMed]
13. Leskó, A.; Hunyadi, R.; Kállay, M.; Antal, E. Magas cukortartalmú mustokból erjesztett borok összetétele. *Borászati Füzetek* **2015**, 45–47.
14. Guymon, J.F.; Crowell, E.A. Direct Gas Chromatographic Determination of Levo- and Meso-2,3-Butanediols in Wines and Factors Affecting Their Formation. *Am. J. Enol. Vitic.* **1967**, *18*, 200. [CrossRef]
15. Tokaji Borvidék Hegyközségi Tanácsa. *A TOKAJ Oltalom Alatt Álló Eredetmegjelölés Termékleírása*; Tokaji Borvidék Hegyközségi Tanácsa: Tokaj, Hungary, 2017.
16. *Commission Regulation (EEC) No 2676/90 of 17 September 1990 Determining Community Methods for the Analysis of Wines*; The European Commission: Brussels, Belgium, 1990.
17. Son, H.-S.; Hwang, G.-S.; Ahn, H.-J.; Park, W.-M.; Lee, C.-H.; Hong, Y.-S. Characterization of Wines from Grape Varieties through Multivariate Statistical Analysis of 1H NMR Spectroscopic Data. *Food Res. Int.* **2009**, *42*, 1483–1491. [CrossRef]
18. Kallitsounakis, G.; Catarino, S. An Overview on Botrytized Wines. *Ciênc. Téc. Vitivinícola* **2020**, *35*, 76–106. [CrossRef]
19. Magyar, I.; Soós, J. Botrytized Wines – Current Perspectives. *Int. J. Wine Res.* **2016**, *8*, 29–39. [CrossRef]

fermentation

Review

The Role of Malt on Beer Flavour Stability

Luis F. Guido * and Inês M. Ferreira

LAQV/REQUIMTE, Faculdade de Ciências da Universidade do Porto, Rua do Campo Alegre 687,
4169-007 Porto, Portugal
* Correspondence: lfguido@fc.up.pt; Tel.: +351-220-402-644

Abstract: Delaying flavour staling has been one of the greatest and most significant challenges for brewers. The choice of suitable raw materials, particularly malting barley, is the critical starting point to delay the risk of beer staling. Malting barley and the malting process can have an impact on beer instability due to the presence of pro-oxidant and antioxidant activities. Malt contains various compounds originating from barley or formed during the malting process, which can play a significant role in the fundamental processes of brewing through their antioxidant properties. This review explores the relationship between malt quality, in terms of antioxidant and pro-oxidant activities, and the flavour stability of beer.

Keywords: beer staling; malt quality; beer stability; beer flavour; antioxidant activity; pro-oxidant activity

Citation: Guido, L.F.; Ferreira, I.M. The Role of Malt on Beer Flavour Stability. *Fermentation* **2023**, *9*, 464. https://doi.org/10.3390/ fermentation9050464

Academic Editors: Maria Dimopoulou, Spiros Paramithiotis, Yorgos Kotseridis and Jayanta Kumar Patra

Received: 14 April 2023
Revised: 10 May 2023
Accepted: 11 May 2023
Published: 13 May 2023

1. Introduction

There has been controversy about the pathways involved in the synthesis of beer staling compounds. Although the filling and packaging processes and storage conditions are widely recognised as playing a central role, attention is being increasingly directed towards the properties of barley malt and the malting technique to better understand the factors affecting beer flavour stability. Malt is responsible for providing the principal colour of beer, and different malt types and malting conditions are responsible for both positive and negative flavours in beer [1]. The scarce knowledge on the significance of the balance between the antioxidant and pro-oxidant potential exhibited by barley and malt has motivated our research team to conduct studies aiming at elucidating the contribution of individual components to the overall antioxidant capacity of malt. A comprehensive review of the overall antioxidant properties of malt and how they are influenced by individual constituents of barley and the malting process may be accessed [2].

Although the staling of beer is a complex phenomenon, volatile carbonyl compounds have attracted most attention in the context of beer flavour instability. However, the myriad of flavour notes changing during staling is due to a much broader range of chemical entities. Unsaturated aldehydes, especially those with very low sensory thresholds having 7 to 10 carbon atoms, are considered to be the most important. Amongst them, *E*-2-nonenal, with an extremely low flavour threshold of 0.11 µg/L (Table 1), has long been the most frequently cited as the cause of a cardboard character typical of aged beer [3]. These aldehydes are potentially formed during malting and mashing by a number of different routes, in which the enzymatic and non-enzymatic degradation of polyunsaturated fatty acids is assumed to be the major source [4].

Acetaldehyde is the predominant carbonyl compound present in beer, making up around 60% of all aldehydes [5]. Despite this, its high flavour threshold (Table 1) means that it does not typically contribute significantly to beer flavour, except in rare cases such as when it causes a "green taste" in young beers [6]. In contrast, the most prominent aroma in a 6-month-old beer sample is often identified as *E*-β-damascenone, a terpenic ketone characterized by a "stewed apple," fruity, and honey-like note (Table 1, [7]). According to some authors, *E*-β-damascenone may be just as significant to the flavour of aged beer as *E*-2-nonenal [8,9]. Furanic aldehydes, such as furfural and 5-hydroxymethyl furfural, are often associated with the development of caramelized stale flavours, but it is widely accepted that these compounds are not directly responsible for such flavours [10].

The Strecker degradation is a chemical reaction between an amino acid and an alpha-dicarbonyl compound, such as those that are intermediates in the Maillard reaction. This reaction converts the amino acid into an aldehyde with one less carbon atom. While the aldehydes produced through the Maillard reaction and Strecker degradation are not typically considered to be the primary beer staling compounds, recent research has shown that aldehydes resulting from the Strecker degradation, such as 2-methylpropanal, 3-methylbutanal, and phenylacetaldehyde, can be found in higher concentrations in malt compared to other types of aldehydes. 3-Methylbutanal was found in the highest concentration (1213–8218 µg kg^{-1}), followed by 2-methylpropanal (612–3469 µg kg^{-1}), and phenylacetaldehyde (198–5105 µg kg^{-1}) [11,12].

In brief, nearly all the aldehydes that are considered aging indicators in beer are already present in malt [13]. Fickert and Schieberle (1998) identified several aldehydes in the volatile fraction of barley malt, including *E*-2-nonenal, which suggests that both lipid peroxidation and the Strecker degradation are important reactions for generating flavours during malt production [14].

Table 1. Sensory descriptors and flavour threshold values of carbonyl compounds in lager beers.

Compound	Flavour Threshold (mg/L)	Sensory Descriptor
Carbonyl Compounds		
Acetaldehyde	25 [a]	Green leaves, fruity
(*E*)-2-nonenal	0.00011 [a]	Papery, cardboard
(*E*)-β-damascenone	0.15 [b]	Stewed apple, honey-like
Furfural	150 [a]	Sweet, bready, caramellic
5-Hydroxymethyl-furfural	1000 [a]	Fatty, waxy, caramellic
Diacetyl	0.1–0.2 [c]	Butterscotch, rancid
2,3-Pentanedione	1 [c]	Fruity, sweet

[a] [15]; [b] [16]; [c] [17].

2. From Barley to Malt

The barley is transformed into malt through the malting process, which is mainly the transformation of insoluble starch into simple sugars. The malting process consists of three technological steps: steeping, germination, and kilning or roasting (final heat treatments).

During the steeping stage, the raw barley is soaked in water to increase the grain's moisture content from approximately 12% up to 46%. The moisture increase induces the beginning of germination and the early growth of the embryo, resulting in the activation and development of diastatic enzymes. The starch is then released from the endosperm and converted into simple sugars. Usually, the steeping step consists in alternating periods where the grains are submersed in water with rest periods or dry periods, with a duration of 36–48 h at a temperature of 15–20 °C [18,19].

The next step consists of the germination of steeped barley to obtain "green malt". The "green malt" is characterized by high moisture contents (up to 47%) and high enzymatic activity, resulting in the hydrolysis of cell walls and starch mediated by α-amilase, β-amylase, and β-glucanases. The germination step lasts for 4–6 days and it is conducted at temperatures around 15–22 °C. The moisture of the grains is kept around 45% and the growth of the rootlets is controlled to evaluate the degree of sprouting [18,19].

After germination, the moisture content of "green malt" must be reduced (to approximately 5%) in order to stop germination and assure optimal conditions for the conservation and storage of the grains. This is achieved by submitting the grains to a final heat treatment step (kilning), which will halt the sprouting process and avoid all the starch reserves that are consumed by deactivating enzymatic activity. The kilning process consists of a first phase called the withering phase, where the grains are exposed to temperatures between 50–65 °C for 12, reducing the moisture content to approximately 12% and halting the germination (above 40 °C). The second phase (curing phase) is carried out at temperatures up to 82–85 °C for 4 h, reducing moisture to values around 4% and ensuring enzyme inactivity. The grains are then cooled down to ensure an ideal temperature for discharge and storage [18,19].

The thermal processing steps have the greatest impact on the colour and flavour of malt. The obtained malted grains are usually called pale malt and are applied in almost all types of beer. Light malt is commonly fast-dried to limit the formation of Maillard reaction products, while for specialty dark malts the moisture is reduced slowly in order to achieve higher grain temperatures, inducing the Maillard reaction [20].

Specialty malts are produced to provide characteristic flavours and colours to beer, but they are typically used in very small amounts (<5%) compared to pale malt. They can be produced by roasting barley, green malt, or kilned pale malt [2,20]. resulting in low enzymatic activity and low levels of fermentable sugars and amino acids. Specialty malts are primarily used to add colour and aroma to beer. Specialty dark malts are generally classified into coloured malts, caramel malts, and roasted malts, which are obtained through roasting at temperatures ranging from 110 to 220 °C. Different degrees of roasting, which are determined by time and temperature, result in various intensities of colour and flavour through the Maillard and caramelization reactions, as well as pyrolysis reactions at high temperatures (>150 °C) [2,20,21].

The primary flavour characteristics of malt are developed during the malting process. However, recent studies have focused on understanding the impact of different barley varieties on the flavour of the beer. Through analysis of various barley varieties, researchers have identified how different metabolites influence the chemistry and flavour of the beer. The presence of various compounds, such as aldehydes, ketones, alcohols, furans, and aromatic compounds, contributes to the unique chemical composition of the final beer and thus, imparts its distinctive sensory attributes [22,23].

3. Malt Antioxidant Activity

Antioxidants can be broadly defined as compounds that inhibit oxidative reactions by decreasing molecular oxygen levels, scavenging chain-initiating and chain-propagating free radicals, chelating metals, or decomposing peroxides [24]. As such, they are believed to play a crucial role in malting and brewing by inhibiting oxidative damage.

Sulphites and ascorbic acid are commonly used as antioxidants during brewing to produce beers with high antioxidant activity [25]. However, due to consumer demand and stricter regulations, there has been a trend toward reducing the use of added antioxidants. Consequently, more attention is now being given to the brewing process and the properties of raw materials. Barley malt already contains various endogenous antioxidants such as phenolic compounds, phytic acid, ascorbic acid, and enzymes [26]. Protecting the endogenous antioxidants present in barley during malting can increase the brew's reduction potential, thus inhibiting oxidative processes harmful to flavour stability and avoiding the use of exogenous antioxidant compounds [27]. Important antioxidant compounds in malting and brewing include:

(i)   Melanoidins and reductones

Maillard reaction products (MRPs) are formed by the reaction between carbonyl groups of reducing sugars and amino groups of amino acids, peptides, or proteins, yielding a complex mixture of compounds with different molecular weights. The polymerization of low molecular weight compounds into high molecular weight compounds, also designated as melanoidins (MLD), may occur in the late stages of the Maillard reaction. MLD (Figure 1) are important for the quality and characteristics of many types of foods and beverages not only due to their colour and aroma, but also due to their health benefits and antioxidant properties [2]. MRPs act as scavengers for reactive oxygen species such as superoxide, peroxide, and hydroxyl radicals. They are known to be highly efficient antioxidants in the production and storage of food. MRPs have been found to have metal chelating properties, to be effective at reducing hydroperoxides to non-radical products, and to break the radical chain by donating a hydrogen atom [26]. Studies have reported that the antioxidant capacity of malt can increase during kilning and roasting as a result of the Maillard reaction, which leads to the development of reductones and MRPs [28–32]. MRPs have been identified as the primary contributors to the antioxidant activity of roasted malts [30,33], with a positive influence on the maintenance and development of malt-reducing properties [34].

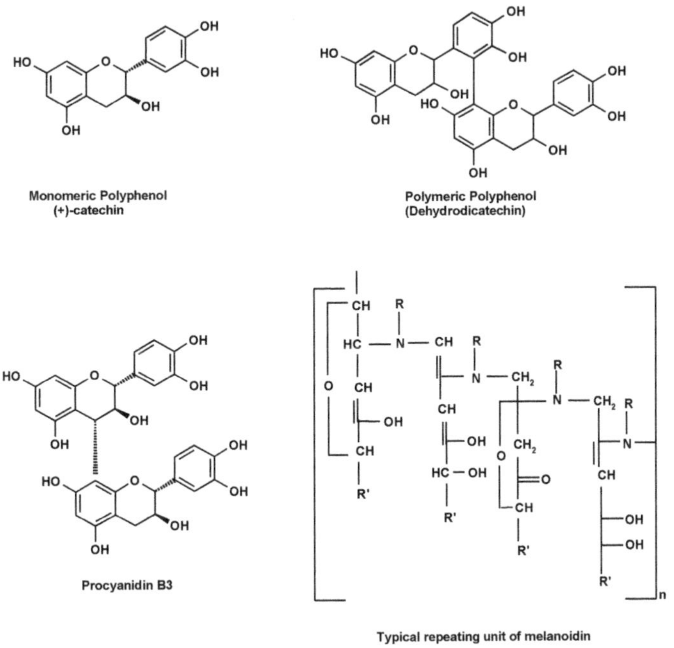

Monomeric Polyphenol
(+)-catechin

Polymeric Polyphenol
(Dehydrodicatechin)

Procyanidin B3

Typical repeating unit of melanoidin

Figure 1. Chemical structures for some important antioxidant species, naturally present in barley malt.

(ii)   Phenolic substances (phenolic acids and polyphenol compounds)

Barley contains 100 to 400 mg/kg of phenolic compounds, consisting of 80% of flavan-3-ols, 13% of flavonols, 5% of phenolic acids, and 2% of apolar compounds. Among flavan-3-ols, the most abundant compounds are the monomer forms, (+)-catechin and (−)-epicatechin, and polymer forms, constituted mainly by units of (+)-catechin and (+)-gallocatechin (Figure 1). Monomeric, dimeric, and trimeric flavan-3-ols accounted for 58% to 68% of the total phenolic content, with a predominance of trimeric flavan-3-ols [35].

Phenolic compounds are effective radical scavengers and can inhibit non-enzymatic lipid peroxidation. They also act as enzymatic lipid peroxidation inhibitors, as discussed below. Malt and hops both contribute these substances to wort and beer, but the majority comes from the malt. Malt-derived polyphenols account for about 70–80% of the polyphenols in the wort, with the remaining 20–30% coming from hops. The properties of the polyphenols from these two sources are likely to be different and highly dependent on the degree of polymerization. Lower molecular weight polyphenols are particularly effective as antioxidants, with the reducing power and solubility of polyphenols decreasing with increasing molecular weight.

The contribution of malt to the redox potential, and its status throughout the brewing process and in the final beer, depend on various factors and the interactions between different reducing agents. It is important to note that in certain situations, reducing agents can become pro-oxidants. Therefore, the oxidation/reduction state plays a crucial role in the deterioration of beer, and an increase in reducing activity is beneficial for enhancing flavour stability.

The antioxidant capacity of barley is primarily attributed to its polyphenols, which can vary depending on the barley variety. Therefore, selecting the appropriate barley variety is the first step in reducing the potential for oxidation. Research has shown that most of the polyphenols found in malt are already present in barley, indicating that the natural antioxidants in barley make a significant contribution to the antioxidant activity of malt [36]. Low-molecular-weight polyphenols (<5 kDa) are responsible for 80% of the antioxidant activity in malt and beer samples [37]. The kilning stage during malt production also has a considerable impact on the antioxidant levels and reduction potential of malt. The formation of melanoidins and reductones during kilning significantly influences the concentration of MRPs in malt, and this is affected by the kilning temperature and time. The malting process, in particular the kilning regimes and roasting temperatures, may have an important impact in terms of the phenolic composition and antioxidant features. Among the malts studied, extracts from light types (kilning temperature $\leq$160 °C) contained higher amounts of total and individual phenolic compounds, ferulic and *p*-coumaric acids in particular, than dark extracts (malt kilning temperature $\geq$200 °C) [38].

Recent studies have revealed that phenolic compounds can inhibit lipoxygenases from germinating barley [39]. The effectiveness of phenolic compounds in inhibiting autoxidation and enzymatic oxidation can vary widely, ranging from 1 to 100 depending on their chemical structure [27]. During malting, there is a reduction in phenolic content, with catechin monomers being the most affected, as shown by Goupy et al. (1999) [33]. The antioxidant (+)-catechin and ferulic acid decreased the rate of formation of some carbonyl compounds during beer forced-ageing in the presence of air, but had no impact during the extended storage of beer at low levels of oxygen [4]. The beer produced from proanthocyanidin-free barley called Caminant was considered to be slightly inferior by a tasting panel compared to reference beers brewed from barley varieties cultivated under comparable conditions. This may be attributed to the deficient polyphenol level, particularly the catechin fraction, in the Caminant barley variety. As a result, this variety poses challenges in terms of flavour stability [40]. Even though the potential benefit of polyphenols to the colloidal stability of beer, by forming non-biological haze with proteins, is now largely recognised, their beneficial role for flavour stability is still an open question. To prove this, it was recently demonstrated that the partial removal of polyphenols by polyvinylpolypyrrolidone has no impact on flavour stability [41]. PCA analysis has revealed that the chemical composition and sensory characteristics of aged beer are affected by the varietal differences in barley [42]. The presence of natural polyphenols in barley has been found to have a positive impact on beer flavour stability. However, it should be noted that technological factors can also significantly influence beer flavour stability, as evidenced by the significant heterogeneity observed in the malt bed during industrial kilning.

Boivin et al. (1993) demonstrated that malt contains compounds that can inhibit the lipoxygenase activity of germinating barley, prevent lipid oxidation, and exhibit reducing power, according to various methods for evaluating antioxidant properties. The malt's reducing compounds are produced during germination and increase during the first stage of kilning, but decrease at higher temperatures. The amount of reducing compounds in malt depends on the barley cultivar and the kilning conditions applied [26]. The kilning regime not only affects the malt's antioxidant properties but also influences the colour and flavour profile of the final beer product [29].

Barley and germinating barley produce enzymes that exhibit antioxidant activity. One such enzyme is superoxide dismutase (SOD, EC 1.15.1.1), which catalyses the conversion of superoxide radicals to hydrogen peroxide, which is subsequently broken down into water and oxygen by catalase (CAT, EC 1.11.1.6). Together, SOD and CAT maintain oxygen in a stable, less reactive state by reducing the levels of superoxide and hydroperoxide. Barley contains both of these enzymes and their activities increase during germination.

They can also survive pale kilning regimes but are destroyed at mashing temperatures exceeding 65 °C [43]. Peroxidase (POD, EC 1.11.1.7) is a primary antioxidant that can protect against medium oxidation by removing hydrogen peroxide. However, malt POD can also oxidize endogenous barley phenolic compounds, such as ferulic acid, (+)-catechin, and (−)-epicatechin, which could have negative effects on beer quality [44]. The residual enzyme activities in malt depend on both the barley cultivar and the malting process. Natural antioxidant compounds in malt, such as phenolic compounds and MRPs, may play a significant role in inhibiting oxidative processes during malting and brewing. These antioxidants can inhibit lipoxygenase action during malting and mashing and decrease the autoxidation reaction during the brewing process and beer storage. While enzyme antioxidants can only act during malting and at the beginning of mashing, phenolic compounds and MRPs can act throughout the process and even after the beer has been stored. Therefore, careful selection of barley cultivars and malting regimes can help maximize the antioxidant potential of malt and improve beer quality.

4. Malt Pro-Oxidant Activity

Pro-oxidant malt compounds are primarily associated with the enzymes that are responsible for the breakdown of lipids. These enzymes include lipase (EC 3.1.1.3), lipoxygenase (LOX, EC 1.13.11.12), and the hydroperoxide-reactive enzyme system (Figure 2).

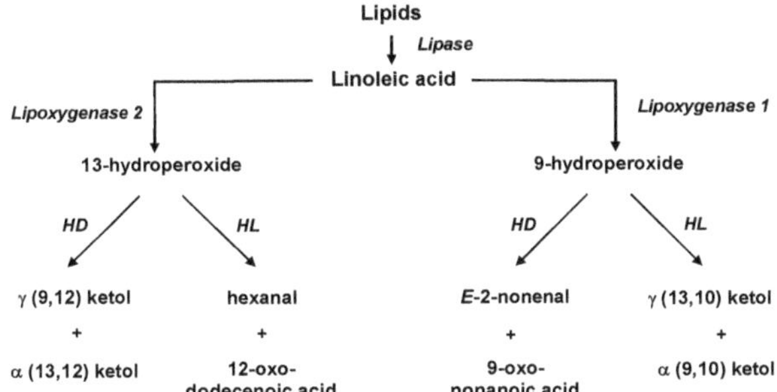

Figure 2. The enzymatic oxidation of lipids: lipase, lipoxygenase, and hydroperoxidase-reactive enzyme system. HL: hydroperoxide lyase. HD: hydroperoxide dehydrogenase (after [43]).

Oxidation of malt phenolic compounds by the catalytic action of polyphenol oxidase (PPO, EC 1.14.18.1) also occurs during the malting process. All these enzymes are found in most cereals, including barley [44], but they may also be synthesized by microflora developing during malting.

Pro-oxidant enzymes are primarily involved in lipid degradation. Lipase is the first enzyme to act on the ester bond between fatty acid and glycerol of triglycerides and diglycerides, releasing free fatty acids from lipids. Lipoxygenase catalyzes the oxidation of polyunsaturated free fatty acids, such as linoleic acid (C18:2), forming hydroperoxides. Lipoxygenase may also be involved in the creation of oxidative cross-linking between thiol-rich proteins via reactions, resulting in macromolecular reticulations that could alter the filterability performance of wort and beer, possibly affecting their quality [43]. The primary oxidation products of lipoxygenase activity, hydroperoxides, are decomposed into off-flavour compounds, such as unsaturated aldehydes, by hydroperoxide reactive enzyme systems, namely, hydroperoxide lyase and hydroperoxide isomerase (EC 4.2.1.92) [26]. High moisture content (above 40%) and low temperatures (below 60 °C) promote lipoxygenase (LOX) activity during the withering phase of kilning, resulting in the synthesis of E-2-nonenal and adduct formation. These nonenal adducts, also known as malt-RNP, which account for approximately 25% of the nonenal potential in the mash, can have a negative impact on the flavour stability of beer [45]. However, Carlsberg Research Laboratory developed a low-lipoxygenase barley cultivar in 2002, which expresses mutant LOX-1 protein. This barley variety can produce beer with significantly enhanced flavour stability and reduced levels of E-2-nonenal. Through mutation breeding, a LOX-1-null barley line was obtained, which can improve the flavour stability of beer without affecting other important beer qualities [46]. The results of a recent study clearly indicated that the LOX-less barley malt showed less nonenal potential than the control, and the beer brewed from the LOX-less barley malt contained much lower concentrations of trans-2-nonenal (T2N) and gamma-nonalactone, especially after the (forced or natural) aging of the beer, compared with the beer brewed under the same conditions using the control malt [47].

Polyphenol oxidase is able to catalyse the oxidation of polyphenols compounds with oxygen in very reactive quinonic compounds (Figure 3). In the oxidized state, they can cross-link and polymerize with proteins or cell-wall polysaccharides, directly influencing the formation of non-biological haze in wort and beer. Polyphenol oxidase is primarily responsible for enzymatic browning in fruits and vegetables. Enzymatic or chemical oxidation of polyphenols typically results in a loss of their antioxidant capacity. However, recent studies suggest that partially oxidized polyphenols may exhibit higher antioxidant activity than non-oxidized phenols [48].

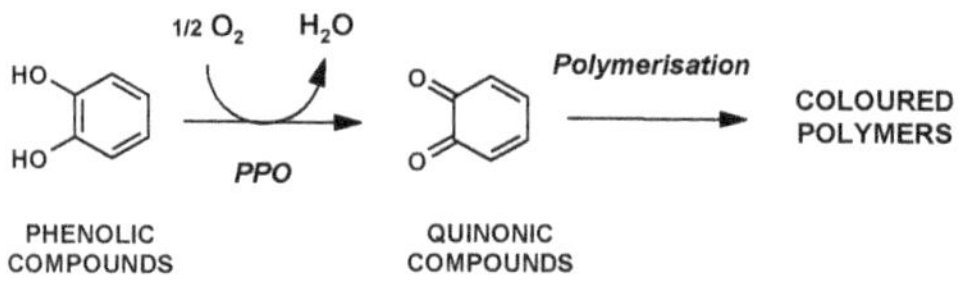

Figure 3. The action of polyphenol oxidase (PPO).

The pro-oxidant activity of malt extracts can also be attributed to flavonoids, procyanidins, and certain MRPs [29]. Many phenolic compounds act as antioxidants only at high concentrations, but at lower levels, they may have pro-oxidant effects [49]. Apart from their well-established antioxidant properties, MRPs may also exhibit pro-oxidant properties. Highly reactive radicals are generated in the initial stages of the Maillard reaction, and their disappearance is accompanied by the gradual development of browning. The level of these radicals depends on the intensity and duration of the heat treatment applied because at low temperatures, the reaction steps contributing to the formation of pro-oxidant compounds last longer than with high-temperature treatments [50]. High molecular weight browning compounds, generated by roasting barley, have been shown to act as pro-oxidants in metal-catalyzed oxidation reactions [51].

Figure 4 illustrates some possible routes of the pro- and antioxidant enzymatic activity in the malting and brewing process.

Through their sequential action, these enzymes are most active during the malting and mashing processes. Enzymatic activity is destroyed during the kilning and mashing steps, except for POD, which is a highly heat-stable enzyme. However, POD, which can oxidize phenolic compounds, appears to have limited action in the finished product due to the extremely low levels of hydrogen peroxide. In contrast, phenolic compounds and MRPs may play a significant role throughout the entire process and even after beer storage. Evidence has been provided for the inhibitory action of malt polyphenols on lipoxygenase (LOX) activity in finished malts. The anti-radical power, which is highly correlated with polyphenolic content, was found to be similar for both malt and barley, highlighting the essential role of barley's endogenous polyphenols on beer flavour stability [52]. The radical scavenging properties of highly polymerized phenolic compounds may also be effective against oxidative reactions during the malting and mashing stages. A deeper understanding of this issue could be achieved by investigating the mechanism by which malt polyphenols exert their protective action throughout the malting, mashing, or filtration steps against oxidative reactions.

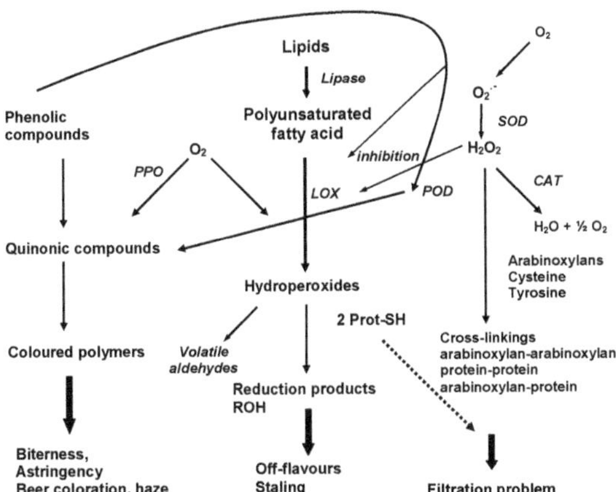

Figure 4. Implication of pro- and antioxidant enzymes in the malting and brewing process.

5. Concluding Remarks

The quality of malt can impact beer instability due to the presence of lipids, oxidative enzymes (such as lipoxygenase, hydroperoxide lyase, and hydroperoxide dehydrogenase), polyphenols, and phenolic acids. Polyphenols and phenolic acids present in malt are natural antioxidants capable of delaying, retarding, or preventing oxidation processes. Therefore, they are considered to have a significant effect on malting and brewing as inhibitors of oxidative damage. As a result, increasing attention is being directed toward the final properties of raw materials. For example, protecting the endogenous antioxidants present in barley during malting could increase the reduction potential of the brew, thus helping to inhibit oxidative processes that are detrimental to flavour stability and avoid the use of exogenous antioxidant compounds. Among the measured variables for malts, malt anti-radical power is the major contributor to beer flavour stability. In conclusion, it is clear that malt quality in terms of pro-oxidant and antioxidant activity plays a central role in beer flavour instability. This can be improved through a suitable choice of barley and malting process. Malt with low lipoxygenase activity, low residual nonenal potential, free phenolic compounds with high antioxidant activity, and a high amount of reducing compounds provide an excellent starting point.

Author Contributions: L.F.G. coordinated the study and contributed to the writing, editing, and correction of the manuscript. He is also the corresponding author of this article. I.M.F. contributed to the writing and editing of the manuscript. All authors have read and agreed to the published version of the manuscript.

Funding: This research was funded by Fundação para a Ciência e Tecnologia, grant number UIDB/50006/2020, UIDP/50006/2020, and PD/BD/135091/2017.

Institutional Review Board Statement: Not applicable.

Informed Consent Statement: Not applicable.

Data Availability Statement: Not applicable.

Conflicts of Interest: The authors declare no conflict of interest.

References

1. Bamforth, C. *Beer: Tap into the Art and Science of Brewing*; Oxford University Press: Oxford, UK, 2003.
2. Carvalho, D.O.; Gonçalves, L.M.; Guido, L.F. Overall antioxidant properties of malt and how they are influenced by the individual constituents of barley and the malting process. *Compr. Rev. Food Sci. Food Saf.* **2016**, *15*, 927–943. [CrossRef]

3. Santos, J.R.; Carneiro, J.R.; Guido, L.F.; Almeida, P.J.; Rodrigues, J.A.; Barros, A.A. Determination of E-2-nonenal by high-performance liquid chromatography with UV detection assay for the evaluationof beer ageing. *J. Chromatogr. A* **2003**, *985*, 395–402. [CrossRef]

4. Walters, M. Natural Antioxidants and flavour stability. *Ferment* **1997**, *10*, 111–119.

5. Guido, L.F.; Fortunato, N.A.; Rodrigues, J.A.; Barros, A.A. Voltammetric assay for the aging of beer. *J. Agric. Food Chem.* **2003**, *51*, 3911–3915. [CrossRef]

6. Stewart, G.G. The production of secondary metabolites with flavour potential during brewing and distilling wort fermentations. *Fermentation* **2017**, *3*, 63. [CrossRef]

7. Murakami, A.A.; Goldstein, H.; Navarro, A.; Seabrooks, J.R.; Ryder, D.S. Investigation of beer flavor by gas chromatography-olfactometry. *J. Am. Soc. Brew. Chem.* **2003**, *61*, 23–32. [CrossRef]

8. Gijs, L.; Chevance, F.; Jerkovic, V.; Collin, S. How low pH can intensify β-damascenone and dimethyl trisulfide production through beer aging. *J. Agric. Food Chem.* **2002**, *50*, 5612–5616. [CrossRef] [PubMed]

9. Guido, L.F.; Carneiro, J.R.; Santos, J.R.; Almeida, P.J.; Rodrigues, J.A.; Barros, A.A. Simultaneous determination of E-2-nonenal and beta-damascenone in beer by reversed-phase liquid chromatography with UV detection. *J. Chromatogr. A* **2004**, *1032*, 17–22. [CrossRef]

10. Carneiro, J.R.; Guido, L.F.; Almeida, P.J.; Rodrigues, J.A.; Barros, A.A. The impact of sulphur dioxide and oxygen on the behaviour of 2-furaldehyde in beer: An industrial approach. *Int. J. Food Sci. Technol.* **2006**, *41*, 545–552. [CrossRef]

11. Ditrych, M.; Filipowska, W.; De Rouck, G.; Jaskula-Goiris, B.; Aerts, G.; Andersen, M.L.; De Cooman, L. Investigating the evolution of free staling aldehydes throughout the wort production process. *Brew. Sci.* **2019**, *72*, 10–17.

12. Reichel, S.; Carvalho, D.O.; Santos, J.R.; Bednar, P.; Rodrigues, J.A.; Guido, L.F. Profiling the volatile carbonyl compounds of barley and malt samples using a low-pressure assisted extraction system. *Food Control* **2021**, *121*, 107568. [CrossRef]

13. Van Waesberghe, J.W. Flavour stability starts with malt and in the brewhouse. *Brauwelt Int.* **2002**, *20*, 375–377.

14. Fickert, B.; Schieberle, P. Identification of the key odorants in barley malt (caramalt) using GC/MS techniques and odour dilution analyses. *Food Nahrung* **1998**, *42*, 371–375. [CrossRef]

15. Meilgaard, M.C. Flavor chemistry of beer. II. Flavor and threshold of 239 aroma volatiles. *Tech. Quart. Master. Brew. Assoc. Am.* **1975**, *12*, 151–168.

16. Guyot-Declerck, C.; François, N.; Ritter, C.; Govaerts, B.; Collin, S. Influence of pH and ageing on beer organoleptic properties. A sensory analysis based on AEDA data. *Food Qual. Prefer.* **2005**, *16*, 157–162. [CrossRef]

17. Wainwright, T. Diacetyl—A review. 1. analytical and biochemical considerations. ii. brewing experience. *J. Inst. Brew.* **1973**, *79*, 451–470. [CrossRef]

18. Briggs, D.E.; Hough, J.S.; Stevens, R.; Young, T.W. *Malting and Brewing Science: Malt and Sweet Wort*; Springer: Berlin/Heidelberg, Germany, 1982; Volume 1.

19. EBC. *Malting Technology: Manual of Good Practice*; Carl, F.H., Ed.; EBC: Nürnberg, Germany, 2000.

20. Belitz, H.D.; Grosch, W.; Schieberle, P. Springer Food chemistry 4th revised and extended edition. *Annu. Rev. Biochem.* **2009**, *79*, 655–681.

21. Mosher, M.; Trantham, K. *Brewing Science: A Multidisciplinary Approach*; Springer: Berlin/Heidelberg, Germany, 2017.

22. Dong, L.; Hou, Y.; Li, F.; Piao, Y.; Zhang, X.; Zhang, X.; Li, C.; Zhao, C. Characterization of volatile aroma compounds in different brewing barley cultivars. *J. Sci. Food Agric.* **2015**, *95*, 915–921. [CrossRef] [PubMed]

23. Bettenhausen, H.M.; Barr, L.; Broeckling, C.D.; Chaparro, J.M.; Holbrook, C.; Sedin, D.; Heuberger, A. Influence of malt source on beer chemistry, flavor, and flavor stability. *J. Food Res. Int.* **2018**, *113*, 487–504. [CrossRef]

24. Halliwell, B.; Gutteridge, J.M.C.; Aruoma, O.I. The deoxyribose method: A simple "test-tube" assay for determination of rate constants for reactions of hydroxyl radicals. *Anal. Biochem.* **1987**, *165*, 215–219. [CrossRef]

25. Van Gheluwe, J.E.A.; Yalyi, Z.; Dadic, M. Oxidation in brewing. *Brew. Dig.* **1970**, *45*, 70–79.

26. Boivin, P.; Clamagirand, V.; Maillard, M.N.; Berset, C.; Malanda, M. Malt quality and oxidation risk in brewing. *Proc. Conv. Int. Brew. Asia Pac. Sect.* **1996**, *24*, 110–115.

27. Boivin, P.; Allain, D.; Clamagirant, V.; Maillard, M.N.; Cuvelier, M.E.; Berset, C.; Richard, H.; Nicolas, J.; Forget-Richard, F. Mesure de l'activité antioxygène de l'orge et du malt: Approche multiple. *Proc. Congr. Eur. Brew. Conv.* **1993**, 397–404.

28. Maillard, M.N.; Soum, M.H.; Boivin, P.; Berset, C. Antioxidant activity of barley and malt: Relationship with phenolic content. *J. Food Sci. Technol.* **1996**, *29*, 238–244. [CrossRef]

29. Woffenden, H.M.; Ames, J.M.; Chandra, S.; Anese, M.; Nicoli, M.C. Effect of kilning on the antioxidant and pro-oxidant activities of pale malts. *J. Agric. Food Chem.* **2002**, *50*, 4925–4933. [CrossRef]

30. Samaras, T.S.; Camburn, P.A.; Chandra, S.X.; Gordon, M.H.; Ames, J.M. Antioxidant properties of kilned and roasted malts. *J. Agric. Food Chem.* **2005**, *53*, 8068–8074. [CrossRef] [PubMed]

31. Vanderhaegen, B.; Neven, H.; Verachtert, H.; Derdelinckx, G. The chemistry of beer aging—A critical review. *Food Chem.* **2006**, *95*, 357–381. [CrossRef]

32. Inns, E.L.; Buggey, L.A.; Booer, C.; Nursten, H.E.; Ames, J.M. Effect of modification of the kilning regimen on levels of free ferulic acid and antioxidant activity in malt. *J. Agric. Food Chem.* **2011**, *59*, 9335–9343. [CrossRef]

33. Coghe, S.; Vanderhaegen, B.; Pelgrims, B.; Basteyns, A.V.; Delvaux, F.R. Characterization of dark specialty malts: New insights in color evaluation and pro-and antioxidative activity. *J. Am. Soc. Brew. Chem.* **2003**, *61*, 125–132. [CrossRef]

34. Čechovská, L.; Konečný, M.; Velíšek, J.; Cejpek, K. Effect of Maillard reaction on reducing power of malts and beers. *Czech J. Food Sci.* **2012**, *30*, 548–556. [CrossRef]
35. Goupy, P.; Hugues, M.; Boivin, P.; Amiot, M.J. Antioxidant composition and activity of barley (*Hordeum vulgare*) and malt extracts and of isolated phenolic compounds. *J. Sci. Food Agric.* **1999**, *79*, 1625–1634. [CrossRef]
36. Chandra, C.S.; Buggey, L.A.; Peters, S.; Cann, C.; Liegeois, C. Factors Affecting the Development of Antioxidant Properties of Malts during the Malting and Roasting Process. *Hgca Proj. Rep.* **2001**. Available online: https://ahdb.org.uk/factors-affecting-the-development-of-antioxidant-properties-of-malts-during-the-malting-and-roasting-process (accessed on 14 April 2023).
37. Pascoe, H.M.; Ames, J.M.; Chandra, S. Critical stages of the brewing process for changes in antioxidant activity and levels of phenolic compounds in ale. *J. Am. Soc. Brew. Chem.* **2003**, *61*, 203–209. [CrossRef]
38. Moreira, M.M.; Morais, S.; Carvalho, D.O.; Barros, A.A.; Delerue-Matos, C.; Guido, L.F. Brewer's spent grain from different types of malt: Evaluation of the antioxidant activity and identification of the major phenolic compounds. *Food Res. Int.* **2013**, *54*, 382–388. [CrossRef]
39. Boivin, P.; Malanda, M.; Clamagirand, V. Evaluation of the organoleptic quality of malt. Evolution during malting and varietal influence. *Proc Congr. Eur. Brew. Conv.* **1995**, 159–168.
40. Back, W.; Forster, C.; Krottenthaler, M.; Lehmann, J.; Sacher, B.; Thum, B. New research findings on improving taste stability. *Brauwelt Int.* **1999**, *17*, 394–405.
41. Bushnell, S.E.; Guinard, J.X.; Bamforth, C.W. Effects of sulfur dioxide and polyvinylpolypyrrolidone on the flavor stability of beer as measured by sensory and chemical analysis. *J. Am. Soc. Brew. Chem.* **2003**, *61*, 133–141. [CrossRef]
42. Guido, L.F.; Curto, A.; Boivin, P.; Benismail, N.; Gonçalves, C.; Barros, A.A. Predicting the organoleptic stability of beer from chemical data using multivariate analysis. *Eur. Food Res. Technol.* **2007**, *226*, 57–62. [CrossRef]
43. Bamforth, C.; Clarkson, S.; Large, P. The relative importance of polyphenol oxidase, lipoxygenase and peroxidases during wort oxidation. *Proc. Congr. Eur. Brew. Conv.* **1991**, *23*, 617–624.
44. Boivin, P. Pro-and anti-oxidant enzymatic activity in malt. *Cerevisia* **2001**, *26*, 109–116.
45. Guido, L.F.; Boivin, P.; Benismail, N.; Gonçalves, C.R.; Barros, A.A. An early development of the nonenal potential in the malting process. *Eur. Food Res. Technol.* **2005**, *220*, 200–206. [CrossRef]
46. Hirota, N.; Kuroda, H.; Takoi, K.; Kaneko, T.; Kaneda, H.; Yoshida, I.; Takashio, M.; Ito, K.; Takeda, K. Brewing performance of malted lipoxygenase-1 null barley and effect on the flavor stability of beer. *Cereal Chem.* **2006**, *83*, 250–254. [CrossRef]
47. Yunhong, Y.; Huang, S.; Dong, J.; Fan, W.; Huang, S.; Liu, J.; Chang, Z.; Tian, Y.; Hao, J.; Hu, S. The influence of LOX-less barley malt on the flavour stability of wort and beer. *J. Inst. Brew.* **2014**, *120*, 93–98.
48. Manzocco, L.; Calligaris, S.; Mastrocola, D.; Nicoli, M.C.; Lerici, C.R. Review of non-enzymatic browning and antioxidant capacity in processed foods. *Trends Food Sci. Technol.* **2000**, *11*, 340–346. [CrossRef]
49. Yen, G.C.; Chen, H.Y.; Peng, H.H. Antioxidant and pro-oxidant effects of various tea extracts. *J. Agric. Food Chem.* **1997**, *45*, 30–34. [CrossRef]
50. Nicoli, M.C.; Anese, M.; Parpinel, M. Influence of processing on the antioxidant properties of fruit and vegetables. *Trends Food Sci. Technol.* **1999**, *10*, 94–100. [CrossRef]
51. Carvalho, D.O.; Øgendal, L.H.; Andersen, M.L.; Guido, L.F. High molecular weight compounds generated by roasting barley malt are pro-oxidants in metal-catalyzed oxidations. *Eur. Food Res. Technol.* **2016**, *242*, 1545–1553. [CrossRef]
52. Guido, L.F.; Curto, A.F.; Boivin, P.; Benismail, N.; Gonçalves, C.R.; Barros, A.A. Correlation of malt quality parameters and beer flavor stability: Multivariate analysis. *J. Agric. Food Chem.* **2007**, *55*, 728–733. [CrossRef] [PubMed]

fermentation

Review

Zero- and Low-Alcohol Fermented Beverages: A Perspective for Non-Conventional Healthy and Sustainable Production from Red Fruits

Marcello Brugnoli [1], Elsa Cantadori [1,2], Mattia Pia Arena [1], Luciana De Vero [1], Andrea Colonello [2] and Maria Gullo [1,3,*]

1   Department of Life Sciences, University of Modena and Reggio Emilia, 42122 Reggio Emilia, Italy; marcello.brugnoli@unimore.it (M.B.); elsa.cantadori@unimore.it (E.C.); mattiapia.arena@unimore.it (M.P.A.); luciana.devero@unimore.it (L.D.V.)
2   Ponti SpA, 28074 Ghemme, Italy; andrea.colonello@ponti.com
3   NBFC, National Biodiversity Future Center, 90133 Palermo, Italy
*   Correspondence: maria.gullo@unimore.it

Abstract: The growing health consciousness among consumers is leading to an increased presence of functional foods and beverages on the market. Red fruits are rich in bioactive compounds such as anthocyanins with high antioxidant activity. In addition, red fruits contain sugars and are rich in phenolic compounds, vitamin C, dietary fibers, and manganese. Due to these characteristics, they are also suitable substrates for fermentation. Indeed, nowadays, microbial transformation of red fruits is based on alcoholic or lactic fermentation, producing alcoholic and non-alcoholic products, respectively. Although products fermented by acetic acid bacteria (AAB) have been thoroughly studied as a model of health benefits for human beings, little evidence is available on the acetic and gluconic fermentation of red fruits for obtaining functional products. Accordingly, this review aims to explore the potential of different red fruits, namely blackberry, raspberry, and blackcurrant, as raw materials for fermentation processes aimed at producing low- and no-alcohol beverages containing bioactive compounds and no added sugars. AAB are treated with a focus on their ability to produce acetic acid, gluconic acid, and bacterial cellulose, which are compounds of interest for developing fruit-based fermented beverages.

Keywords: red fruits; acetic acid bacteria; vinegar; non-alcoholic beverages; gluconic acid; acetic acid

Citation: Brugnoli, M.; Cantadori, E.; Arena, M.P.; De Vero, L.; Colonello, A.; Gullo, M. Zero- and Low-Alcohol Fermented Beverages: A Perspective for Non-Conventional Healthy and Sustainable Production from Red Fruits. *Fermentation* **2023**, *9*, 457. https://doi.org/10.3390/fermentation9050457

Academic Editors: Maria Dimopoulou, Spiros Paramithiotis, Yorgos Kotseridis and Jayanta Kumar Patra

Received: 21 April 2023
Revised: 4 May 2023
Accepted: 8 May 2023
Published: 10 May 2023

1. Introduction

Nowadays, the consumption of functional fermented foods and beverages is a well-established habit as consumers are strongly interested in products with health claims. As a matter of fact, since ancient times, fermented products have been a part of human nutrition. Originally, their production was performed to improve the shelf life of perishable raw materials from agriculture and animal husbandry. Subsequently, many different microorganisms have been selected in order to obtain disparate fermented products with favorite quality characteristics, mostly regarding shelf life, taste, texture, mouthfeel, flavor, and color [1–6]. More recently, in addition to researching certain sensory and technological characteristics, the challenge goes so far to obtain products with the added quality of beneficial influencing human health. In this frame, zero- and low-ethanol beverages production is an expanding globally promoted market, although availability, acceptability, and affordability are still issues and current gaps that need to be filled [7].

Thus, a diversity of fermented products is obtained starting from disparate raw materials, depending on their availability and diffusion in the territories of origin. Therefore, milk, meat, cereals, vegetables, and fruits are widely used in both traditional and modern food manufacturing processes.

In general, fruits are suitable raw materials due to their fermentative aptitudes, such as high content of sugars, in particular glucose and fructose. It has been reported that fermentation enhances the nutritive value of the final products, which are generally characterized by lower amounts of glucose and great antioxidant content [8]. Among the different kinds of fruits, red fruits are extremely beneficial for human health and, with their intense flavor, are appreciated by consumers. Several studies have stated their phytochemical composition, which includes minerals, fibers, and antioxidant compounds able to exert a protective effect against many chronic diseases [9–12] (Figure 1).

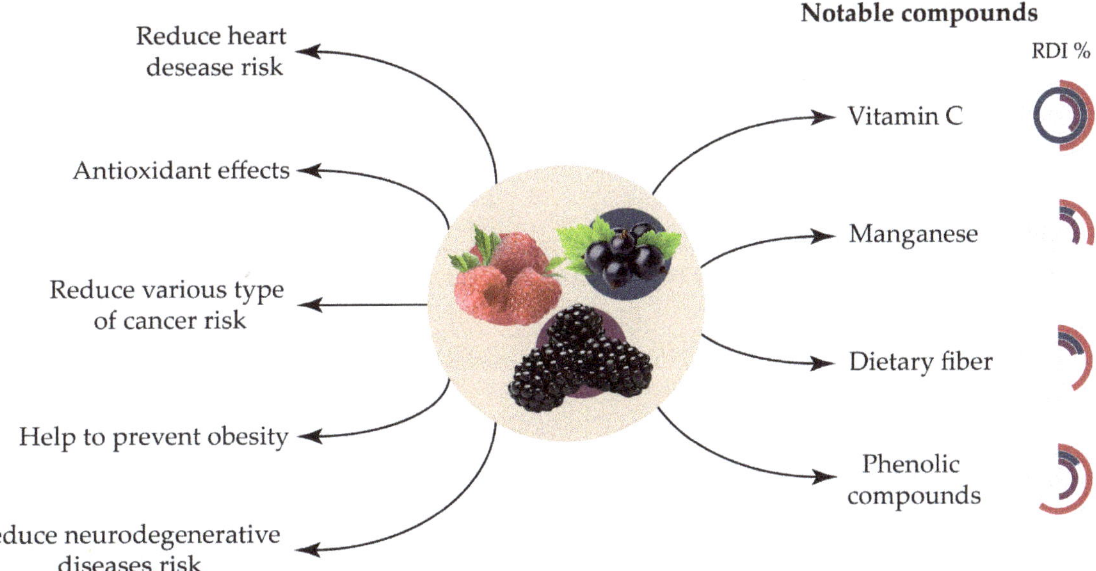

Figure 1. Raspberry, blackcurrant, and blackberry are the main health benefits reported in the literature. Notable compounds are listed, and the recommended daily intake (RDI) is reported as a percentage covered by a 100 g portion of raspberry (dark red), blackcurrant (blue), and blackberry (violet).

Polyphenols, carotenoids, and anthocyanins are the main kinds of antioxidant phytochemicals found in red fruits.

Anthocyanins are responsible for the red/dark color of fruits and are the major flavonoid family molecules present in raspberry, blackberry, and blackcurrant. The most common naturally occurring anthocyanins are the 3-O-glycosides or 3,5-di-O-glycosides of cyanidin, delphinidin, peonidin, petunidin, pelargonidin, and malvidin. Numerous studies have shown that they can have various biological activities, including antimicrobial and anti-inflammatory activities, protective action against various degenerative diseases, and an important role in decreasing the invasiveness of tumor cells [13–17]. Nevertheless, the profile and concentration of anthocyanins are different depending on the fruit. Some red fruits, for example, strawberries, have lower concentrations of anthocyanins, and others, such as black currants, have higher concentrations [18].

Other than anthocyanins, red fruits also contain vitamin C, ellagitannins, and several minerals such as manganese, giving berries and derivative products outstanding health benefits for human beings [19]. Amongst health benefits, the antioxidant effect is a major characteristic of red fruits [20].

However, health-based recommendations include reducing alcohol consumption and calories, thus promoting zero/low-ethanol and no added sugars beverages [21].

For these reasons, owing to outstanding functional properties, red fruits represent valid non-conventional raw materials for producing low- and no-alcohol fermented beverages with health benefits. This is in line with the recent increasing trend to exploit non-conventional raw materials containing fermentable sugars for producing new functional beverages. Date palm fruits, for instance, were proven to be a valid starting substrate for the formulation of functional foods and beverages, as they show a high amount of sugars, which makes them highly fermentable and dietary fibers, minerals, vitamins, and phenolic compounds, which confer functional features to end-products [22].

Moreover, legumes, single-cell protein, bee pollen, and tropical fruits assume a greater role in the food market as non-conventional matrices suitable to produce dairy-free functional products [23]. The recognition of the beneficial effects of consuming functional products on a daily basis led to the scientific interest in developing new products; in this perspective, quince, kiwifruit, prickly pear, and pomegranate juices have been explored as fermentable substrates to developing new non-dairy fermented beverages [24].

In addition to health reasons, sometimes using alternatives to traditional raw materials can also have an economic return since the resources existing in a territory can be exploited in the best possible way. This is the case, for example, of non-conventional edible plants that are spontaneous, wild, or cultivated vegetable species, e.g., wax mallow, used in certain regions and cultures as therapeutic herbs and can be opportunely used to produce fermented beverages showing beneficial bioactivity [25,26]. In this frame, fermented beverages, such as vinegar with different content of acetic acid, gluconic beverages, and kombucha tea, fit as functional products with health benefits for human beings. These beverages are the result of the fermentative activity of different microbial groups, including yeasts, acetic acid bacteria (AAB), and lactic acid bacteria (LAB).

Vinegars and vinegar-based beverages are produced via a double fermentation: an alcoholic fermentation performed by yeasts and then an acetic acid fermentation by AAB. Yeast hydrolyzes sucrose into glucose and fructose, which are used to produce ethanol. AAB oxidizes ethanol into acetic acid, which is the main organic acid that characterizes vinegar-based beverages, even though the concentration is significantly lower compared to vinegar [27].

Gluconic beverages are produced by a single-step fermentation in which glucose is oxidized into gluconic acid. The latter is a weak organic acid exploited to improve the sensorial complexity of foods and beverages. Gluconic acid can be further oxidized into 2,5-diketo-D-gluconate acid and 5-keto-D-gluconic acid [28].

Kombucha tea is produced by the activity of a consortium of microorganisms growing on sugared tea. AAB and yeasts are the most present, whereas LAB occurs less frequently [29]. Kombucha fermentation starts with the hydrolyzation of sucrose into glucose and fructose by yeasts, then bacteria start oxidative and fermentative processes of glucose, fructose, and ethanol [30]. The result is a beverage containing a mix of organic acids, mainly acetic and gluconic acids, ethanol in small amounts, CO_2, and a floating layer of bacterial cellulose produced by AAB [31]. In addition, kombucha is reported to have beneficial effects on human health, such as anticancer, antimicrobial, antioxidant properties, and anti-aging activity [32].

In this review, raspberry, blackcurrant, and blackberry are focused on their composition and aptitude to develop fermentations. Alcoholic and lactic fermentations are described as the more applied microbial transformations. Acetic acid fermentation for producing low- and non-alcoholic products, namely vinegar and acetic and gluconic beverages (Figure 2), are discussed, highlighting their potential as well as the limited availability of existing marketed products.

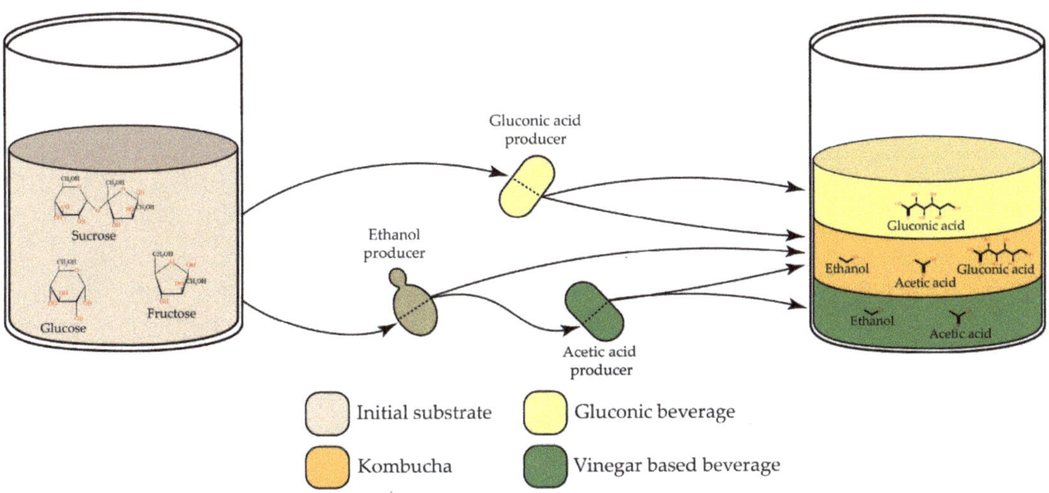

Figure 2. Microbial transformations and interactions in the production of gluconic beverages, kombucha, and vinegar-based beverages starting from a sugared substrate.

2. Red Fruits Features

2.1. Blackberry Fruit (Rubus Subg. Rubus)

Blackberry is a worldwide consumed fruit, mostly produced in North America, Europe, and Asia [33]. The major producers are North America and Europe, producing 65,000 and 45,000 tons/year, respectively [34]. Amongst European countries, Serbia and Hungary lead the production, representing almost all European annual production.

At the industrial level, blackberries are used for different productions, such as dietary supplements or jams. However, they are mostly consumed as fresh fruits or sold as individually quick-frozen packs. Although blackberries' chemical composition is strictly dependent on several factors, such as the cultivar or the stage of ripeness [35], generally, they are rich in sugars, minerals, and phenolic compounds [33,36]. Total sugars, soluble solids, and total anthocyanin increase as the fruit ripens. On the other hand, protein content and total phenolic compounds significantly decrease along the maturation steps [37]. Glucose and fructose are the main sugars, with sucrose present in traces. Potassium and magnesium are the main minerals detected in blackberries, followed by calcium and manganese [37] (Table 1).

Table 1. Chemical composition of fresh blackberries per 100 g at ripe stage (adapted from [33,38–43]).

Blackberry Composition	Lowest Reported	Highest Reported
	Content [g/100 g]	
Water	85.8	90.3
Protein	1.00	1.49
Total lipids	0.42	0.53
Ash	0.21	1.20
Total fiber	0.80	6.6
Total sugars	4.88	10.22
Sucrose	0.07	1.08
Glucose	2.31	2.61
Fructose	2.40	3.38
Maltose	-	0.07
Galactose	-	0.03

Table 1. *Cont.*

Minerals	Content [mg/100 g]	
Calcium	7.25	29.0
Iron	0.62	4.70
Magnesium	10.7	21.4
Phosphorus	7.25	22.0
Potassium	79.7	185.5
Sodium	0.30	1.00
Zinc	0.18	0.31
Copper	0.05	0.17
Manganese	0.42	1.47
Vitamins	**Content [mg/100 g]**	
Total ascorbic acid	1.50	44.0
Thiamin	-	0.02
Riboflavin	-	0.03
Niacin	-	0.65
Pantothenic acid	-	0.28
Vitamin B6	-	0.03
A-tocopherol	-	1.17
B-tocopherol	-	0.04
Γ-tocopherol	-	1.34

Blackberry contains a high amount of citric and malic acid. In addition, various studies reported the presence of shikimic, fumaric, and succinic acids [44,45]. Organic acid content is of fundamental importance for evaluating fruits' quality levels since they act as a stabilizer for anthocyanins. The health benefits of blackberries are associated with anthocyanins and other phenolic compounds such as ellagitannins, flavonols, and procyanidins [33,46]. Anthocyanins are responsible for the characteristic color of blackberry and are strong antioxidant compounds with potential antidiabetic, anticancer, anti-inflammatory, antimicrobial, and anti-obesity effects, as well as prevention of cardiovascular diseases [47,48].

2.2. Raspberries (Rubus idaeus)

Raspberries hold a special position due to their culinary versatility, the ideal nutritional profile of low calories, high fiber, mineral, potassium, sodium, and vitamins, the presence of several essential micronutrients, and phytochemical composition (Table 2).

Table 2. Composition of fresh red raspberries per 100 g (adapted from [9,42,49–52]).

Raspberry Composition	Lowest Reported	Highest Reported
	Content [g/100 g]	
Water	85.7	88.6
Protein	1.00	1.80
Total lipids	0.10	0.65
Carbohydrate	10.1	11.90
Dietary fiber	6.50	11.94
Total sugars	3.60	6.50
Sucrose	0.20	4.20
Glucose	1.86	2.50
Fructose	2.35	3.65

Table 2. *Cont.*

Minerals	Content [mg/100 g]	
Calcium	24.0	35.6
Iron	0.55	0.80
Magnesium	9.00	23.0
Phosphorus	30.0	35.0
Potassium	133	184
Sodium	0.02	4.00
Zinc	0.30	0.42
Copper	-	0.09
Manganese	0.11	0.67
Vitamins	**Content [mg/100 g]**	
Total ascorbic acid	13.4	43.9
Thiamin	0.03	0.10
Riboflavin	0.04	0.10
Niacin	0.03	0.70
Pantothenic acid	0.01	0.50
Vitamin B6	0.06	0.30
Total folate (µg)	21.0	36.0
Choline	-	12.3
Vitamin A, RAE (µg)	-	2.00
Lutein—zeaxanthin (µg)	136	360
Vitamin E	0.30	1.60
A-tocopherol	0.30	1.60
Vitamin K (µg)	-	7.38

Raspberries are consumed as fresh or frozen fruits or as processed products such as juices, jams, or jellies [49,53].

Worldwide, 822,493 tonnes of raspberry are produced per year. Mexico, Serbia, and the Russian Federation altogether produce more than 50% of the world's total raspberry. The major raspberry producer in the world is the Russian Federation, with 174,000 tonnes of production per year; Mexico comes second with 128,848 tonnes of yearly production, and the third largest producer of raspberry with 120,058 tonnes of production per year is Serbia [9].

Raspberries are a good source of phenolic compounds and many nutrients, such as vitamins, minerals, and fatty acids [54]. Raspberries are also rich in fructose and contain small amounts of glucose and sucrose. Among vitamins, vitamin C is the most abundant, followed by riboflavin, folic acid, and niacin. It is worth noting that a 100 g portion of raspberries provides 50% of the recommended intake of vitamin C [49]. Raspberries are also a good source of manganese, potassium, copper, and iron. The nutrient profile of raspberry potentially helps regulate blood sugar levels by slowing digestion and contributes to a satiety effect (given by the high fiber content) [12].

Likewise, other red fruits, such as raspberries, contain high levels of anthocyanins and ellagitannins. Ellagic acid and ellagitannins exhibit a wide range of beneficial effects on human health, such as antioxidant, antimutagenic, anticarcinogenic, and antiviral. Besides anthocyanins and ellagitannins, raspberries contain other phenolic compounds, including quercetin, kaempferol, and gallic acid, reaching a total phenolic content between 160–645 mg/100 g of fresh fruit [55–57].

2.3. Blackcurrants (Ribes nigrum)

Black currants represent an important cultivation among small fruits, with an annual production of 185,000 tonnes. Most of the world's production is concentrated in Europe, which represents the largest world producer with 160,000 tonnes per year [34]. Globally, Germany, Poland, and the United Kingdom contribute to about 80% of the total production of blackcurrant.

Blackcurrant is widely recognized for containing high concentrations of phenolic compounds (125–151 mg/100 g fresh weight), especially of proanthocyanins and anthocyanins, which together constitute 80% of total phenolics [55,58]. Furthermore, blackcurrants contain high levels of vitamin C, about five times more than oranges [47,54], high minerals (potassium, calcium, magnesium, and sodium), and monosaccharides [33] (Table 3).

Table 3. Nutrient composition of fresh blackcurrant per 100 g (adapted from [20,59–64]).

Blackcurrant Composition	Lowest Reported	Highest Reported
	Content [g/100 g]	
Water	77.0	83.0
Dietary fiber	5.30	6.20
Total sugars	7.10	14.0
Sucrose	0.10	1.30
Glucose	1.71	3.42
Fructose	0.85	1.52
Minerals	Content [mg/100 g]	
Calcium	31.3	64.2
Iron	1.13	6.36
Magnesium	17.0	65.9
Phosphorus	35.0	40.0
Potassium	251	320
Sodium	0.98	2.50
Zinc	0.16	0.36
Copper	0.15	0.20
Manganese	0.002	0.52
Vitamins	Content [mg/100 g]	
Total ascorbic acid	98.0	284
Thiamin	0.08	0.11
Riboflavin	0.08	0.11
Niacin	37.6	41.1
Vitamin B6	0.10	0.50
Vitamin A	17.8	20.0
A-tocopherol	0.50	0.90

Vitamin C levels range from 98 to 284 mg/100 g of fresh fruit, covering 100% of recommended daily intake with a portion of just 25 g. High levels of vitamin C, anthocyanins, and phenolic compounds suggest that blackcurrant can be used as a potential nutraceutical ingredient. Therefore, the phytochemicals present in blackcurrant have been extensively studied for their antioxidant activity [65], anti-inflammation activity [66], neuroprotective actions [67], anti-obesity properties [66,68], and anti-cancer properties [69].

Blackcurrants could be consumed as fresh fruit or as juices obtained from frozen processed berries. However, when berries are frozen, chemical changes can occur, including the concentration of solutes and chemicals, oxidative reactions, and enzyme activity. Contrarily, total phenolic and anthocyanin contents decrease during the processing of berry fruits into juices. Djordjević and co-workers [63] reported a strong reduction in anthocyanins, varying from 12% to 80%, and a slight decrease in total phenolics during the processing of berry fruits into juice. However, the content of total phenolics increased by 46.09–171.76% when berries were frozen and stored for 1 year, while in juices, total phenolics increased by 107.58%. Contrarily, the content of total anthocyanins in berries and juices after 1 year of storage decreased by 5.63–52.76% and 13.04–36.82%, respectively.

3. Red Fruits' Conventional Fermentation through Lactic Acid Bacteria and Yeasts

LAB are among the most used microorganisms to transform vegetables and fruits into more stable products. During fermentation, microbial enzymes produce newly derived compounds impacting aroma and functionality, reduce sugar content, improve nutritional value, and extend the shelf life of products.

Frozen fruits, juices, or smoothies can be fermented by LAB, obtaining healthy and functional products rich in bioactive compounds. To produce low-alcohol or non-alcoholic berry beverages, various LAB species (*Lactiplantibacillus plantarum*, *Levilactobacillus brevis*, *Lactocaseibacillus rhamnosus*, *Lactobacillus acidophilus*, *Lactocaseibacillus casei*) strains have been used in berry fermentation [59,70,71].

During fermentation, microbial enzymes produce new derived compounds impacting aroma and functionality (e.g., vitamins, phenolic compounds, or bioactive peptides), reduce sugar content, improve nutritional value, and extend the shelf life of fruit-based beverages [72,73]. Hence, depending on the juice and starter culture mix, fermented beverages with outstanding beneficial effects related to phenolic content, bioactive compounds presence, vitamin content, and probiotic activity could be obtained.

L. casei showed good adaptation when inoculated in blackberry juice, providing a beverage with functional characteristics [74].

Wu and co-workers [69] fermented blueberry and blackberry juices using three potential probiotic strains, reaching the recommended level for probiotic effects in both juices with each strain. In addition, fermentation improved the overall acceptability of both juices. Authors observed an increasing trend in syringic acid, ferulic acid, gallic acid, and lactic acid during fermentation. On the other hand, *p*-coumaric acid, protocatechuic acid, chlorogenic acid, and anthocyanins decreased during fermentation, with cyanindin-3-glucoside and peonidin-3-glucoside being the most affected, with a reduction of over 30%.

Low-ethanol blackcurrant beverage was obtained via sequential fermentation with *Metschnikowia* yeasts, showing promising future prospects for the development of low-ethanol content beverages [75]. The feasibility of efficient alcoholic fermentation using red fruits as raw material to produce fruit wine is well documented and is a common practice worldwide. Raspberry wines are characterized by high anthocyanins content and aromatic descriptors associated with volatile compounds such as ethyl pyruvate and ethyl butyrate [76,77]. In addition, raspberry wine is a rich source of phenolic compounds and represents a traditional product in Asian countries, especially Korea [78]. Comparing the proanthocyanidin content of grape wine with raspberry wine, the latter's results are three times richer, which explains the characteristic bitterness and astringency of raspberry wine [79]. Blackberry, raspberry, and blackcurrant wines are established products in the USA, mainly due to the peculiar flavor which distinguishes them from grape wine. In addition, blackberry wine showed higher total phenolic and total anthocyanin contents compared to grape wine and other fruit wines, as reported by Johnson and de Mejia [80].

4. Acetic Acid Bacteria-Based Beverages

4.1. Vinegar and Vinegar-Based Beverages

Vinegar is widespread in the world, and it is mainly known as a condiment and a preservative of foods. However, more recently, vinegars with healthy attributes were rising in the market. Some benefits to consuming vinegar include enhanced immunity, reduced risk factors for cardiovascular diseases, improved digestion, appetite suppression, and reduced fasting blood glucose, blood pressure, and serum cholesterol [81].

Conventionally, vinegar is produced from several raw materials, such as grapes, apples, rice, and diluted ethanol, according to established practices [82,83]. Moreover, several works highlight the feasibility of vinegar production from other fermentable raw materials, mainly fruits such as dates, oranges, strawberries, pineapples, and prickly pears [18,84–87].

Basically, in order to obtain vinegar both in submerged and static fermentation regimes, mixed AAB starter cultures are used [88,89]. In vinegar, AAB drives the production of acetic acid, thereby preventing the growth of microbial competitors. Some AAB also produce a cellulosic layer when growing in media containing sugars. Bacterial cellulose is a biopolymer of interest in the biotechnological industry, as well as a compound occurring in some fermented beverages [90,91]. Cunha and co-workers [92] produced blackberry vinegars through successive acetification cycles and evaluated bioactive compound variation from raw material to the final product. Vinegars were characterized by an average acetic acid content of 51.6 g/L and considerable quantities of phenolic compounds. Interestingly, phenolic compounds, antioxidant potential, and anthocyanins content were observed to be stable along several acetic fermentation cycles. On the other hand, a slight decrease in phenolic compounds, antioxidant potential, and anthocyanins content was observed when comparing blackberry wine and vinegar after acetic fermentation. However, the anthocyanins content in blackberry vinegar was appreciable (32.78 mg cyanidin 3-glucoside/L).

Su and Chien [93] obtained a blueberry vinegar with an anthocyanins content of 3.22 mg/100 mL via fermenting fruits and barks. In addition, Dogaru and co-workers [94] reported a higher antioxidant capacity of raspberry and blackberry vinegars compared to bilberry and apple vinegars, reaching a total antioxidant capacity of 16.0 and 15.2 mM Fe^{2+}/L, respectively.

The high phenolic compounds content and high antioxidant capacity of red fruit vinegars, along with the presence of organic acids and amino acids, could have health-promoting effects on human beings. Indeed, bioactive compounds of red fruit vinegars have been correlated with positive effects, such as an increase in digestion absorption and decreases in cardiovascular disease, serum cholesterol level, arterial stiffening, and blood pressure, by various studies [18,95–99]. In addition, the bioactive compounds in vinegars can be produced and/or increased through the overall vinegar fermentation process, where phenolic compounds are transformed into new antioxidative molecules.

Next to the production of vinegar, a range of non-dairy products consists of low-alcoholic and non-alcoholic fermented beverages. These products, although they could be an alternative to dairy-based beverages in terms of texture, flavor, as well as nutritional value, have received little attention from consumers and industry, especially in Western society [100]. Instead, the production and consumption of vinegar-based beverages obtained from fruits are more developed in Asian countries such as China or Korea [101].

Kim and co-workers [102] extensively investigated the physicochemical properties of various commercially available vinegar-based beverages at low- or non-alcohol content consumed in the Korean market, including their pH, acidity, sugar, total soluble sugar, total acid, and total amino acid content. Acetic acid content ranged between 0.84 to 1.91 g/L, representing more than 50% of total acid content. Oxalic, citric, malic, succinic, and lactic acids were also present. The authors also evaluated total phenolic compounds, anthocyanins, flavonoid content, and antioxidant activity. Specifically, blackberry vinegar-based beverages had high total anthocyanin content (13.21 mg/100 mL), antioxidants activity (10.98 %), total polyphenol content (87.25 mg/100 mL), and flavonoid value (51.12 mg/100 mL).

4.2. Kombucha and Gluconic Beverages

Kombucha is a slightly sweet and sparkling beverage obtained from fermented green or black tea with sugar via a microbial consortium composed of several AAB, yeasts, and LAB [103]. This microbial consortium forms a powerful symbiosis capable of inhibiting the growth of potentially contaminating microorganisms. The fermentation process also leads to the formation of a cellulose pellicle due to the activity of AAB, mainly belonging to the *Komagataeibacter* genus [104]. Actual food trends toward minimally processed products, without additives, with high nutritional value and health benefits, have increased with consumer awareness. In this context, traditional Kombucha tea has recently captured the attention of researchers and consumers. In addition, Kombucha consumption has been associated with a wide range of health functions, such as anti-inflammatory and hypoglycemic effects, and antioxidant, antimicrobial, and antiproliferative properties, and mainly related to the presence of organic acids, vitamins, minerals, and phenolic compounds [105].

Although studies [106] confirmed that the fermentation process breaks down larger compounds already present in the liquid into small molecules with greater bioavailability, initial substrate composition in terms of bioactive compounds is fundamental. Indeed, recently, different substrates have been tested in the fermentation of kombucha tea, resulting in new kombucha beverages with different sensorial properties and functional properties [107].

Salak, commonly known as snake fruit in Indonesia, was used to produce Salak kombucha, fermenting salak juice over 14 days. The obtained functional fermented beverage, rich in antioxidants, such as tannins and polyphenols, and organic acids, such as acetic, citric, and lactic, was demonstrated to have anti-hyperglycemia activity [108].

Moreover, kombucha prepared using black tea, sugar, and different berry fruits (blackberry, raspberry, and red goji berry) resulted in much richer in mineral and phenolic contents compared to standard kombucha [109]. Blackberry kombucha was the most appreciated and was characterized by the highest catechins content (92.38 mg/100 g) and good contents in potassium (1487.52 mg/kg), calcium (271.45 mg/kg), and magnesium (236.41 mg/kg). On the other hand, raspberry kombucha had higher potassium (1486.323 mg/kg) content but was lower in calcium (236.47 mg/kg) and phosphorus (197.52 mg/kg).

Likewise, in red fruit vinegar production, the total phenol content of kombucha increases due to the release of small molecules with higher antioxidant activities caused by enzyme activity or acidity increase during the fermentation process. On the other hand, as observed by Ulusoy and co-workers [110], extended storage time could affect total phenol content in black carrot kombucha, blackthorn kombucha, and raspberry kombucha. However, even though the reduction in total phenol content, raspberry kombucha showed high antioxidant capacity primarily constituted by anthocyanins and ellagitannins.

Other than antimicrobial and anti-proliferative properties, kombucha produced from red fruits could have a potential gastroprotective effect given by the high phenolic content, as reported by Barbosa and co-workers [111]. Indeed, phenolic compounds have been reported to stimulate e Prostaglandin E2 and to improve the status of different oxidative stress biomarkers [111,112]. By choosing initial substrates rich in phenolic compounds, the beneficial effects of kombucha could be further improved.

An innovative trend of the last years is acidic beverages containing mainly gluconic acid as an acidifier. These beverages are based on the fermentation by *Gluconobacter* sp., which transforms the glucose present in the fruit juice into gluconic acid. However, few examples of beverages from fruits containing no ethanol obtained by AAB exist on the market.

The study of Hornedo-Ortega and co-workers [113] demonstrated that alcoholic fermentation of strawberry purees decreased the anthocyanin content, while gluconic fermentation preserved these compounds, which is an advantage of this last process. Following these results, the authors reported that the chemical composition and antioxidant activity of strawberry gluconic beverage are stable for 60 days of storage at 4 °C [113].

However, the use of kombucha microbial consortium to ferment alternative raw matrices is becoming even more popular in trying to achieve the aim of producing novel pro-healthy and eco-friendly products. Moreover, in the agricultural and food field, there are numerous by-products coming from other food productions that still have an exploitable potential use. As a matter of fact, many times, the waste products of food processes still contain nutritional valuable compounds, such as proteins, sugars and polysaccharides, minerals, and secondary metabolites. Thus, matrices such as soybean whey and banana peel extract have been shown to be suitable for microbial fermentation, leading to beverages containing bioactive compounds with antioxidant and antimicrobial features [114,115].

5. Opportunities and Challenges

Currently, zero- and low-ethanol fermented beverages available on the market are produced mainly from fermentation processes by LAB. Most of these products are milk-based beverages [100]. Nowadays, there is an expanding market of zero- and low-alcohol beverages due to major awareness about the long-term effects of ethanol intake, and societal and individual vulnerability factors on alcohol consumption, in conjunction with a greater propensity of consumers to purchase healthier foods [116–118]. However, the habit of consuming alcoholic beverages is still very widespread despite the health and religious aspects, being linked to deeply rooted cultural factors and food styles. The consumption of zero- and low-alcohol beverages could be enlarged and supported by the introduction of new fermentation processes.

In this light, exploring AAB for producing non-conventional beverages from red fruits meets different needs which cross the consumer health and acceptance, as well as sustainable principles, reducing food wastes and recovering seasoning surplus. Moreover, considering fruits or leaves (as in the case of kombucha-based beverages), it is possible to set up bioprocesses for obtaining beverages at zero- and low-ethanol content and with no added sugars.

Numerous raw materials could be suitable as a substrate for fermentation by AAB, but few examples of marketed beverages, except for vinegar and, more recently, kombucha tea, are available. In particular, to obtain functional beverages, red fruits could represent suitable substrates bringing a positive impact on human health [119]. Eventually, even red fruit waste or by-products, such as leaves or seeds, could be used as raw materials rich in bioactive compounds. Ziemlewska and co-workers [120] utilized the kombucha microbial community as a starter culture to ferment red fruit leaves, obtaining an extract rich in bioactive compounds. Fermented and raw extracts were compared in terms of antioxidant potential and anti-aging properties. Results showed higher effects in fermented extracts, highlighting the positive impact of AAB on bioactive compounds.

In this frame, the know-how acquired in vinegar production can be the starting point for vinegar-based, zero- and low-ethanol, and gluconic beverages. Indeed, established techniques, such as static and submerged methods, could be suitable for developing new healthy products [121–123]. Future challenges that need to be considered include the optimization of vinegar fermentation methods for non-conventional raw materials and the increase in consumer awareness for new healthy products.

However, AAB fermentation could play a key role in the emerging market of zero-and low-alcohol beverages, contributing to more sustainable productions of beverages and positively impacting public health.

Author Contributions: M.B. writing and editing; E.C. original draft preparation; M.P.A. writing and editing, L.D.V. editing; A.C. editing; M.G. conceptualization, supervision, and resources. All authors have read and agreed to the published version of the manuscript.

Funding: Part of this work was granted by the European Commission—NextGenerationEU, Project "Strengthening the MIRRI Italian Research Infrastructure for Sustainable Bioscience and Bioeconomy", code n. IR0000005.

Institutional Review Board Statement: Not applicable.

Informed Consent Statement: Not applicable.

Data Availability Statement: Not applicable.

Conflicts of Interest: The authors declare no conflict of interest.

References

1. De Roos, J.; De Vuyst, L. Acetic acid bacteria in fermented foods and beverages. *Curr. Opin. Biotechnol.* **2018**, *49*, 115–119. [CrossRef]
2. Gullo, M.; Giudici, P. Acetic acid bacteria in traditional balsamic vinegar: Phenotypic traits relevant for starter cultures selection. *Int. J. Food Microbiol.* **2008**, *125*, 46–53. [CrossRef]
3. Bassi, D.; Puglisi, E.; Cocconcelli, P.S. Comparing natural and selected starter cultures in meat and cheese fermentations. *Curr. Opin. Food Sci.* **2015**, *2*, 118–122. [CrossRef]
4. Pereira, G.V.M.; De Carvalho Neto, D.P.; Junqueira, A.C.D.O.; Karp, S.G.; Letti, L.A.; Magalhães Júnior, A.I.; Soccol, C.R. A Review of Selection Criteria for Starter Culture Development in the Food Fermentation Industry. *Food Rev. Int.* **2019**, *36*, 135–167. [CrossRef]
5. Calvert, M.D.; Madden, A.A.; Nichols, L.M.; Haddad, N.M.; Lahne, J.; Dunn, R.R.; McKenney, E.A. A review of sourdough starters: Ecology, practices, and sensory quality with applications for baking and recommendations for future research. *PeerJ* **2021**, *9*, e11389. [CrossRef] [PubMed]

6. Yousseef, M.; Lafarge, C.; Valentin, D.; Lubbers, S.; Husson, F. Fermentation of cow milk and/or pea milk mixtures by different starter cultures: Physico-chemical and sensorial properties. *LWT* **2016**, *69*, 430–437. [CrossRef]
7. World Health Organization. *A Public Health Perspective on Zero-and Low-Alcohol Beverages*; WHO: Geneva, Switzerland, 2023.
8. Anagnostopoulos, D.A.; Tsaltas, D. *Chapter 10–Fermented Foods and Beverages*; Woodhead Publishing: Cambridge, UK, 2019.
9. Zhang, X.; Ahuja, J.K.; Burton-Freeman, B.M. Characterization of the nutrient profile of processed red raspberries for use in nutrition labeling and promoting healthy food choices. *Nutr. Health Aging* **2019**, *5*, 225–236. [CrossRef]
10. Xue, B.; Hui, X.; Chen, X.; Luo, S.; Dilrukshi, H.; Wu, G.; Chen, C. Application, emerging health benefits, and dosage effects of blackcurrant food formats. *J. Funct. Foods* **2022**, *95*, 105147. [CrossRef]
11. Derosa, G.; Maffioli, P.; Sahebkar, A. Ellagic Acid and Its Role in Chronic Diseases. *Adv. Exp. Med. Biol.* **2016**, *928*, 473–479. [CrossRef] [PubMed]
12. Burton-Freeman, B.M.; Sandhu, A.K.; Edirisinghe, I. Red Raspberries and Their Bioactive Polyphenols: Cardiometabolic and Neuronal Health Links. *Adv. Nutr. Int. Rev. J.* **2016**, *7*, 44–65. [CrossRef]
13. Jakobek, L.; Seruga, M.; Medvidovic-Kosanovic, M.; Novak, I. Anthocyanin content and antioxidant activity of various red fruit juices. *Dtsch. Lebensm.* **2007**, *103*, 58.
14. Ma, Y.; Ding, S.; Fei, Y.; Liu, G.; Jang, H.; Fang, J. Antimicrobial activity of anthocyanins and catechins against foodborne pathogens Escherichia coli and Salmonella. *Food Control.* **2019**, *106*, 106712. [CrossRef]
15. Ma, Z.; Du, B.; Li, J.; Yang, Y.; Zhu, F. An Insight into Anti-Inflammatory Activities and Inflammation Related Diseases of Anthocyanins: A Review of Both In Vivo and In Vitro Investigations. *Int. J. Mol. Sci.* **2021**, *22*, 11076. [CrossRef]
16. Zhong, H.; Xu, J.; Yang, M.; Hussain, M.; Liu, X.; Feng, F.; Guan, R. Protective Effect of Anthocyanins against Neurodegenerative Diseases through the Microbial-Intestinal-Brain Axis: A Critical Review. *Nutrients* **2023**, *15*, 496. [CrossRef]
17. Rabelo, A.C.S.; Guerreiro, C.d.A.; Shinzato, V.I.; Ong, T.P.; Noratto, G. Anthocyanins Reduce Cell Invasion and Migration through Akt/mTOR Downregulation and Apoptosis Activation in Triple-Negative Breast Cancer Cells: A Systematic Review and Meta-Analysis. *Cancers* **2023**, *15*, 2300. [CrossRef]
18. Vilela, A.; Cosme, F. Drink Red: Phenolic Composition of Red Fruit Juices and Their Sensorial Acceptance. *Beverages* **2016**, *2*, 29. [CrossRef]
19. Skrovankova, S.; Sumczynski, D.; Mlcek, J.; Jurikova, T.; Sochor, J. Bioactive Compounds and Antioxidant Activity in Different Types of Berries. *Int. J. Mol. Sci.* **2015**, *16*, 24673–24706. [CrossRef] [PubMed]
20. Nour, V.; Trandafir, I.; Ionica, M.E. Ascorbic acid, anthocyanins, organic acids and mineral content of some black and red currant cultivars. *Fruits* **2011**, *66*, 353–362. [CrossRef]
21. Who, J.; Consultation, F.E. Diet, nutrition and the prevention of chronic diseases. *World Health Organ. Technol. Rep. Ser.* **2003**, *916*, 1–149.
22. Cantadori, E.; Brugnoli, M.; Centola, M.; Uffredi, E.; Colonello, A.; Gullo, M. Date Fruits as Raw Material for Vinegar and Non-Alcoholic Fermented Beverages. *Foods* **2022**, *11*, 1972. [CrossRef] [PubMed]
23. Pontonio, E.; Rizzello, C.G. Editorial: Ad-Hoc Selection of Lactic Acid Bacteria for Non-conventional Food Matrices Fermentations: Agri-Food Perspectives. *Front. Microbiol.* **2021**, *12*, 681830. [CrossRef] [PubMed]
24. Randazzo, W.; Corona, O.; Guarcello, R.; Francesca, N.; Germanà, M.A.; Erten, H.; Moschetti, G.; Settanni, L. Development of new non-dairy beverages from Mediterranean fruit juices fermented with water kefir microorganisms. *Food Microbiol.* **2016**, *54*, 40–51. [CrossRef]
25. Silva, K.A.; Uekane, T.M.; de Miranda, J.F.; Ruiz, L.F.; da Motta, J.C.B.; Silva, C.B.; Pitangui, N.D.S.; Gonzalez, A.G.M.; Fernandes, F.F.; Lima, A.R. Kombucha beverage from non-conventional edible plant infusion and green tea: Characterization, toxicity, antioxidant activities and antimicrobial properties. *Biocatal. Agric. Biotechnol.* **2021**, *34*, 102032. [CrossRef]
26. Gadhoumi, H.; Gullo, M.; De Vero, L.; Martinez-Rojas, E.; Tounsi, M.S.; Hayouni, E.A. Design of a New Fermented Beverage from Medicinal Plants and Organic Sugarcane Molasses via Lactic Fermentation. *Appl. Sci.* **2021**, *11*, 6089. [CrossRef]
27. Ou, A.S.; Chang, R.C. Taiwan fruit vinegar. In *Vinegars of the World*; Solieri, L., Giudici, P., Eds.; Springer: Milano, Italy, 2009; pp. 223–242.
28. Cañete-Rodríguez, A.; Santos-Dueñas, I.; Jiménez-Hornero, J.; Ehrenreich, A.; Liebl, W.; García-García, I. Gluconic acid: Prop-erties, production methods and applications—An excellent opportunity for agro-industrial by-products and waste bio-valorization. *Process Biochem.* **2016**, *51*, 1891–1903. [CrossRef]
29. Landis, E.A.; Fogarty, E.; Edwards, J.C.; Popa, O.; Eren, A.M.; Wolfe, B.E. Microbial Diversity and Interaction Specificity in Kombucha Tea Fermentations. *Msystems* **2022**, *7*, e00157-22. [CrossRef]
30. May, A.; Narayanan, S.; Alcock, J.; Varsani, A.; Maley, C.; Aktipis, A. Kombucha: A novel model system for cooperation and conflict in a complex multi-species microbial ecosystem. *PeerJ* **2019**, *7*, e7565. [CrossRef]
31. Bishop, P.; Pitts, E.R.; Budner, D.; Thompson-Witrick, K.A. Kombucha: Biochemical and microbiological impacts on the chemical and flavor profile. *Food Chem. Adv.* **2022**, *1*, 100025. [CrossRef]
32. Chakravorty, S.; Bhattacharya, S.; Chatzinotas, A.; Chakraborty, W.; Bhattacharya, D.; Gachhui, R.; Paul, S.K. Kombucha Drink: Production, quality, and safety aspects. *Sci. Beverages* **2019**, *220*, 259–288. [CrossRef]
33. Strik, B.; Finn, C.; Clark, J.; Bañados, M.P. Worldwide Production of Blackberries. *Acta Hortic.* **2008**, *777*, 209–218. [CrossRef]
34. FAOSTAT, F. Forestry Database. Available online: https://www.fao.org/forestry/statistics/84922/en/ (accessed on 15 April 2023).

35. Siriwoharn, T.; Wrolstad, R.E.; Finn, C.E.; Pereira, C.B. Influence of Cultivar, Maturity, and Sampling on Blackberry (*Rubus* L. Hybrids) Anthocyanins, Polyphenolics, and Antioxidant Properties. *J. Agric. Food Chem.* **2004**, *52*, 8021–8030. [CrossRef]
36. Schulz, M.; Seraglio, S.K.T.; Della Betta, F.; Nehring, P.; Valese, A.C.; Daguer, H.; Gonzaga, L.V.; Costa, A.C.O.; Fett, R. Blackberry (*Rubus ulmifolius* Schott): Chemical composition, phenolic compounds and antioxidant capacity in two edible stages. *Food Res. Int.* **2019**, *122*, 627–634. [CrossRef] [PubMed]
37. Tosun, I.; Ustun, N.S.; Tekguler, B. Physical and chemical changes during ripening of blackberry fruits. *Sci. Agric.* **2008**, *65*, 87–90. [CrossRef]
38. Guedes, M.N.S.; De Abreu, C.M.P.; Maro, L.A.C.; Pio, R.; De Abreu, J.R.; De Oliveira, J.O. Chemical characterization and mineral levels in the fruits of blackberry cultivars grown in a tropical climate at an elevation. *Acta Sci. Agron.* **2013**, *35*, 191–196. [CrossRef]
39. Kafkas, E.; Koşar, M.; Türemiş, N.; Başer, K. Analysis of sugars, organic acids and vitamin C contents of blackberry genotypes from Turkey. *Food Chem.* **2005**, *97*, 732–736. [CrossRef]
40. Izadyar, A.B.; Wang, S.Y. Changes of lipid components during dormancy in 'Hull Thornless' and 'Triple Crown Thornless' blackberry cultivars. *Sci. Hortic.* **1999**, *82*, 243–254. [CrossRef]
41. Zia-Ul-Haq, M.; Riaz, M.; De Feo, V.; Jaafar, H.Z.; Moga, M. *Rubus Fruticosus* L.: Constituents, Biological Activities and Health Related Uses. *Molecules* **2014**, *19*, 10998–11029. [CrossRef] [PubMed]
42. de Souza, V.R.; Pereira, P.A.P.; da Silva, T.L.T.; De Oliveira Lima, L.C.; Pio, R.; Queiroz, F. Determination of the bioactive compounds, antioxidant activity and chemical composition of Brazilian blackberry, red raspberry, strawberry, blueberry and sweet cherry fruits. *Food Chem.* **2014**, *156*, 362–368. [CrossRef]
43. Moraes, D.P.; Lozano-Sánchez, J.; Machado, M.L.; Vizzotto, M.; Lazzaretti, M.; Leyva-Jimenez, F.J.J.; da Silveira, T.L.; Ries, E.F.; Barcia, M.T. Characterization of a new blackberry cultivar BRS Xingu: Chemical composition, phenolic compounds, and antioxidant capacity in vitro and in vivo. *Food Chem.* **2020**, *322*, 126783. [CrossRef]
44. Fan-Chiang, H.-J.; Wrolstad, R.E. Sugar and Nonvolatile Acid Composition of Blackberries. *J. AOAC Int.* **2010**, *93*, 956–965. [CrossRef]
45. Mikulic-Petkovsek, M.; Schmitzer, V.; Slatnar, A.; Stampar, F.; Veberic, R. Composition of Sugars, Organic Acids, and Total Phenolics in 25 Wild or Cultivated Berry Species. *J. Food Sci.* **2012**, *77*, C1064–C1070. [CrossRef]
46. Wang, S.Y.; Lin, H.S. Antioxidant activity in fruits and leaves of blackberry, raspberry. *J. Agric. Food Chem.* **2000**, *48*, 140–146. [CrossRef]
47. Khoo, G.M.; Clausen, M.R.; Pedersen, H.L.; Larsen, E. Bioactivity and chemical composition of blackcurrant (*Ribes nigrum*) cultivars with and without pesticide treatment. *Food Chem.* **2012**, *132*, 1214–1220. [CrossRef]
48. He, K.; Li, X.; Chen, X.; Ye, X.; Huang, J.; Jin, Y.; Li, P.; Deng, Y.; Jin, Q.; Shi, Q.; et al. Evaluation of antidiabetic potential of selected traditional Chinese medicines in STZ-induced diabetic mice. *J. Ethnopharmacol.* **2011**, *137*, 1135–1142. [CrossRef]
49. Rao, A.V.; Snyder, D.M. Raspberries and Human Health: A Review. *J. Agric. Food Chem.* **2010**, *58*, 3871–3883. [CrossRef]
50. Alibabić, V.; Skender, A.; Bajramović, M.; Šertović, E.; Bajrić, E. Evaluation of morphological, chemical, and sensory characteristic-sof raspberry cultivars grown in Bosnia and Herzegovina. *Turk. J. Agric. For.* **2018**, *42*, 67–74. [CrossRef]
51. Rambaran, T.F.; Bowen-Forbes, C.S. Chemical and sensory characterisation of two *Rubus rosifolius* (red raspberry) varieties. *Int. J. Food Sci.* **2020**, *2020*, 8. [CrossRef]
52. Wang, S.Y.; Zheng, W. Preharvest application of methyl jasmonate increases fruit quality and antioxidant capacity in raspberries. *Int. J. Food Sci. Technol.* **2005**, *40*, 187–195. [CrossRef]
53. Bowen-Forbes, C.S.; Zhang, Y.; Nair, M.G. Anthocyanin content, antioxidant, anti-inflammatory and anticancer properties of blackberry and raspberry fruits. *J. Food Compos. Anal.* **2009**, *23*, 554–560. [CrossRef]
54. Bobinaite, R.; Viškelis, P.; Venskutonis, P.R. Chemical Composition of Raspberry (*Rubus* spp.) Cultivars. In *Nutritional Composition of Fruit Cultivars*; Simmonds, M.S.J., Preedy, V.R., Eds.; Academic Press: Cambridge, MA, USA, 2016; pp. 713–731. ISBN 9780124081178.
55. Lugasi, A.; Hóvári, J.; Kádár, G.; Denes, F. Phenolics in raspberry, blackberry and currant cultivars grown in Hungary. *Acta Aliment.* **2011**, *40*, 52–64. [CrossRef]
56. Krivokapić, S.; Vlaović, M.; Vratnica, B.D.; Perović, A.; Perović, S. Biowaste as a Potential Source of Bioactive Compounds—A Case Study of Raspberry Fruit Pomace. *Foods* **2021**, *10*, 706. [CrossRef]
57. Biesalski, H.-K.; Dragsted, L.O.; Elmadfa, I.; Grossklaus, R.; Müller, M.; Schrenk, D.; Walter, P.; Weber, P. Bioactive compounds: Definition and assessment of activity. *Nutrition* **2009**, *25*, 1202–1205. [CrossRef]
58. Cortez, R.E.; de Mejia, E.G. Blackcurrants (*Ribes nigrum*): A Review on Chemistry, Processing, and Health Benefits. *J. Food Sci.* **2019**, *84*, 2387–2401. [CrossRef]
59. Pinto, T.; Vilela, A.; Cosme, F. Chemical and Sensory Characteristics of Fruit Juice and Fruit Fermented Beverages and Their Consumer Acceptance. *Beverages* **2022**, *8*, 33. [CrossRef]
60. Raudsepp, P.; Kaldmäe, H.; Kikas, A.; Libek, A.-V.; Püssa, T. Nutritional quality of berries and bioactive compounds in the leaves of black currant (*Ribes nigrum* L.) cultivars evaluated in Estonia. *J. Berry Res.* **2010**, *1*, 53–59. [CrossRef]
61. Haeknel, H.; Wegner, R. Some B vitamins in different varieties of black currants. *Ernahrungsforschung* **1957**, *2*, 801–802.
62. Paunović, S.M.; Nikolić, M.; Miletić, R.; Mašković, P. Vitamin and mineral content in black currant (*Ribes nigrum* L.) fruits as affected by soil management system. *Acta Sci. Pol. Hortorum Cultus* **2017**, *16*, 135–144. [CrossRef]

63. Djordjević, B.; Šavikin, K.; Zdunić, G.; Janković, T.; Vulić, T.; Pljevljakušić, D.; Oparnica, C. Biochemical Properties of the Fresh and Frozen Black Currants and Juices. *J. Med. Food* **2013**, *16*, 73–81. [CrossRef]
64. Ersoy, N.; Kupe, M.; Gundogdu, M.; Ilhan, G.; Ercisli, S. Phytochemical and Antioxidant Diversity in Fruits of Currant (*Ribes* spp.). *Not. Bot. Horti Agrobot. Cluj-Napoca* **2018**, *46*, 381–387. [CrossRef]
65. Michalska, A.; Wojdyło, A.; Łysiak, G.P.; Figiel, A. Chemical Composition and Antioxidant Properties of Powders Obtained from Different Plum Juice Formulations. *Int. J. Mol. Sci.* **2017**, *18*, 176. [CrossRef] [PubMed]
66. Cao, L.; Park, Y.; Lee, S.; Kim, D.-O. Extraction, Identification, and Health Benefits of Anthocyanins in Blackcurrants (*Ribes nigrum* L.). *Appl. Sci.* **2021**, *11*, 1863. [CrossRef]
67. Subash, S.; Essa, M.M.; Al-Asmi, A.; Al-Adawi, S.; Vaishnav, R.; Guillemin, G.J. Effect of dietary supplementation of dates in Alzheimer's disease APPsw/2576 transgenic mice on oxidative stress and antioxidant status. *Nutr. Neurosci.* **2014**, *18*, 281–288. [CrossRef]
68. Tsuda, T. Recent Progress in Anti-Obesity and Anti-Diabetes Effect of Berries. *Antioxidants* **2016**, *5*, 13. [CrossRef]
69. Wu, Y.; Li, S.; Tao, Y.; Li, D.; Han, Y.; Show, P.L.; Wen, G.; Zhou, J. Fermentation of blueberry and blackberry juices using Lactobacillus plantarum, Streptococcus thermophilus and Bifidobacterium bifidum: Growth of probiotics, metabolism of phenolics, antioxidant capacity in vitro and sensory evaluation. *Food Chem.* **2021**, *348*, 129083. [CrossRef]
70. Samtiya, M.; Aluko, R.E.; Dhewa, T.; Moreno-Rojas, J. Potential Health Benefits of Plant Food-Derived Bioactive Components: An Overview. *Foods* **2021**, *10*, 839. [CrossRef]
71. Castellone, V.; Bancalari, E.; Rubert, J.; Gatti, M.; Neviani, E.; Bottari, B. Eating Fermented: Health Benefits of LAB-Fermented Foods. *Foods* **2021**, *10*, 2639. [CrossRef]
72. Rodríguez, L.G.R.; Gasga, V.M.Z.; Pescuma, M.; Van Nieuwenhove, C.; Mozzi, F.; Burgos, J.A.S. Fruits and fruit by-products as sources of bioactive compounds. Benefits and trends of lactic acid fermentation in the development of novel fruit-based functional beverages. *Food Res. Int.* **2020**, *140*, 109854. [CrossRef] [PubMed]
73. Paramithiotis, S.; Das, G.; Shin, H.-S.; Patra, J.K. Fate of Bioactive Compounds during Lactic Acid Fermentation of Fruits and Vegetables. *Foods* **2022**, *11*, 733. [CrossRef]
74. Bernal-Castro, C.A.; Díaz-Moreno, C.; Gutiérrez-Cortés, C. Inclusion of prebiotics on the viability of a commercial *Lactobacillus casei* subsp. rhamnosus culture in a tropical fruit beverage. *J. Food Sci. Technol.* **2019**, *56*, 987–994. [CrossRef]
75. Kelanne, N.M.; Siegmund, B.; Metz, T.; Yang, B.; Laaksonen, O. Comparison of volatile compounds and sensory profiles of alcoholic black currant (*Ribes nigrum*) beverages produced with Saccharomyces, Torulaspora, and Metschnikowia yeasts. *Food Chem.* **2021**, *370*, 131049. [CrossRef]
76. Aguirre, M.J.; Chen, Y.Y.; Isaacs, M.; Matsuhiro, B.; Mendoza, L.; Torres, S. Electrochemical behaviour and antioxidant capacity of anthocyanins from Chilean red wine, grape and raspberry. *Food Chem.* **2010**, *121*, 44–48. [CrossRef]
77. Duarte, W.F.; Dias, D.R.; Oliveira, J.M.; Vilanova, M.; Teixeira, J.A.; e Silva, J.B.A.; Schwan, R.F. Raspberry (*Rubus idaeus* L.) wine: Yeast selection, sensory evaluation and instrumental analysis of volatile and other compounds. *Food Res. Int.* **2010**, *43*, 2303–2314. [CrossRef]
78. Jagtap, U.B.; Bapat, V.A. Wines from fruits other than grapes: Current status and future prospectus. *Food Biosci.* **2015**, *9*, 80–96. [CrossRef]
79. Lim, J.W.; Jeong, J.T.; Shin, C.S. Component analysis and sensory evaluation of Korean black raspberry (Rubus coreanus Mique) wines. *Int. J. Food Sci. Technol.* **2012**, *47*, 918–926. [CrossRef]
80. Johnson, M.; DE Mejia, E. Comparison of Chemical Composition and Antioxidant Capacity of Commercially Available Blueberry and Blackberry Wines in Illinois. *J. Food Sci.* **2011**, *77*, C141–C148. [CrossRef]
81. Xia, T.; Zhang, B.; Duan, W.; Zhang, J.; Wang, M. Nutrients and bioactive components from vinegar: A fermented and functional food. *J. Funct. Foods* **2019**, *64*, 103681. [CrossRef]
82. Giudici, P.; De Vero, L.; Gullo, M. Vinegars. In *Acetic Acid Bacteria: Fundamentals and Food Applications*, 1st ed.; Sengun, I.Y., Ed.; CRC Press, Taylor & Francis Group: Boca Raton, FL, USA, 2017; pp. 261–287.
83. Pothimon, R.; Gullo, M.; La China, S.; Thompson, A.K.; Krusong, W. Conducting High acetic acid and temperature acetification processes by Acetobacter pasteurianus UMCC 2951. *Process Biochem.* **2020**, *98*, 41–50. [CrossRef]
84. Di Donna, L.; Bartella, L.; De Vero, L.; Gullo, M.; Giuffrè, A.M.; Zappia, C.; Capocasale, M.; Poiana, M.; D'urso, S.; Caridi, A. Vinegar production from Citrus bergamia by-products and preservation of bioactive compounds. *Eur. Food Res. Technol.* **2020**, *246*, 1981–1990. [CrossRef]
85. Hidalgo, C.; Torija, M.; Mas, A.; Mateo, E. Effect of inoculation on strawberry fermentation and acetification processes using native strains of yeast and acetic acid bacteria. *Food Microbiol.* **2013**, *34*, 88–94. [CrossRef]
86. Roda, A.; Lucini, L.; Torchio, F.; Dordoni, R.; De Faveri, D.M.; Lambri, M. Metabolite profiling and volatiles of pineapple wine and vinegar obtained from pineapple waste. *Food Chem.* **2017**, *229*, 734–742. [CrossRef]
87. Ben Hammouda, M.; Castro, R.; Durán-Guerrero, E.; Attia, H.; Azabou, S. Vinegar production via spontaneous fermentation of different prickly pear fruit matrices: Changes in chemical composition and biological activities. *J. Sci. Food Agric.* **2023**. *ahead of print*. [CrossRef]
88. Gullo, M.; Zanichelli, G.; Verzelloni, E.; Lemmetti, F.; Giudici, P. Feasible acetic acid fermentations of alcoholic and sugary substrates in combined operation mode. *Process Biochem.* **2016**, *51*, 1129–1139. [CrossRef]

89. Gullo, M.; Verzelloni, E.; Canonico, M. Aerobic submerged fermentation by acetic acid bacteria for vinegar production: Process and biotechnological aspects. *Process Biochem.* **2014**, *49*, 1571–1579. [CrossRef]

90. La China, S.; Bezzecchi, A.; Moya, F.; Petroni, G.; Di Gregorio, S.; Gullo, M. Genome sequencing and phylogenetic analysis of K1G4: A new Komagataeibacter strain producing bacterial cellulose from different carbon sources. *Biotechnol. Lett.* **2020**, *42*, 807–818. [CrossRef]

91. Barbi, S.; Taurino, C.; La China, S.; Anguluri, K.; Gullo, M.; Montorsi, M. Mechanical and structural properties of environmental green composites based on functionalized bacterial cellulose. *Cellulose* **2021**, *28*, 1431–1442. [CrossRef]

92. Da Cunha, M.A.A.; De Lima, K.P.; Santos, V.A.Q.; Heinz, O.L.; Schmidt, C.A.P. Blackberry Vinegar Produced By Successive Acetification Cycles: Production, Characterization And Bioactivity Parameters. *Braz. Arch. Biol. Technol.* **2016**, *59*, e16150136. [CrossRef]

93. Su, M.-S.; Chien, P.-J. Antioxidant activity, anthocyanins, and phenolics of rabbiteye blueberry (*Vaccinium ashei*) fluid products as affected by fermentation. *Food Chem.* **2007**, *104*, 182–187. [CrossRef]

94. Dogaru, D.V.; Hădărugă, N.; Traşcă, T.; Jianu, C.; Jianu, I. Researches regarding the antioxidant capacity of some fruits vinegar. *J. Agroaliment. Process. Technol.* **2009**, *15*, 506–510.

95. Bortolini, D.G.; Maciel, G.M.; Fernandes, I.D.A.A.; Rossetto, R.; Brugnari, T.; Ribeiro, V.R.; Haminiuk, C.W.I. Biological potential and technological applications of red fruits: An overview. *Food Chem. Adv.* **2022**, *1*, 100014. [CrossRef]

96. Udani, J.K.; Singh, B.B.; Singh, V.J.; Barrett, M.L. Effects of Açai (*Euterpe oleracea* Mart.) berry preparation on metabolic parameters in a healthy overweight population: A pilot study. *Nutr. J.* **2011**, *10*, 45. [CrossRef]

97. De Oliveira, P.R.B.; da Costa, C.A.; de Bem, G.; De Cavalho, L.C.R.M.; De Souza, M.A.V.; Neto, M.D.L.; Sousa, P.J.D.C.; De Moura, R.S.; Resende, A.C. Effects of an Extract Obtained from Fruits of Euterpe oleracea Mart. in the Components of Metabolic Syndrome Induced in C57BL/6J Mice Fed a High-fat Diet. *J. Cardiovasc. Pharmacol.* **2010**, *56*, 619–626. [CrossRef]

98. Lamas, C.; Lenquiste, S.; Baseggio, A.; Cuquetto-Leite, L.; Kido, L.; Aguiar, A.; Erbelin, M.; Collares-Buzato, C.; Maróstica, M.; Cagnon, V. Jaboticaba extract prevents prediabetes and liver steatosis in high-fat-fed aging mice. *J. Funct. Foods* **2018**, *47*, 434–446. [CrossRef]

99. Gale, A.M.; Kaur, R.; Baker, W.L. Hemodynamic and electrocardiographic effects of açaí berry in healthy volunteers: A randomized controlled trial. *Int. J. Cardiol.* **2014**, *174*, 421–423. [CrossRef]

100. Baschali, A.; Tsakalidou, E.; Kyriacou, A.; Karavasiloglou, N.; Matalas, A.-L. Traditional Low-Alcoholic and Non-alcoholic Fermented Beverages Consumed in European Countries: A Neglected Food Group. *Nutr. Res. Rev.* **2017**, *30*, 1–24. [CrossRef]

101. Li, X.; Cao, W.; Shen, Y.; Li, N.; Dong, X.-P.; Wang, K.-J.; Cheng, Y.-X. Antioxidant compounds from Rosalaevigata fruits. *Food Chem.* **2012**, *130*, 575–580. [CrossRef]

102. Kim, S.-H.; Cho, H.-K.; Shin, H.-S. Physicochemical properties and antioxidant activities of commercial vinegar drinks in Korea. *Food Sci. Biotechnol.* **2012**, *21*, 1729–1734. [CrossRef]

103. Villarreal-Soto, S.A.; Beaufort, S.; Bouajila, J.; Souchard, J.-P.; Taillandier, P. Understanding Kombucha Tea Fermentation: A Review. *J. Food Sci.* **2018**, *83*, 580–588. [CrossRef]

104. La China, S.; De Vero, L.; Anguluri, K.; Brugnoli, M.; Mamlouk, D.; Gullo, M. Kombucha Tea as a Reservoir of Cellulose Producing Bacteria: Assessing Diversity among *Komagataeibacter* Isolates. *Appl. Sci.* **2021**, *11*, 1595. [CrossRef]

105. Jayabalan, R.; Malbaša, R.V.; Lončar, E.S.; Vitas, J.S.; Sathishkumar, M. A Review on Kombucha Tea-Microbiology, Composition, Fermentation, Beneficial Effects, Toxicity, and Tea Fungus. *Compr. Rev. Food Sci. Food Saf.* **2014**, *13*, 538–550. [CrossRef]

106. Ziemlewska, A.; Zagórska-Dziok, M.; Nizioł-Łukaszewska, Z.; Kielar, P.; Młoń, M.; Szczepanek, D.; Sowa, I.; Wójciak, M. In Vitro Evaluation of Antioxidant and Protective Potential of Kombucha-Fermented Black Berry Extracts against H_2O_2-Induced Oxidative Stress in Human Skin Cells and Yeast Model. *Int. J. Mol. Sci.* **2023**, *24*, 4388. [CrossRef]

107. Liu, Y.; Zheng, Y.; Yang, T.; Mac Regenstein, J.; Zhou, P. Functional properties and sensory characteristics of kombucha analogs prepared with alternative materials. *Trends Food Sci. Technol.* **2022**, *129*, 608–616. [CrossRef]

108. Zubaidah, E.; Apriyadi, T.E.; Kalsum, U.; Widyastuti, E.; Estiasih, T.; Srianta, I.; Blanc, P.J. In vivo evaluation of snake fruit Kombucha as hyperglycemia therapeutic agent. *Int. Food Res. J.* **2018**, *25*, 453–457.

109. Akarca, G. Determination of Potential Antimicrobial Activities of some Local Berries Fruits in Kombucha Tea Production. *Braz. Arch. Biol. Technol.* **2021**, *64*, e21210023. [CrossRef]

110. Ulusoy, A.; Tamer, C.E. Determination of suitability of black carrot (*Daucus carota* L. spp. sativus var. *atrorubens Alef.) juice concentrate, cherry laurel (Prunus laurocerasus), blackthorn (Prunus spinosa) and red raspberry (Rubus ideaus) for kombucha beverage production.* *J. Food Meas. Charact.* **2019**, *13*, 1524–1536. [CrossRef]

111. Barbosa, E.L.; Netto, M.C.; Junior, L.B.; de Moura, L.F.; Brasil, G.A.; Bertolazi, A.A.; de Lima, E.M.; Vasconcelos, C.M. Kombucha fermentation in blueberry (*Vaccinium myrtillus*) beverage and its in vivo gastroprotective effect: Preliminary study. *Futur. Foods* **2022**, *5*, 100129. [CrossRef]

112. Alanko, J.; Riutta, A.; Holm, P.; Mucha, I.; Vapaatalo, H.; Metsä-Ketelä, T. Modulation of arachidonic acid metabolism by phenols: Relation to their structure and antioxidant/prooxidant properties. *Free. Radic. Biol. Med.* **1998**, *26*, 193–201. [CrossRef]

113. Hornedo-Ortega, R.; Krisa, S.; García-Parrilla, M.C.; Richard, T. Effects of gluconic and alcoholic fermentation on anthocyanin composition and antioxidant activity of beverages made from strawberry. *LWT* **2016**, *69*, 382–389. [CrossRef]

114. Tu, C.; Tang, S.; Azi, F.; Hu, W.; Dong, M. Use of kombucha consortium to transform soy whey into a novel functional beverage. *J. Funct. Foods* **2018**, *52*, 81–89. [CrossRef]

115. Pure, A.E.; Pure, M.E. Antioxidant and Antibacterial Activity of Kombucha Beverages Prepared using Banana Peel, Common Nettles and Black Tea Infusions. *Appl. Food Biotechnol.* **2016**, *3*, 125–130. [CrossRef]
116. Liguori, L.; Russo, P.; Albanese, D.; Di Matteo, M. Production of Low-Alcohol Beverages: Current Status and Perspectives. In *Food Processing for Increased Quality and Consumption*; Academic Press: Cambridge, MA, USA, 2018; pp. 347–382.
117. Moushmoush, B.; Abi-Mansour, P. Alcohol and the heart: The long-term effects of alcohol on the cardiovascular system. *Arch. Intern. Med.* **1991**, *151*, 36–42. [CrossRef]
118. Brown, S.A.; Vik, P.W.; Patterson, T.L.; Grant, I.; Schuckit, M.A. Stress, vulnerability and adult alcohol relapse. *J. Stud. Alcohol* **1995**, *56*, 538–545. [CrossRef]
119. Costa, A.G.V.; Garcia-Diaz, D.F.; Jimenez, P.; Silva, P.I. Bioactive compounds and health benefits of exotic tropical red–black berries. *J. Funct. Foods* **2013**, *5*, 539–549. [CrossRef]
120. Ziemlewska, A.; Nizioł-Łukaszewska, Z.; Zagórska-Dziok, M.; Bujak, T.; Wójciak, M.; Sowa, I. Evaluation of Cosmetic and Dermatological Properties of Kombucha-Fermented Berry Leaf Extracts Considered to Be By-Products. *Molecules* **2022**, *27*, 2345. [CrossRef] [PubMed]
121. Mizzi, J.; Gaggìa, F.; Cionci, N.B.; Di Gioia, D.; Attard, E. Selection of Acetic Acid Bacterial Strains and Vinegar Production from Local Maltese Food Sources. *Front. Microbiol.* **2022**, *13*, 897825. [CrossRef] [PubMed]
122. Sokollek, S.J.; Hammes, W.P. Description of a Starter Culture Preparation for Vinegar Fermentation. *Syst. Appl. Microbiol.* **1997**, *20*, 481–491. [CrossRef]
123. Vegas, C.; González, Á.; Mateo, E.; Mas, A.; Poblet, M.; Torija, M.J. Evaluation of representativity of the acetic acid bacteria species identified by culture-dependent method during a traditional wine vinegar production. *Food Res. Int.* **2013**, *51*, 404–411. [CrossRef]

Article

Total Lipids and Fatty Acids in Major New Zealand Grape Varieties during Ripening, Prolonged Pomace Contacts and Ethanolic Extractions Mimicking Fermentation

Emma Sherman [1], Muriel Yvon [2], Franzi Grab [2], Erica Zarate [3], Saras Green [3], Kyung Whan Bang [3] and Farhana R. Pinu [1,*]

[1] Biological Chemistry and Bioactives Group, The New Zealand Institute for Plant and Food Research Limited, Auckland 1025, New Zealand; emma.sherman@plantandfood.co.nz
[2] Viticulture and Oenology Group, The New Zealand Institute for Plant and Food Research Limited, Blenheim 7201, New Zealand; muriel.yvon@plantandfood.co.nz (M.Y.); franziska.grab@plantandfood.co.nz (F.G.)
[3] School of Biological Sciences, University of Auckland, Auckland 1010, New Zealand; s.green@auckland.ac.nz (S.G.); kban028@live.com (K.W.B.)
* Correspondence: farhana.pinu@plantandfood.co.nz

Abstract: Despite the important roles of lipids in winemaking, changes in lipids during grape ripening are largely unknown for New Zealand (NZ) varieties. Therefore, we aimed to determine the fatty acid profiles and total lipid content in two of NZ's major grape varieties. Using gas chromatography–mass spectrometry, absolute quantification of 45 fatty acids was determined in Sauvignon blanc (SB) and Pinot noir (PN) grapes harvested at two different stages of ripeness. Lipid concentrations were as high as 0.4 g/g in seeds of both varieties, while pulp contained the least amount. Many unsaturated fatty acids were present, particularly in grape seeds, while skin contained relatively higher amounts of saturated fatty acids that increased throughout ripening. For both varieties, a significant increase in lipid concentration was observed in grapes harvested at the later stage of ripeness, indicating an association between lipids and grape maturity, and providing a novel insight about the use of total lipids as another parameter of grape ripeness. A variety-specific trend in the development and extraction of grape lipids was found from the analysis of the must and ethanolic extracts. Lipid extraction increased linearly with the ethanol concentration and with the extended pomace contact time. More lipids were extracted from the SB pomace to the must than PN within 144 h, suggesting a must matrix effect on lipid extraction. The knowledge generated here is relevant to both industry and academia and can be used to develop lipid diversification strategies to produce different wine styles.

Keywords: GC-MS; wine; ripening indices; lipid extraction; varietal differences; juice matrix

Citation: Sherman, E.; Yvon, M.; Grab, F.; Zarate, E.; Green, S.; Bang, K.W.; Pinu, F.R. Total Lipids and Fatty Acids in Major New Zealand Grape Varieties during Ripening, Prolonged Pomace Contacts and Ethanolic Extractions Mimicking Fermentation. *Fermentation* **2023**, *9*, 357. https://doi.org/10.3390/fermentation9040357

Academic Editors: Maria Dimopoulou, Spiros Paramithiotis, Yorgos Kotseridis and Jayanta Kumar Patra

Received: 9 March 2023
Revised: 22 March 2023
Accepted: 30 March 2023
Published: 4 April 2023

1. Introduction

Lipids are an important group of molecules that directly contribute to wine aroma development, and recent work has suggested they may also play a role in wine mouthfeel [1,2]. Lipids, including fatty acids, phospholipids, sterols and others have important biological functions in all cell types such as energy storage, cellular communication, biological process regulation and maintenance of cell membrane structure [3,4]. This group of metabolites is distributed within different grape tissues, particularly concentrating in the skin and seeds. A study of six different grape varieties revealed that lipid concentration ranged from 0.15% to 0.24% of fresh berry weight [5]. Grape seeds usually contain the highest proportion of lipids and are rich predominantly in mono- and polyunsaturated fatty acids. Grape skins contain a range of lipids that act as the main protective barrier to prevent evaporation of water and cellular contents [6].

Very little research has been carried out thus far to characterize lipids in different grape varieties compared with other primary metabolites (e.g., sugars, organic and amino acids). As the grape juice matrix mostly consists of water, there was a misconception among scientists that lipids are not extracted from the pomace to the juice, especially in the production of white wines as no, or very little, skin contact is used in this process. However, a pilot study on New Zealand Sauvignon blanc (SB) juices showed that the total lipid concentration of grape juices can be as high as 2.8 g/L, with <15% available as free fatty acids [7]. Moreover, the SB juices sampled during commercial processing were shown to contain a diverse range of lipid species, including odd-numbered and hydroxy fatty acids, as well as other common saturated and unsaturated free and bound fatty acids. Quantitative data from a recently published study also showed that a variety of free fatty acids ranging from C6 to C24 are present in New Zealand SB juices [8]. Palmitic, stearic, linoleic, and γ-linolenic acids were the four most abundant free fatty acids detected in the 380 juice samples analyzed [7,8]. In another study, Arita, et al. [9] carried out comprehensive lipidome analyses of Pinot noir (PN) and Koshu grape berries and found clear differences in fatty acids and other lipid components. For example, at least 36 of 49 lipid components were significantly higher in PN skins than Koshu skins. PN skins also contained more lipids that had alkyl chains with >18 carbons, and a loss of C18:3 fatty acids during the ripening of Koshu grapes was observed, which may have been converted into (Z)-hex-3-enal, the precursor of C6-aroma compounds.

As grape pulp contains comparatively less lipid and fatty acids than grape skins and seeds, most of these compounds are extracted by the grape juice from the pressing of grapes and through prolonged skin/pomace contact. Studies have shown that extended pomace or skin contact and different pressing conditions alter the fatty acid composition of the grape juices either by increasing specific fatty acids (C18:2 and C18:3 fatty acids) or by reducing the amount of C6 compounds [10]. Comparison among different white and red varieties of grapes indicated a variety and grape tissue–specific differences in lipid composition [4]. For example, grape seeds usually contain ~60% unsaturated fatty acids, with linoleic acid being the predominant, and a high concentration of glycerophospholipids. Moreover, lignoceric acid was one of the main free saturated fatty acids in grape skin along with palmitic and stearic acids [4,11]. These results clearly highlight the diversity of grape lipids across grape varieties and their localization in different grape tissues.

Lipids and fatty acids play a significant role in wine yeast metabolism, particularly under anaerobic winemaking conditions. *Saccharomyces cerevisiae* and many other wine yeast strains are not able to produce fatty acids while growing in the absence of oxygen [12]. Under winemaking conditions, yeast cells are subjected to various stresses arising from osmotic pressure, ethanol toxicity and anaerobiosis. Lipids, specifically fatty acids, present in fermentation media then become an important source of nutrients that ensure optimum growth and fermentation performance of wine yeasts [13]. The literature evidence shows that lipids and fatty acids can modulate wine yeast metabolism, thus influencing the production of wine aroma compounds [14–18]. Pre-fermentative supplementation of common saturated and unsaturated fatty acids usually presents in grape juices significantly affected the growth and metabolism of wine yeasts, albeit in different ways [15,18]. Therefore, the availability of different fatty acids, even during the pre-inoculum preparation, causes significant variations in the yeast cells, thus modifying their metabolism and overall aroma production during winemaking.

In addition to modifying their metabolic activity, availability of fatty acids and other lipids changes the membrane composition of *S. cerevisiae* cells, thereby providing better protection against different stresses [19,20]. Supplemented fatty acids (e.g., palmitic and palmitoleic acids or mixture of fatty acids) are promptly consumed by wine yeasts and incorporated into the cell, allowing the yeast cells to be more viable in different fermentation conditions, which resulted in better survival and fermentation performance [21,22]. In addition to providing protection at low temperature, different lipids and fatty acids present in the yeast cell membrane proved to provide protection against ethanol toxicity during fermentation [23]. Modification and rearrangement of yeast cell lipid composition by changing the availability of lipid molecules in the exogenous media could be used to produce different styles of wines from grape juices with modulated lipid and fatty acid contents. Additionally, recent research by Phan and Tomasino [24] found that the PN lipidome could be used to predict wine origin, showing that lipids persist in wines after fermentation, albeit at very low concentrations (<0.1%). Investigations into potential sensory impacts of lipids in finished wines found that phospholipids could induce a detectable increase in perceived viscosity of model wine [1]. However, attempts to increase the concentrations of lipids in real wines by adding yeast product were unsuccessful, and therefore the implications for mouthfeel perception unclear [25].

Most of the studies carried out on New Zealand wines investigated the influence of fatty acids on wine yeast metabolism and aroma production. The only study completed on the comprehensive lipidome of grape juices in New Zealand was on SB juices [7]. Therefore, there is a lack of knowledge about lipid composition in different grape varieties and how these molecules contribute to the wine quality and sensory properties. This project aimed to fill the gap in knowledge, mainly by determining total lipid content and fatty acid composition of two main New Zealand grape varieties. Here, we investigated the differences in lipid and fatty acid composition at different stages of ripening of SB and PN grapes while determining the effect of skin (pomace) contact time on lipid extraction into the must. We also explored the influence of alcohol concentration (mimicking wine fermentation) on lipid extraction from grape pomace.

2. Materials and Methods

2.1. Experimental Design-Collection of Grapes and Sampling Protocols

Approximately 30 kg of grapes was harvested from the same blocks at two different ripening stages (harvest 1 = unripe grapes and harvest 2 = ripe grapes). SB was sourced from the Marlborough Research Centre's Rowley Vineyard (harvest 1 on 23 March 2019 and harvest 2 on 30 March 2019) while PN grapes were harvested from Omaka Settlement Vineyard of Dog Point situated in Marlborough wine region, New Zealand (harvest 1 on 15 March 2019 and harvest 2 on 22 March 2019). These vineyards follow the standard seasonal canopy management used in New Zealand.

After harvest, a representative sample of 100 grapes was collected from approximately 25 different bunches from each harvest to separate and collect the different grape tissue types (skin, seeds, and pulp). Then, 25 kg of grapes was weighed, crushed, and destemmed using dry ice and 100 ppm potassium metabisulfite (PMS) was added. Using a hydro-bladder press (Fratelli Marchisio & C.S.p.A., Pieve di Teco, Italy), crushed grapes were immediately pressed following protocols for SB developed in our research winery and the juice collected under CO_2 cover. Resulting juices and pomaces were weighed separately and the juice-to-pomace ratio was determined for the reconstitution of the musts and pomaces. A further 60 ppm PMS was added to each treatment to minimize oxidation. Figure 1 shows a schematic diagram of the sample preparation protocol (total n = 122 including grape skin, seeds, and pulp). Approximately 100 mL fresh juice after each press was collected in triplicate for various analyses. Juices and pomaces were reconstituted with the same juice and pomace ratio determined after pressing (final volume ~2 L) in triplicate to simulate extended pomace contact under standard cold soak conditions at 6 °C. Pomace contact time varied depending on grape variety to make it relatable with commercial winery practices: SB juices were kept in contact with pomace for 12 and 24 h while PN juices were in contact with pomace for 72 and 144 h. Another set of samples was prepared by reconstituting pomace using aqueous ethanol solutions in triplicate (final volume ~2 L) using the same juice and pomace ratio after pressing at three different ethanol concentrations (0%, 9% and 13% ethanol). Moreover, 0% mimicked the beginning of fermentation, while 9% and 13% represented mid and late stages of fermentation. Pomace contact was maintained for 72 and 168 h; however, the temperature regime was different for the grape varieties. SB ethanolic extractions of pomaces were carried out at 15 °C resembling commercial SB winemaking in New Zealand and PN ethanolic extractions of pomaces were performed at 24 °C, similar to commercial PN production (Figure 1).

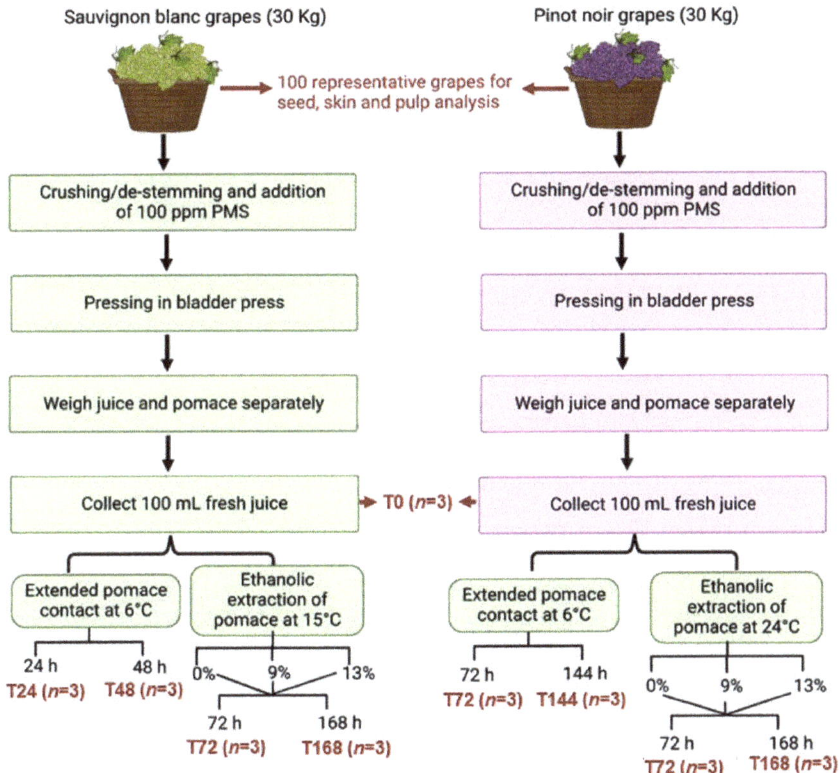

Figure 1. Sample preparation protocol for the project. Here, PMS = potassium metabisulphite, T0 = fresh juices without pomace contact. T denotes the time of pomace contact, number of samples from each stage is shown in red. Green represents Sauvignon blanc while purple shows Pinot noir sampling method. The total number of samples was 122.

2.2. Determination of Major Oenological Parameters

Total soluble solids (TSS) content of all samples was determined using a handheld digital refractometer (Atago, Tokyo, Japan). Titratable acidity and pH were determined using a Mettler Toledo (Columbus, OH, USA) T70 autotitrator with an end-point titration to pH 8.2 and calculated in tartaric acid equivalents (g/L) [26]. Aqueous sodium hydroxide (0.1 M) was used as the titrant. Wine samples were degassed prior to analysis.

Primary amino acid (PAA) concentrations were measured using the nitrogen by o-phthaldialdehyde (NOPA) assay adapted for small volumes [27]. The reaction was measured using a Molecular Devices (San Jose, CA, USA) Spectramax 384 Plus plate reader with a 1-cm pathlength cuvette reference correction. Sample PAA concentrations were quantified in duplicate against a five-point isoleucine standard curve ($R^2 > 0.98$). Ammonium concentrations were measured by an enzymatic assay monitoring the deprotonation of NADPH at 340 nm using the plate reader. Enzymes were purchased from Megazyme (Bray, Ireland); ketoglutaric acid was purchased from Sigma-Aldrich (St. Louis, MO, USA). Samples were appropriately diluted (usually two-fold) and quantified in duplicate against a five-point standard curve ($R^2 > 0.98$). Yeast available nitrogen (YAN) was calculated as the sum of PAA plus ammonium expressed in mg/L of nitrogen.

Glucose and fructose were quantified by enzymatic assay based on the reduction in NADP to NADPH, and the reaction was monitored at 340 nm using the plate reader.

Enzymes and cofactors were purchased from Megazyme (Bray, Ireland). Samples were appropriately diluted and quantified in duplicate against an eight-point standard curve ($R^2 > 0.98$). This method was adapted from the Compendium of International Methods of Analysis OIV-MA-AS311-02.

The optical density of the juice was determined in duplicate directly in a UV-transparent 96-well microplate at 280, 320, and 420 nm based on the method described in Martin, et al. [26]. Absorbance at 280 nm was used to quantify total phenolics against a five-point gallic acid standard curve (five-point, $R^2 > 0.98$).

2.3. Quantification of Major Organic and Amino Acids

A Shimadzu Prominence high performance liquid chromatography (HPLC) (Shimadzu Corporation, Kyoto, Japan) system with a diode array detector (DAD) equipped with an Allure Organic Acids Restek column (Bellefonte, PE, United States; 5 μm, 240 × 4.6 mm) was used to quantify organic acids (tartaric, malic, ascorbic, citric and succinic acids) in samples based on the method reported by Shi et al., 2011 [28] and validated in our laboratory. Briefly, samples were diluted five-fold and filtered through a 0.22 μm syringe filter before injection. A 25-min isocratic method using phosphate buffer (25 mM, pH 2.3) was run at 0.6 mL/min and 30 °C. Samples were run in duplicate and quantified on a six-point standard curve ($R^2 > 0.98$).

Quantification of amino acids in grape juices was performed on an Agilent 1200 series HPLC (Santa Clara, CA, USA) equipped with a Thermo Fisher Scientific Hypersil Gold C18 column (Waltham, MA, USA; 5 μm, 250 × 3.0 mm) using gradient elution with a phosphate/borate buffer (10 mM each, pH 8.2) containing 0.1 $v/v\%$ tetrahydrofuran (THF) and a combination of methanol, acetonitrile and water as the organic solvent (45:45:10), run at 1.5 mL/min and 40 °C. The method was adapted from an application note by Henderson and Brooks [29], validated in the laboratory and described in Martin et al. [26,27]. Briefly, online derivatization of primary amino acids was carried out with o-phthalaldehyde and 3-mercaptopropionic acid and detected by a DAD at 340 nm excitation and 450 nm emission. Samples were treated with iodoacetic acid to encourage the reduction in cysteine. Secondary amino acids were derivatized online with 9-fluorenylmethyl chlo-roformate and detected by fluorescence (260 nm excitation, 315 nm emission). A standard mix of 17 amino acids was purchased from Agilent (Santa Clara, CA, USA). Internal standards sarcosine (100 mg/L) and α-aminobutyric acid (100 mg/L) were added to all standards and samples to account for potential injection volume variability. Samples were analyzed both undiluted and diluted ten-fold in water for the quantification of low and high abundance amino acids respectively and filtered through a 0.45-μm syringe filter before injection. All samples were analyzed in duplicate and quantified on a six-point standard curve ($R^2 > 0.98$) [29].

2.4. Lipid Extraction and Transesterification

Lipid extraction of the samples was performed using a modified protocol published by Díaz de Vivar, et al. [30]. As different types of samples were generated during this study, optimization of the extraction method was carried out to determine the amount and volume of sample required for extracting lipids. For juice and alcoholic extracts, 2 mL of the sample provided optimum results while 10 mg of the sample was required for analyzing seeds, skin, and pulp. Grape seeds, skins and pulp were freeze-dried and ground into a powder, which was then used for the extraction of lipids and fatty acids. Lipids were extracted by adding 250 μL of distilled water to 2 mL of juice and alcoholic extracts and 10 mg powder of seeds, skins and pulp. After this, 125 μL of chloroform with internal standard (C23: 0.48 mg in 50 mL of chloroform) and 250 μL of methanol were added and mixed thoroughly. The samples were centrifuged at 2500 rpm for 5 min and transferred to new glass vials. Another 250 μL of distilled water and 250 μL of chloroform were added to the samples, mixed, and again centrifuged at 2500 rpm for 5 min. The top layer was discarded and the bottom layer containing the lipids in chloroform was transferred to a gas-chromatography

(GC) vial. Using a SpeedVac (Savant SP5121P, ThermoFisher, Waltham, MA, USA) the chloroform layer was concentrated prior to transesterification.

Transesterification was performed using a modified protocol published by Díaz de Vivar, et al. [30]. Lipid extract (1 mL) was transferred to screw-cap glass culture tubes and 2 mL of internal standard (C19:0) solution dissolved in methanol: toluene 4:1 (*v/v*) was added. For the seed samples, a 1:10 dilution was performed using toluene. A magnetic stir bar was placed inside the tube and 200 μL of acetyl chloride was added to the tube over a period of 1 min. The tube was tightly closed with the Teflon cap prior to determining the weight and it was then placed in a heating/stirring dry block at 100 °C for 1 h. After this, the tube was cooled in water, dried and the weight was determined again to check for any leakage. To stop the reaction and neutralize the mixture, 5 mL/sample of 6% K_2CO_3 solution was added very slowly and mixed by vortexing prior to centrifuging at 2500 rpm for 5 min. Approximately 200 μL of the upper toluene phase was transferred to a vial for analysis by gas chromatography–mass spectrometry (GC-MS).

2.5. GC-MS Analysis

The transesterified lipid extracts were injected on to a GC-MS (Agilent GC 7890 coupled to a MSD 5975, Agilent Technologies, Santa Clara, CA, USA) with a quadrupole mass selective detector (EI) operated at 70 eV using helium as the carrier gas. Instrument analytical parameters were based on those developed by Kramer et al. [31]. Column selection was based on the recommendations from the Official Methods for the determination of trans fat (American Oil Chemists Society). The column was a fused silica Rtx-2330 100 m long, 0.25 mm internal diameter, 0.2 μm highly polar stationary phase (90% biscyanopropyl 10% cyanopropylphenyl polysiloxane, Shimadzu, Kyoto, Japan). The carrier gas was instrument grade helium (99.99%, BOC). One microlitre of the sample was injected using a CTC PAL autosampler into a glass 4-mm ID straight inlet liner packed with deactivated glass wool (Restek Sky®, Bellefonte, PA, USA). The inlet temperature was 250 °C, in splitless mode, and the column flow was set at 1 mL/min, with a column head pressure of 9 psi, giving an average linear velocity of 19 cm/s. Purge flow was set to 50 mL/min 1 min after injection. After injection at 60 °C, the oven temperature was raised to 150 °C at a rate of 40 °C min^{-1}, and then to 230 °C at 3 °C min^{-1} and finally held constant for 30 min. The GC oven temperature programming started isothermally at 45 °C for 2 min, increased 10 °C/min to 215 °C, held 35 min, increased 40 °C/min to 250 °C and held 10 min. The transfer line to the mass spectrometric detector (MSD) was maintained at 250 °C, the MSD source at 230 °C and the MSD quadrupole at 150 °C. The detector was turned on 14.5 min into the run. The detector was run in positive-ion, electron-impact ionization mode, at 70 eV, with the electron multiplier set with no additional voltage relative to the autotune value. Data were acquired at 1463 amu/s in scan mode from 41 to 420 atomic mass units, with a detection threshold of 100 ion counts.

Fatty acid methyl ester (FAME) peaks were identified by comparing their retention times with those of 52 authentic FAME standards (Nu-Chek-Prep, Inc., Elysian, MN, USA-52 component FAME mix; GLC reference standard 674) and based on the in-house MS library. A five-point calibration curve was prepared in order to quantify the fatty acids positively identified in the samples.

2.6. GC-MS Data Mining

Data analysis was automated and performed with in-house R package developed at the University of Auckland metabolomics laboratory [7,8]. The raw data output from the GC-MS was converted to AIA format (.cdf) and analyzed using automated mass spectral deconvolution and identification software (AMDIS, http://www.amdis.net/, accessed on 4 April 2020) and an in-house MS library of 52 fatty acids derivatized compounds and one extra standard (C19:0). The reference ion used as a measure of abundance for each compound is usually the most abundant fragment and is not the molecular ion. As the output from AMDIS returns zero values that are not suitable for statistical analysis, an in-house R-script MassOmics was used in conjunction with the AMDIS output to produce data that include trace levels of metabolites normally excluded by AMDIS. The values are generated from the maximum height of the reference ion for the compound peak. Unlike peak area, peak height is affected by chromatographic disturbances such as column contamination, and as a result early eluting peaks may sometimes be under-represented. Data were checked against negative controls and obvious contamination, or artifacts were highlighted in the uncorrected results and removed in the corrected results. Coeluting peaks were highlighted, checked and corrected. Where two identifications were equally likely for one peak, both identifications have been reported. The resulting dataset was then normalized by the internal standard "nonadecanoic acid" and average peak responses from experimental "blank" samples were deducted from experimental samples to account for baseline response. Quantification was performed using calibration curves obtained from the analysis of standard mix samples. Lastly, sample biomass/volume normalization and dilution correction (if required) were performed to obtain the final quantified data. Approximate total lipid content was calculated by summing the concentrations of all detected free fatty acids in the samples according to Van Wychen and Laurens [32].

2.7. Statistical Analysis

Data were log-transformed prior to performing any statistical analyses. Independent Student's *t*-tests were performed to compare the changes in the total lipids, fatty acids and oenological parameters (e.g., alcohol content, volatile thiol concentrations, amino and organic acids) of each treatment compared with their respective controls using an in-house R script. One-way analysis of variance (ANOVA) was also performed to compare different treatments. Microsoft Excel 2010 was used to determine mean and standard deviation of triplicate controls and treatments. A web-based platform Metaboanalyst 4.0 (http://www.metaboanalyst.ca, accessed on 4 April 2020) was used to perform different unsupervised and supervised statistical analyses including principal component analysis (PCA), partial least square-discriminant analysis (PLS-DA), variance in projection (VIP) and also to generate the heatmap (distance measure: Euclidean and clustering algorithm: Ward). A machine learning algorithm in Metaboanalyst 4.0 "pattern searching" was also used to determine the top 25 features correlated (using Pearson correlation method) with the total lipids and major fatty acids [33].

3. Results and Discussion

3.1. Chemical Composition of the Sauvignon Blanc (SB) and Pinot Noir (PN) Juices and Alcoholic Extracts

We determined the chemical composition of both the SB and PN juices just after pressing, juices with prolonged pomace contact and ethanolic extracts of the grape pomaces. Table 1 details the main oenological measurements and organic and amino acids present in the SB and PN juices. Although we intended to collect the SB grapes at 18 °Brix for harvest 1 and 21.5 °Brix for harvest 2, the TSS in our final samples was 19.9 °Brix and 21.8 °Brix, respectively. Similarly, the TSS in the PN grapes was 22.1 °Brix (instead of 20 °Brix) and 24.7 °Brix (instead of 23.5 °Brix) for harvest 1 and harvest 2, respectively. As the Marlborough region experienced a much warmer summer in 2019, the grape ripening was faster than previous vintages. However, there was at least a 2 °Brix difference in the TSS between the harvests, and titratable acidity and YAN data also reflected the different fruit maturities at each harvest (Table 1).

Most of the chemical parameters shown in Table 1 varied between harvests, depending on the grape variety. We observed no significant variation in pH and ammonium ($p > 0.05$) between harvest 1 and 2 in both the SB and PN juices. In the PN juices, both tartaric and malic acid concentrations were significantly lower and sugar concentrations were much higher in the harvest 2 juices. Tartaric acid and total reducing sugar concentrations changed little between harvests in the SB juices, but a significant reduction in malic acid concentration and titratable acidity was observed in the harvest 2 juices, suggesting a more advanced ripeness stage compared to with harvest 1. SB and PN amino acids showed varietal differences, and most of their concentrations increased in the harvest 2 samples (Table 1). Additionally, total polyphenols increased 3.5-fold in the PN harvest 2 grape juices compared with harvest 1, suggesting that the harvest 2 grapes were indeed more mature.

Table 1. Chemical properties of harvest 1 and 2 Sauvignon blanc and Pinot noir juices prior to pomace contact.

	Sauvignon Blanc			Pinot Noir		
	Harvest 1	**Harvest 2**	*p*-Value	**Harvest 1**	**Harvest 2**	*p*-Value
Total soluble solids (°Brix)	19.9 (0.1)	21.8 (0.0)	<0.05	22.1 (0.1)	24.7 (0.1)	<0.05
Titratable acidity (g/L)	11.7 (0.04)	9.77 (0.02)	<0.05	9.09 (0.06)	6.95 (0.05)	<0.05
pH	3.02 (0.01)	3.05 (0.00)	>0.05	3.12 (0.01)	3.22 (0.01)	>0.05
Tartaric acid (g/L)	7.66 (0.14)	7.09 (0.02)	>0.05	6.30 (0.07)	3.30 (0.08)	<0.01
Malic acid (g/L)	5.74 (0.11)	4.70 (0.01)	<0.05	4.68 (0.05)	3.62 (0.01)	<0.05
YAN (mg N/L)	155 (14)	179 (4)	<0.05	191 (3)	219 (0)	<0.05
Ammonium (mg N/L)	61 (14)	67 (2)	>0.05	82 (0)	81 (1)	>0.05
PAA (mg N/L)	94 (0)	112 (2)	<0.05	109 (2)	139 (1)	<0.03
Total reducing sugar (g/L)	221 (19)	237 (4)	>0.05	231 (1)	293 (11)	<0.05
Glucose (g/L)	122 (7)	129 (3)	>0.05	121 (2)	155 (7)	>0.05
Fructose (g/L)	99 (11)	108 (7)	<0.05	110 (3)	138 (4)	<0.01
Total Phenolics (mg gallic acid/L)	*	252 (2)	ND	424 (5)	1517 (69)	<0.01
Aspartic acid (µmol/L)	237 (7)	397 (3)	<0.01	254 (3)	161 (4)	<0.01
Glutamic acid (µmol/L)	827 (14)	861 (14)	<0.05	751 (6)	786 (15)	>0.05
Serine (µmol/L)	281 (5)	357 (6)	>0.05	377 (5)	504 (12)	<0.01
Arginine (µmol/L)	1243 (22)	1476 (37)	<0.05	1473 (20)	2143 (51)	<0.01
Alanine (µmol/L)	682 (11)	806 (7)	<0.01	807 (8)	1181 (21)	<0.01
Histidine (µmol/L)	810 (12)	901 (17)	<0.05	767 (32)	1102 (29)	<0.01
Threonine (µmol/L)	454 (6)	446 (4)	>0.05	551 (9)	682 (10)	<0.05
Valine (µmol/L)	143 (4)	158 (6)	>0.05	153 (8)	171 (4)	>0.05
Proline (µmol/L)	873 (107)	1032 (93)	<0.05	681 (92)	636 (54)	>0.05
Methionine (µmol/L)	77 (6)	106 (2)	<0.05	142 (2)	213 (4)	<0.01
Isoleucine (µmol/L)	48 (2)	59 (1)	>0.05	89 (1)	107 (2)	>0.05
Leucine (µmol/L)	51 (6)	72 (1)	>0.05	124 (0)	168 (3)	<0.05
Phenylalanine (µmol/L)	39 (0)	64 (1)	<0.05	45 (0)	59 (1)	>0.05

* indicates missing data; ND = not determined; PAA = primary amino acids; YAN = yeast available nitrogen. *p*-values were determined by comparing harvest 1 and harvest 2 values using independent *t*-test. Standard deviations are shown within brackets.

Chemical analyses of the musts from extended pomace contact (24 and 48 h contact for SB; 72 and 144 h contact for PN) revealed that the composition of the musts changed because of pomace contact, with the duration of contact an important factor in these variations. There was a significant reduction in total reducing sugars after 24 h of pomace contact in the harvest 1 SB must, which was not expected. This may be due to fermentation starting despite the higher SO_2 addition rates and cold temperature, or insufficient precision of analysis of high sugar concentration samples by the plate reader assay [34].

Concentrations of polyphenols, tannins and monomeric anthocyanins increased significantly ($p < 0.05$) in the PN musts because of extended pomace contact. As winemakers are aware, these secondary metabolites are extracted better with time during pomace contact, which is also widely reported in the literature, for example by [35].

3.2. Overview of Fatty Acids Present in Grapes and Their Extracts Harvested at Different Time Points

The main aim of our project was to determine the lipid content and composition in two main New Zealand grape varieties, and to explore how pomace contact and ethanolic extraction (mimicking winemaking conditions) influence lipid and fatty acid composition.

This is the first study of this kind in New Zealand, which has allowed us to generate unique datasets on SB and PN grape tissues, juices, and extracts. The method we used here provided an overview of fatty acids that are found in both bound and free form. Table 2 shows the complete list of fatty acids detected in all the samples. Using an in-house MS library, 45 fatty acids were positively identified and quantified in grape juices and extracts while 44 fatty acids were determined in grape seeds, skin, and pulp (Table 2). These fatty acids ranged from C8 to C24, the majority of which were unsaturated. This was in accordance with previously published data [4,7]. At least two medium-chain fatty acids, undecanoic and tridecanoic acids, were unique to grape skin and seeds, while four other long-chain fatty acids including oleic, trans- vaccenic, trans-elaidic and 11-cis-eicosenoic acids were only found in the grape juices and extracts. Therefore, either these long-chain fatty acids were too low in concentration to be detected in the grape tissues, or somehow, they were produced as a result of the extraction process. It is possible that some lipid and fatty acid oxidation occurred during the extraction process despite our measures to exclude oxygen (i.e., using N_2 gas prior to closing the vessels and adding 60 ppm PMS).

Table 2. List of saturated and unsaturated fatty acids detected and identified in grape juices, extracts and tissues.

	Fatty Acids	Other Known Names	No of Carbons and Double Bonds	Type of Fatty Acid
1	Octanoic acid	Caprylic acid	C8:0	Saturated
2	Decanoic acid	Capric acid	C10:0	Saturated
3	Undecanoic acid *	Undecylic acid	C11:0	Saturated
4	Dodecanoic acid	Lauric acid	C12:0	Saturated
5	Tridecanoic acid *	Tridecylic acid	C13:0	Saturated
6	Tetradecanoic acid	Myristic acid	C14:0	Saturated
7	Pentadecanoic acid		C15:0	Saturated
8	Hexadecanoic acid	Palmitic acid	C16:0	Saturated
9	Heptadecanoic acid	Margaric acid	C17:0	Saturated
10	Octadecanoic acid	Stearic acid	C18:0	Saturated
11	Eicosanoic acid	Arachidic acid	C20:0	Saturated
12	Heneicosanoic acid	Heneicosylic acid	C21:0	Saturated
13	Docosanoic acid	Behenic acid	C22:0	Saturated
14	Tetracosanoic acid	Lignoceric acid	C24:0	Saturated
15	9-cis-Tetradecenoic acid	Myristoleic acid	C14:1	Unsaturated
16	9-trans-Tetradecenoic acid	Myristelaidic acid	C14:1	Unsaturated
17	10-cis-Pentadecenoic acid		C15:1	Unsaturated
18	10-trans-Pentadecenoic acid		C15:1	Unsaturated
19	9-cis-Hexadecenoic acid	Palmitoleic acid (cis)	C16:1	Unsaturated
20	(E)-9-hexadecenoic acid	Palmitoleic acid (trans)	C16:1	Unsaturated
21	10-cis-Heptadecenoic acid		C17:1	Unsaturated
22	10-trans-Heptadecenoic acid		C17:1	Unsaturated
23	9-trans-Octadecenoic acid+	Elaidic acid (trans)	C18:1	Unsaturated
24	9-cis-Octadecenoic acid+	Oleic acid	C18:1	Unsaturated
25	11-trans-Octadecenoic acid+	trans-Vaccenic acid	C18:1	Unsaturated
26	11-cis-Octadecenoic acid	cis-Vaccenic acid	C18:1	Unsaturated
27	9,12,15-cis-Octadecatrienoic acid	alpha-Linolenic acid	C18:3	Unsaturated
28	9,12-cis-Octadecadienoic acid	Linoleic acid	C18:2	Unsaturated
29	9,12-trans-Octadecadienoic acid	Linolelaidic acid	C18:2	Unsaturated
30	cis-6,9,12-octadecatrienoic acid	gamma-Linolenic acid	C18:3	Unsaturated
31	10-trans-Nonadecenoic acid	Nonadecylic acid	C19:1	Unsaturated
32	7-trans-Nonadecenoic acid		C19:1	Unsaturated
33	11,14,17-cis-Eicosatrienoic acid	Eicosatrienoic acid	C20:3	Unsaturated
34	11,14-cis-Eicosadienoic acid	Eicosadienoic acid	C20:2	Unsaturated
35	11-trans-Eicosenoic acid	Eicosenoic acid	C20:1	Unsaturated

Table 2. *Cont.*

	Fatty Acids	Other Known Names	No of Carbons and Double Bonds	Type of Fatty Acid
36	11-cis-Eicosenoic acid+		C20:1	Unsaturated
37	8,11,14-cis-Eicosatrienoic acid		C20:3	Unsaturated
38	5,8,11,14,17-cis-Eicosapentaenoic acid		C20:5	Unsaturated
39	5,8,11,14-cis-Eicosatetraenoic acid		C20:4	Unsaturated
40	13,16-cis-Docosadienoic acid		C22:2	Unsaturated
41	13-cis-Docosenoic acid		C22:1	Unsaturated
42	13-trans-Docosenoic acid		C22:1	Unsaturated
43	7,10,13,16,19-docosapentaenoaic acid		C22:5	Unsaturated
44	4,7,10,13,16,19-Docosahexaenoic acid		C22:6	Unsaturated
45	7,10,13,16-cis-Docosatetraenoic acid	Docosapentaenoic acid	C22:4	Unsaturated
46	4,7,10,13,16-docosapentaenoaic acid		C22:5	Unsaturated
47	15-cis-Tetracosenoic acid		C24:1	Unsaturated

* indicates the fatty acids were only present in grape skin and seed; + indicates the fatty acids were only found in grape juices and extracts.

3.3. Total Lipids and Fatty Acids in Sauvignon Blanc and Pinot Noir Grape Tissues

Determination of the total lipids from transesterification of lipids to fatty acid methyl esters is a common and widely used analytical practice [36,37]. In grapes, different classes of lipids are distributed among the different tissue types. Therefore, we used this transesterification protocol to accurately quantify 45 fatty acids in the various grape tissues. The summation of all the fatty acids presents in the sample provided us an indication of the total lipid content of the samples [7]. As shown in Table 3, grape pulp contained the least amount of lipid as expected. Total lipid content was significantly higher in the PN pulp ($p < 0.05$) and skins ($p < 0.01$) than the SB from both harvests, while seeds from both varieties had similar amounts of lipids. However, grape lipids were mostly concentrated in seeds for both grape varieties, which was expected [4]. Lipids and fatty acids are usually extracted to the grape must during pressing and from subsequent skin/pomace contact [10].

Table 3. Saturated and unsaturated fatty acids detected and identified in Sauvignon blanc (SB) and Pinot noir (PN) juices, extracts and tissues. Standard deviations are shown within brackets.

	Pulp (g/g)	Seeds (g/g)	Skin (g/g)
SB harvest 1	0.009 (0.001)	0.311 (0.075)	0.013 (0.004)
SB harvest 2	0.016 (0.008)	0.486 (0.089)	0.023 (0.001)
Comparison (fold-change, SB harvest 1 vs. harvest 2)	1.84	1.83	1.56
PN harvest 1	0.016 (0.005)	0.221 (0.025)	0.025(0.007)
PN harvest 2	0.029 (0.008)	0.438 (0.077)	0.072 (0.009)
Comparison (fold-change, PN harvest 1 vs. harvest 2)	1.78	1.99	2.92

Our data show an interesting trend of increasing lipids (1.5- to 3-fold) in the harvest 2 grapes for both varieties, indicating that lipid concentrations increase as grapes mature (Table 3). Many studies discuss the role of sugars and organic and amino acids in determining the ripeness of grapes [37–41], but the development of lipids and fatty acids during grape ripening is largely unknown. Our data provide novel information on how lipids can be strongly related to grape maturity and ripeness. It is noteworthy that traditionally used ripeness parameters including total soluble solids and titratable acidity are not necessarily linked with improving wine aroma and quality. However, lipids, specifically fatty acids, play a vital role during wine fermentation and particularly can contribute to the productions of various aroma compounds [15–18]. Therefore, if the total lipids or a specific lipid (fatty acid) marker are directly related with grape ripeness, this will provide researchers or even industries an option to make winemaking decisions, allowing them to diversify wine styles.

As expected, seeds from both varieties contained much higher amounts of unsaturated fatty acids than saturated ones [4]. However, the SB skins had comparatively higher concentrations of saturated fatty acids for both harvests. The PN skins from harvest 1 contained more unsaturated fatty acids while saturated fatty acids increased significantly in the harvest 2 grape skins. Therefore, there might be an increased rate of saturation in fatty acids in the PN grape skin during the ripening process. Interestingly, the SB pulp from the harvest 1 grapes had higher unsaturated fatty acids, while the harvest 2 grape pulps had more saturated fatty acids. However, the PN grape pulp showed an opposite trend. Therefore, the evolution of fatty acid saturation during grape ripening might vary between varieties.

In addition to determining the total lipid content, we also investigated the specific fatty acids that are predominantly present in different grape tissues of both SB and PN. Figure 2 presents the abundance of major fatty acids found in the pulp, seeds and skin of both grape varieties. Stearic acid was the most abundant fatty acid both in the SB and PN grape pulp and skin (Figure 2), while the seeds contained a high amount of linoleic acid (at least >0.1 g/g). This observation was in accordance with previously published studies [4,5]. Among other fatty acids, grape seeds and skin also contained a high concentration of oleic and palmitic acids (>0.05 g/g). Concentrations of other fatty acids present in different grape tissues and varieties were <0.01 g/g. However, even in such small concentrations, these fatty acids play important biological roles during grape development [6].

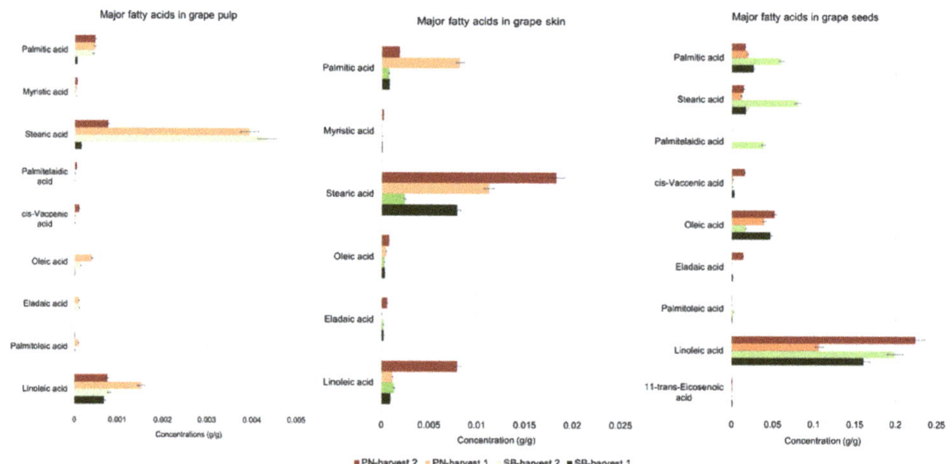

Figure 2. Concentrations of major fatty acids in Pinot noir (PN) and Sauvignon blanc (SB) grape pulp, seed and skin. Fatty acid profiling was completed using 10 mg of each sample type.

3.4. Total Lipid and Fatty Acid Contents in Sauvignon Blanc and Pinot Noir Grape Juices and Extracts

3.4.1. Effect of Harvest Time on Total Lipids and Fatty Acids in Grape Juices

Table 4 shows the ranges of the total lipids in the SB and PN grapes just after pressing and without any pomace contact. A comparison between the harvest 1 and 2 data shows a significant increase ($p < 0.01$; >30% increase) in lipids in harvest 2 compared with harvest 1 for the SB juices. However, lipid concentrations were not significantly different in the PN grapes harvested at the two time points ($p > 0.05$). Therefore, lipid development in grapes may vary depending on variety, and we indeed found a varietal difference in total lipid contents ($p > 0.05$) as the PN juices from both harvests contained more lipid than the SB juices. A more comprehensive multi-seasonal study on lipids is required to confirm these observations as a seasonal variation is largely observed in grape metabolites [8].

Table 4. Total lipids present in Sauvignon blanc and Pinot noir grape juices and the influence of pomace contact on extraction of lipids in the must.

Sample	Pomace Contact Time (h)	Total Lipids-Harvest 1 (g/L)	Total Lipids-Harvest 2 (g/L)	Change in Lipid Level (%)
Sauvignon blanc	0	0.14 (0.01)	0.22 (0.03)	36.13
	24	0.41 (0.03)	0.61 (0.05)	32.55
	48	0.72 (0.04)	1.12 (0.08)	35.84
Pinot noir	0	0.27 (0.03)	0.33 (0.01)	16.24
	72	0.23 (0.02)	0.59 (0.04)	60.95
	144	0.39 (0.05)	0.93 (0.06)	58.02

Standard deviations are shown within brackets.

Absolute quantification of 45 different fatty acids ranging from C8 to C24 allowed us to determine the differences observed in the SB and PN grapes harvested at different time points, thus representing variation caused by the different degree of ripeness of the grapes. Not all the fatty acids present in the grape juices were responsible for the variation. In the SB grape juices, at least 30 fatty acids were significantly different ($p < 0.05$) between harvest 1 and 2, while 23 fatty acids were found to be considerably different ($p < 0.05$) in the PN juices. Therefore, we used the 23 most significant ($p < 0.05$) and common fatty acids found in SB and PN to perform a PCA (Figure 3) where principal component (PC) 1 and PC 2 explained more than 57% of the variation. We observed a clear difference in fatty acid composition between the SB and PN juices. In particular, long-chain polyunsaturated fatty acids such as linoleic acid and gamma-linolenic acid were more abundant in the PN juices than SB.

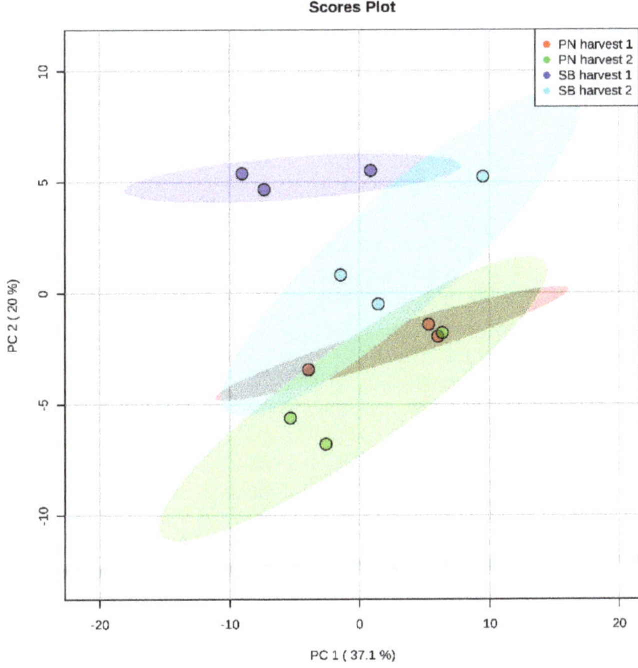

Figure 3. Two-dimensional representation of principal component analysis (PCA) using the 23 most significant fatty acids ($p < 0.05$) present in Sauvignon blanc (SB) and Pinot noir (PN) grape juices harvested at two different maturity levels.

While comparing the fatty acid profiles of the two different harvests, we saw a visible difference between SB harvest 1 and 2 that was not prominent for the PN juices (Figure 3). In SB, levels of some saturated fatty acids including palmitic, stearic and pentadecanoic acids increased significantly in the harvest 2 juices. The PN juice profiles showed no such trend except that the concentration of lignoceric acid was higher in harvest 2 than harvest 1 juices. These data indicate that there is a varietal difference in fatty acid developments in grape juices. Moreover, variation in fatty acids between harvests/maturity also may occur for some varieties, but not for all.

3.4.2. Effect of Pomace Contact Time on Lipid Extraction and Fatty Acids in Grape Must

Our next objective was to explore how pomace contact influences lipid extraction in grape juices. Although pomace contact is an uncommon practice for commercial SB wine production, a short contact time (24–48 h) was applied. As shown in Table 3, lipid concentrations increase 2.5- to 5-fold at 24 and 48 h when compared with juices without any pomace contract for both harvests. In addition, a substantial increase in total lipid in the SB must was observed when comparing harvest 1 and 2 after the pomace contact. As the grape juice matrix can be significantly different at different maturity levels, harvest 2 most probably provided more suitable lipid extraction conditions.

To make this study relatable to commercial practice, we used an extended pomace contact for the PN juices. The PN must data showed a different trend, particularly for harvest 1 (Table 3). There was a slight decrease in the total lipids after 72 h pomace contact in the harvest 1 samples, and the reasons for this are unknown. However, lipid concentration increased 1.5-fold after 144 h when compared with the juices without pomace contact (Table 3). Similar to the SB must, comparison between the harvest 1 and 2 data show an increased rate of extraction of lipids after both 72 and 144 h pomace contact, which reinforces our theory on the role of the juice matrix.

Lipid extraction in the SB must was comparatively higher than in PN even though the PN pomace contact time was much longer. We can again relate this observation to the differences in the juice matrices. The SB juice is generally more acidic, which might have influenced the lipid extraction. Additionally, the SB seeds are usually larger than the PN seeds. Moreover, the PN skins contain a larger number of polyphenols and anthocyanins than SB, and thus extracted more of these secondary metabolites than lipids. Therefore, we speculate that the SB cold soak conditions may allow a better lipid extraction than PN. These data confirm that lipids in commercially produced grape musts (especially for white varieties) can be increased by extending the duration of pomace contact depending on the style of wines to be produced.

As shown in Table 3, lipid contents significantly increase in the must due to the prolonged skin/pomace contact. We observed a similar trend in fatty acids. Figure 4 presents the heatmaps that show those fatty acids that changed significantly ($p < 0.05$) due to the pomace contact. Particularly for SB, levels of more fatty acids (15) increased with time during pomace contact than PN. Only two fatty acids, 7-trans-nonadecanoic and myristoleic acids, were more abundant in juices with no pomace contact. There is a possibility that these fatty acids may have undergone some degradation or transformation process during pomace contact, thus contributing to the development of other fatty acids [7]. For instance, the concentration of myristelaidic acid increased over the time of pomace contact while myristoleic acid decreased. These two fatty acids are closely related to each other (C14:1) and might have just transformed during pomace contact. Some of fatty acids also could be consumed and/or produced by natural microorganisms present in the pomaces and musts [15,42].

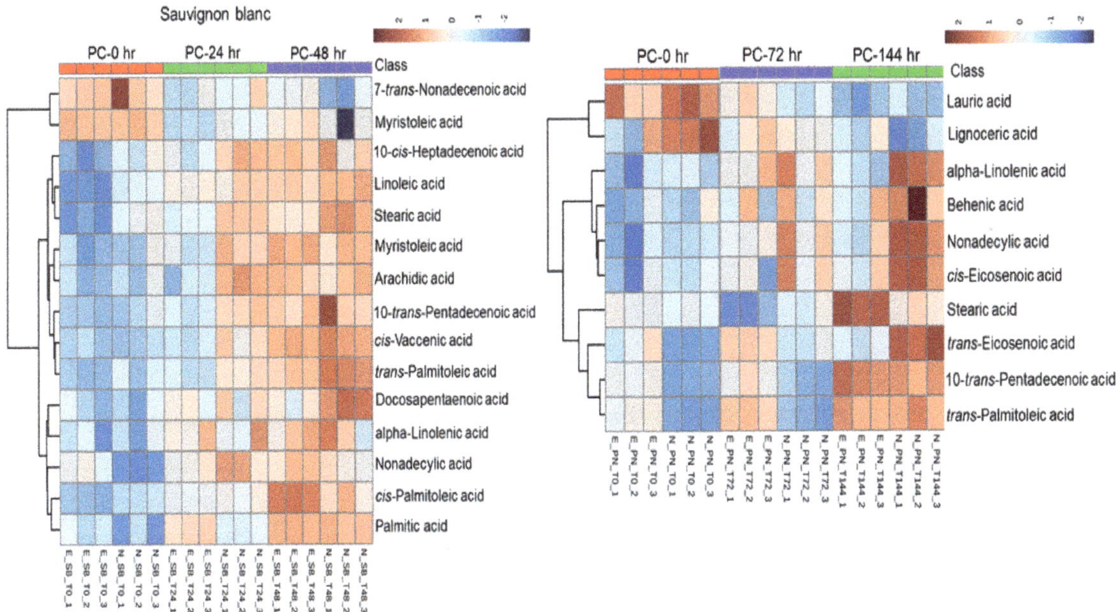

Figure 4. Heat map showing the abundance of most significant fatty acids ($p < 0.05$) that were affected by pomace contact (PC) in Sauvignon blanc (SB) and Pinot noir (PN) must. Here: E = Harvest 1, N = Harvest 2, T = time of pomace contact (PC).

3.5. Effect of Ethanol Concentration Relevant to Fermentation on Lipid Extraction from Grape Pomace

This part of the experiment was designed to simulate different stages of fermentation to observe how different ethanol concentrations affect lipid and fatty acids extraction. We replaced the must juices with two different aqueous ethanol solutions (9 and 13% v/v) to contrast lipid extraction in the absence of ethanol (0%, water only) at 72 and 144 h. Different temperatures were applied to the extractions to reflect commercial winemaking conditions, with the SB extracts incubated at 15 °C and PN at 24 °C (Figure 1). Table 5 shows the summarized data and indicates that lipid extraction increased linearly with the ethanol concentration and also with time. Similar to our observation shown in Table 3, extraction of lipids was greater from the SB pomaces (as high as 3.11 g/L at 144 h) than PN (1.85 g/L at 144 h). Although lipid extraction was expected be better at the higher temperature used for PN, our data proved it otherwise. However, more work is needed to determine the exact reasons behind this observation. As indicated earlier, spontaneous fermentation occurred in some of the extraction conditions, particularly when we used 0 and 9% ethanol, suggesting that some of the lipids may have been originated from the native microbes. This is also evident in commercial winemaking, where winery microbes can contribute to extraction of different metabolites (including flavor precursors) during the maceration process [43,44].

Table 5. Effect of ethanol concentrations on lipid extraction from Sauvignon blanc and Pinot noir grape pomace. Standard deviations are shown within brackets.

Variety	Sauvignon Blanc								
Ethanol (% *v/v*)	0	0	Increase (%)	9	9	Increase (%)	13	13	Increase (%)
Pomace contact time (h)	72	168		72	168		72	168	
Total lipids-harvest 1 (g/L)	0.70 (0.04)	1.09 (0.05)	35.51	0.86 (0.16)	1.80 (0.19)	52.02	1.04 (0.18)	1.75 (0.14)	40.43
Total lipids-harvest 2 (g/L)	0.69 (0.04)	1.14 (0.10)	39.45	0.99 (0.10)	1.84 (0.12)	45.96	1.53 (0.09)	3.11 (0.30)	50.93
Variety	Pinot Noir								
Ethanol (% *v/v*)	0	0	Increase (%)	9	9	Increase (%)	13	13	Increase (%)
Pomace contact time (h)	72	168		72	168		72	168	
Total lipids-harvest 1 (g/L)	0.38 (0.02)	0.85 (0.06)	54.62	0.48 (0.08)	1.06 (0.13)	54.15	1.05 (0.08)	1.60 (0.17)	34.65
Total lipids-harvest 2 (g/L)	0.54 (0.03)	0.89 (0.07)	38.60	0.89 (0.05)	1.24 (0.04)	28.51	1.44 (0.02)	1.85 (0.14)	22.09

We also monitored the changes in the fatty acid profiles during ethanolic extractions of the SB and PN pomaces. For both the SB and PN pomaces, most of the fatty acids were better extracted when ethanol was present, which was expected. However, some of the fatty acids were extracted better when there was no ethanol including linoleic acid for both SB and PN and octanoic, trans-vaccenic acids only for SB. Therefore, we assume that the fatty acid composition of grape must can not only be changed through prolonged pomace contact, but also can be manipulated during the winemaking condition, particularly when ethanol starts to be produced during the fermentation. These results also indicate that different types of fatty acids are available at the different stages of fermentation with the progress of ethanol production. Moreover, fatty acids and lipid components from yeast cells would also contribute to the lipid availability from mid- and late-fermentation, particularly when exogenous lipid sources are all consumed [15]. Therefore, this knowledge would assist us in developing strategies to manipulate fatty acids and lipids during winemaking, thus influencing the final aroma bouquet of wines as many of these fatty acids serve as a pre-cursor for the formation of ethyl and acetate esters [10].

3.6. Correlation of Lipids and Fatty Acids with Major Oenological Parameters

We performed a correlation analysis to investigate if lipids and major fatty acids found in the SB and PN juices have any relationship with major amino acids and other oenological parameters. Figure 5 presents the top 25 features that positively or negatively correlated with the total lipids in the juices. While primary amino acids, YAN and ammonium show strong positive correlation with the total lipids, individual amino acids exhibit negative correlation. The reasons for this remain unknown at this point and warrant further investigation.

Among fatty acids, stearic and linoleic acids also show positive correlation with these ripeness parameters (Figure 6). These two fatty acids are also major fatty acids present in different NZ grape varieties [8]. Although stearic acid strongly correlated with the total lipids (Figures 5 and 6), linoleic acid did not show such trend. We assume that as a polyunsaturated fatty acid, linoleic acid is more prone to oxidation and other chemical changes during the lipid extraction time, while stearic acid is more stable. Therefore, stearic acid has more potential to be a ripeness indicator and a simple user-friendly test to determine stearic acid can be developed to be applied for the determination of berry ripeness.

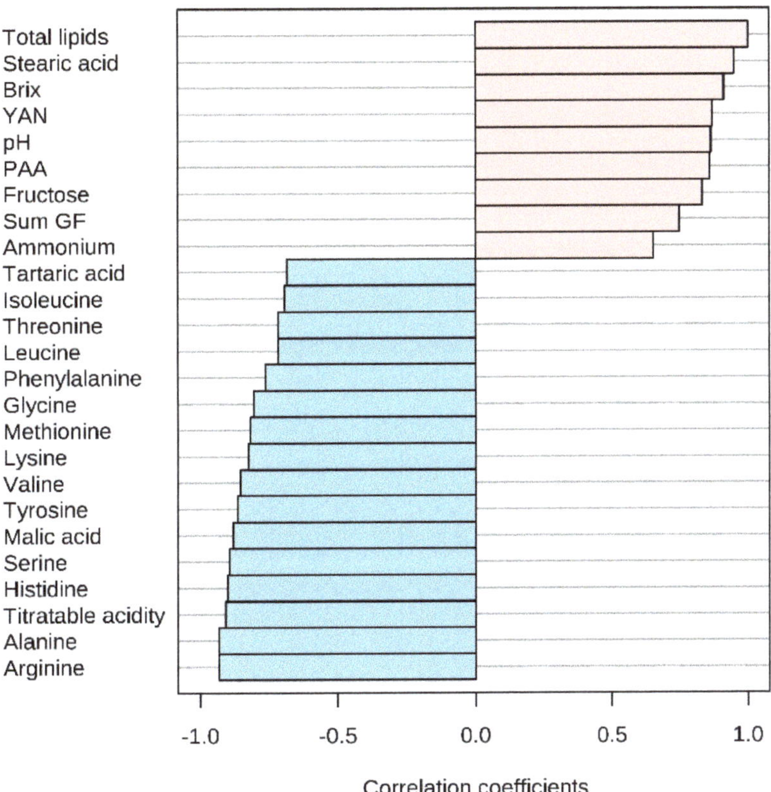

Figure 5. A pattern hunter plot showing the correlation of total lipids with different oenological parameters and nitrogenous compounds analyzed in the Sauvignon blanc and Pinot noir grape juices. Pink bars represent the positive correlation (Pearson r > +0.50) while blue bars show the negative correlation (Pearson r > −0.50). YAN = yeast available nitrogen, PAA = primary amino acids, GF = glucose + fructose.

The relationships between lipids/fatty acids and different ripeness parameters have largely been overlooked and while searching the literature we found them only in a few studies. For instance, Bauman et al. [45] investigated the lipid composition and fatty acid distribution of Concord grapes over four different stages of maturation. They reported that crude lipid content was highest at *véraison* (0.23%) while neutral lipids increased and polar lipids decreased during maturation, indicating that lipid composition of grapes evolves at different stages of grape development. Another study published by Le Fur et al. [46] investigated the changes in phytosterols (ß-sitosterol, campesterol, stigmasterol and lanosterol) in grape skins during the last stages of ripening of Chardonnay grape variety in Burgundy. Barron and Santa-María [47] investigated the relationship between triglycerides and different ripeness and energy indices. Their results indicated a strong correlation of different triglycerides species with energy indices while showing a moderate correlation with different ripeness indices. In a more recently published study using comprehensive lipidomics approach, Masuero et al. [11] reported positive and negative correlation with certain lipid classes with total soluble solids, indicating a relationship among grape ripeness with lipids. However, they did not determine the correlation between the total lipids or major fatty acids with grape ripeness. Our data, therefore, provide a novel insight and we hypothesize that lipid concentration increases with grape maturation, which is related to the different stages of ripening. Thus, lipids could be another parameter for determining grape ripeness alongside sugars/acids. Using this information, further research could be undertaken to develop either a colorometric industry friendly test or non-destructive near infrared spectroscopy (NIRS) method to determine the total lipids or a specific lipid biomarker linked with ripeness. This would provide the industry an important tool to determine grape quality prior to harvest. This in turn would provide vital knowledge for winemakers to select appropriate yeast strains to produce wines with good aromatic quality.

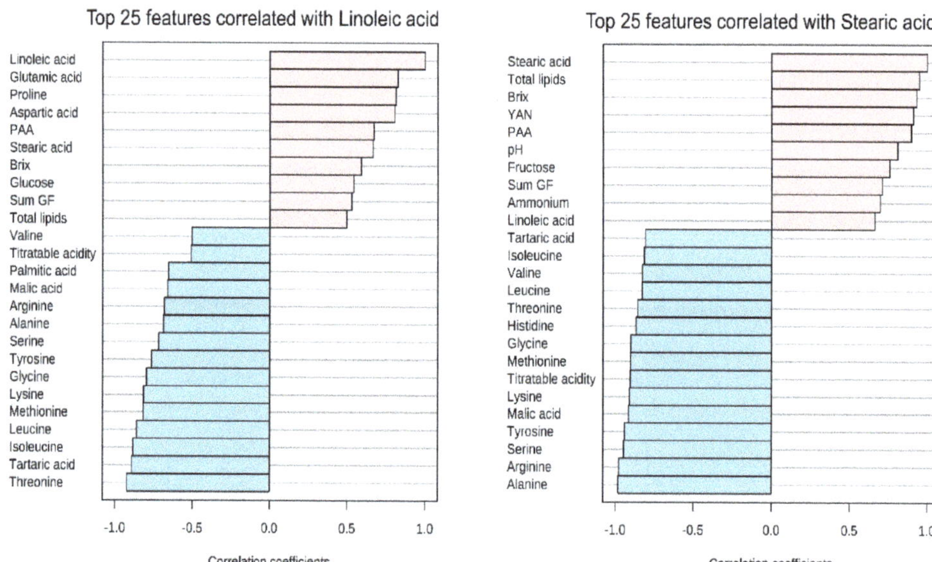

Figure 6. Pattern hunter plots showing the correlation of linoleic and stearic acids with different oenological parameters and nitrogenous compounds analyzed in the Sauvignon blanc and Pinot noir grape juices. Pink bars represent the positive correlation (Pearson r > +0.50) while blue bars show the negative correlation (Pearson r > −0.50). YAN = yeast available nitrogen, PAA = primary amino acids, GF = glucose + fructose.

As is evident from the available published data, lipids and fatty acids are less studied classes of compounds in oenology. Most of the research in this space has focused on the role of fatty acids in aroma development and yeast metabolism [12,13,15,19,21,48–51]. Research by Sherman [2] already demonstrated the prospective role of unsaturated fatty acids as predictors of perceived PN wine body as lipids persist in wines after fermentation, thus contributing to wine sensory properties. Therefore, more research is needed to extend different aspects of utilization of lipid molecules in winemaking and also in grape growing. There is a large knowledge gap and we found no multi-season study that focused on the lipid developments during grape growing and how lipids/fatty acids evolve during different stages of winemaking.

4. Conclusions

In this study, we generated some novel insights on the evolution of lipid and fatty acids in grapes at different stages of maturity by analyzing grape juices, grape tissues and ethanolic extracts. We found that the total lipids and fatty acid composition vary between harvests at different stages of ripeness depending on the grape variety. We observed a strong correlation between lipids/major fatty acids and other commonly used ripeness parameters including total soluble solids, sugars, titratable acidity and YAN, thus indicating a potential role of lipids as another ripeness parameter. Our data also showed that lipid concentrations increased significantly because of prolonged skin/pomace contact depending on the grape variety. Moreover, lipid and fatty acid extraction increased linearly with the ethanol concentration and time of pomace contact while extraction of lipids and fatty acids was greater in SB maceration conditions than PN, suggesting a matrix effect. As our data are based on only one season, observations might not be conclusive and there is a need for a comprehensive multi-season study to confirm these data-generated hypotheses. Research should also be carried out to determine the effect of different vineyard management practices on the developments of lipids in grapes. If this type of research is successfully undertaken, the wine industry will ultimately benefit from the knowledge generated to produce diverse styles of wines via manipulating different lipid classes and fatty acids during grape growing and winemaking.

Author Contributions: Conceptualization, E.S. and F.R.P.; methodology, E.S., M.Y., F.G., E.Z., S.G. and F.R.P.; formal analysis, E.S., K.W.B. and F.R.P.; data curation, E.S. and F.R.P.; writing—original draft preparation, E.S. and F.R.P.; writing—review and editing, E.S., M.Y. and F.R.P.; project administration, F.R.P.; funding acquisition, F.R.P. All authors have read and agreed to the published version of the manuscript.

Funding: The authors declare that this study received funding from New Zealand Lighter Wines Programme (P/471777/04) co-funded by New Zealand Winegrowers, and the Ministry of Primary Industries' Primary Growth Partnership (PGP) venture. The funder was not involved in the study design, collection, analysis, interpretation of data, the writing of this article or the decision to submit it for publication.

Institutional Review Board Statement: Not applicable.

Informed Consent Statement: Not applicable.

Data Availability Statement: Access to data and database can be requested by contacting the corresponding author.

Acknowledgments: We are thankful to Gerald Hope of the Marlborough Research Centre for kindly providing Sauvignon blanc grapes from their Rowley Vineyard and to Nigel Sowman from Dog Point for generously providing Pinot noir grapes. Acknowledgment is due to Victoria Raw, Rob Agnew, Claire Grose, Lily Stuart, Tanya Rutan and Rachel Bishell for their help during the study. We also thank Grant Morris, Megan Jones, Kevin Sutton and Plant and Food's Science Publication Unit for their help with manuscript.

Conflicts of Interest: The authors declare no conflict of interest.

References

1. Phan, Q.; Hoffman, S.; Tomasino, E. Contribution of Lipids to Taste and Mouthfeel Perception in a Model Wine Solution. *ACS Food Sci. Technol.* **2021**, *1*, 1561–1566. [CrossRef]
2. Sherman, E. A Flavoromics Approach to the Characterisation of Pinot Noir Wine Sensory Properties. Ph.D. Thesis, University of Auckland, Auckland, New Zealand, 2017.
3. Kehelpannala, C.; Rupasinghe, T.; Hennessy, T.; Bradley, D.; Ebert, B.; Roessner, U. The state of the art in plant lipidomics. *Mol. Omics* **2021**, *17*, 894–910. [CrossRef]
4. Pérez-Navarro, J.; Da Ros, A.; Masuero, D.; Izquierdo-Cañas, P.M.; Hermosín-Gutiérrez, I.; Gómez-Alonso, S.; Mattivi, F.; Vrhovsek, U. LC-MS/MS analysis of free fatty acid composition and other lipids in skins and seeds of Vitis vinifera grape cultivars. *Food Res. Int.* **2019**, *125*, 108556. [CrossRef]
5. Gallander, J.F.; Peng, A.C. Lipid and Fatty Acid Compositions of Different Grape Types. *Am. J. Enol. Vitic.* **1980**, *31*, 24. [CrossRef]
6. Grncarevic, M.; Radler, F. A Review of the Surface Lipids of Grapes and Their Importance in the Drying Process. *Am. J. Enol. Vitic.* **1971**, *22*, 80.
7. Tumanov, S.; Zubenko, Y.; Greven, M.; Greenwood, D.R.; Shmanai, V.; Villas-Boas, S.G. Comprehensive lipidome profiling of Sauvignon blanc grape juice. *Food Chem.* **2015**, *180*, 249–256. [CrossRef]

8. Pinu, F.R.; Tumanov, S.; Grose, C.; Raw, V.; Albright, A.; Stuart, L.; Villas-Boas, S.G.; Martin, D.; Harker, R.; Greven, M. Juice Index: An integrated Sauvignon blanc grape and wine metabolomics database shows mainly seasonal differences. *Metabolomics* **2019**, *15*, 3. [CrossRef]

9. Arita, K.; Honma, T.; Suzuki, S. Comprehensive and comparative lipidome analysis of *Vitis vinifera* L. cv. Pinot Noir and Japanese indigenous *V. vinifera* L. cv. Koshu grape berries. *PLoS ONE* **2017**, *12*, e0186952. [CrossRef] [PubMed]

10. Ferreira, B.; Hory, C.; Bard, M.H.; Taisant, C.; Olsson, A.; Le Fur, Y. Effects of skin contact and settling on the level of the C18:2, C18:3 fatty acids and C6 compounds in burgundy chardonnay musts and wines. *Food Qual. Prefer.* **1995**, *6*, 35–41. [CrossRef]

11. Masuero, D.; Škrab, D.; Chitarrini, G.; Garcia-Aloy, M.; Franceschi, P.; Sivilotti, P.; Guella, G.; Vrhovsek, U. Grape Lipidomics: An Extensive Profiling thorough UHPLC-MS/MS Method. *Metabolites* **2021**, *11*, 827. [CrossRef] [PubMed]

12. Casu, F.; Pinu, F.R.; Fedrizzi, B.; Greenwood, D.R.; Villas-Boas, S.G. The effect of linoleic acid on the Sauvignon blanc fermentation by different wine yeast strains. *FEMS Yeast Res.* **2016**, *16*, fow050. [CrossRef] [PubMed]

13. Varela, C.; Torrea, D.; Schmidt, S.A.; Ancin-Azpilicueta, C.; Henschke, P.A. Effect of oxygen and lipid supplementation on the volatile composition of chemically defined medium and Chardonnay wine fermented with Saccharomyces cerevisiae. *Food Chem.* **2012**, *135*, 2863–2871. [CrossRef] [PubMed]

14. Deroite, A.; Legras, J.-L.; Rigou, P.; Ortiz-Julien, A.; Dequin, S. Lipids modulate acetic acid and thiol final concentrations in wine during fermentation by Saccharomyces cerevisiae × Saccharomyces kudriavzevii hybrids. *AMB Express* **2018**, *8*, 130. [CrossRef] [PubMed]

15. Pinu, F.R.; Villas-Boas, S.G.; Martin, D. Pre-fermentative supplementation of fatty acids alters the metabolic activity of wine yeasts. *Food Res. Int.* **2019**, *121*, 835–844. [CrossRef] [PubMed]

16. Pinu, F.R.; Edwards, P.J.B.; Jouanneau, S.; Kilmartin, P.A.; Gardner, R.C.; Villas-Boas, S.G. Sauvignon blanc metabolomics: Grape juice metabolites affecting the development of varietal thiols and other aroma compounds in wines. *Metabolomics* **2014**, *10*, 556–573. [CrossRef]

17. Guittin, C.; Maçna, F.; Sanchez, I.; Poitou, X.; Sablayrolles, J.-M.; Mouret, J.-R.; Farines, V. Impact of high lipid contents on the production of fermentative aromas during white wine fermentation. *Appl. Microbiol. Biotechnol.* **2021**, *105*, 6435–6449. [CrossRef]

18. Liu, P.; Ivanova-Petropulos, V.; Duan, C.; Yan, G. Effect of Unsaturated Fatty Acids on Intra-Metabolites and Aroma Compounds of Saccharomyces cerevisiae in Wine Fermentation. *Foods* **2021**, *10*, 277. [CrossRef]

19. Landolfo, S.; Zara, G.; Zara, S.; Budroni, M.; Ciani, M.; Mannazzu, I. Oleic acid and ergosterol supplementation mitigates oxidative stress in wine strains of Saccharomyces cerevisiae. *Int. J. Food Microbiol.* **2010**, *141*, 229–235. [CrossRef]

20. Rupčić, J.; Jurešić, G.C.; Blagović, B. Influence of stressful fermentation conditions on neutral lipids of a Saccharomyces cerevisiae brewing strain. *World J. Microbiol. Biotechnol.* **2010**, *26*, 1331–1336. [CrossRef]

21. Redón, M.; Guillamón, J.M.; Mas, A.; Rozès, N. Effect of lipid supplementation upon Saccharomyces cerevisiae lipid composition and fermentation performance at low temperature. *Eur. Food Res. Technol.* **2009**, *228*, 833–840. [CrossRef]

22. Mannazzu, I.; Angelozzi, D.; Belviso, S.; Budroni, M.; Farris, G.A.; Goffrini, P.; Lodi, T.; Marzona, M.; Bardi, L. Behaviour of Saccharomyces cerevisiae wine strains during adaptation to unfavourable conditions of fermentation on synthetic medium: Cell lipid composition, membrane integrity, viability and fermentative activity. *Int. J. Food Microbiol.* **2008**, *121*, 84–91. [CrossRef]

23. Girardi-Piva, G.; Casalta, E.; Legras, J.-L.; Nidelet, T.; Pradal, M.; Macna, F.; Ferreira, D.; Ortiz-Julien, A.; Tesnière, C.; Galeote, V.; et al. Influence of ergosterol and phytosterols on wine alcoholic fermentation with Saccharomyces cerevisiae strains. *Front. Microbiol.* **2022**, *13*. [CrossRef]

24. Phan, Q.; Tomasino, E. Untargeted lipidomic approach in studying pinot noir wine lipids and predicting wine origin. *Food Chem.* **2021**, *355*, 129409. [CrossRef] [PubMed]

25. Phan, Q.; DuBois, A.; Osborne, J.; Tomasino, E. Effects of Yeast Product Addition and Fermentation Temperature on Lipid Composition, Taste and Mouthfeel Characteristics of Pinot Noir Wine. *Horticulturae* **2022**, *8*, 52. [CrossRef]

26. Martin, D.; Grose, C.; Fedrizzi, B.; Stuart, L.; Albright, A.; McLachlan, A. Grape cluster microclimate influences the aroma composition of Sauvignon blanc wine. *Food Chem.* **2016**, *210*, 640–647. [CrossRef] [PubMed]

27. Martin, D.; Lindsay, M.; Kilmartin, P.; Dias Araujo, L.; Rutan, T.; Yvon, M.; Stuart, L.; Grab, F.; Moore, T.; Scofield, C.; et al. Grape berry size is a key factor in determining New Zealand Pinot noir wine composition. *OENO One* **2022**, *56*, 389–411. [CrossRef]

28. Shi, S.; Condron, L.; Larsen, S.; Richardson, A.E.; Jones, E.; Jiao, J.; O'Callaghan, M.; Stewart, A. In situ sampling of low molecular weight organic anions from rhizosphere of radiata pine (Pinus radiata) grown in a rhizotron system. *Environ. Exp. Bot.* **2011**, *70*, 131–142. [CrossRef]

29. Henderson, J.W.J.; Brooks, A. *Improved Amino Acid Methods using Agilent ZORBAX Eclipse Plus C18 Columns for a Variety of Agilent LC Instrumentation and Separation Goals*; Agilent Application Note 5990–4547EN; Agilent Technologies, Inc.: Wilmington, NC, USA, 2010.

30. De Vivar, M.E.D.; Zárate, E.V.; Rubilar, T.; Epherra, L.; Avaro, M.G.; Sewell, M.A. Lipid and fatty acid profiles of gametes and spawned gonads of Arbacia dufresnii (Echinodermata: Echinoidea): Sexual differences in the lipids of nutritive phagocytes. *Mar. Biol.* **2019**, *166*, 96. [CrossRef]

31. Kramer, J.K.; Hernandez, M.; Cruz-Hernandez, C.; Kraft, J.; Dugan, M.E. Combining results of two GC separations partly achieves determination of all cis and trans 16:1, 18:1, 18:2 and 18:3 except CLA isomers of milk fat as demonstrated using Ag-ion SPE fractionation. *Lipids* **2008**, *43*, 259–273. [CrossRef]

32. Van Wychen, S.; Laurens, L.M.L. *Determination of Total Lipids as Fatty Acid Methyl Esters (FAME) by In Situ Transesterification*; Laboratory Analytical Procedure (LAP); National Renewable Energy Laboratory: Golden, CO, USA, 2013.

33. Chong, J.; Soufan, O.; Li, C.; Caraus, I.; Li, S.; Bourque, G.; Wishart, D.S.; Xia, J. MetaboAnalyst 4.0: Towards more transparent and integrative metabolomics analysis. *Nucleic Acids Res.* **2018**, *46*, W486–W494. [CrossRef]

34. Sherman, E.; Yvon, M.; Martin, D. Influence of surface area to volume ratio on ethanol yield of commercial scale pinot noir ferments. In *A Report from Plant and Food Research Ltd. to New Zealand Winegrowers*; Milestone No. 76003. Contract No. 30998 var5. SPTS No. 17017; New Zealand Wine: Auckland, New Zealand, 2018.

35. Ivanova, V.; Vojnoski, B.; Stefova, M. Effect of winemaking treatment and wine aging on phenolic content in Vranec wines. *J. Food Sci. Technol.* **2012**, *49*, 161–172. [CrossRef] [PubMed]

36. Laurens, L.M.L.; Quinn, M.; Van Wychen, S.; Templeton, D.W.; Wolfrum, E.J. Accurate and reliable quantification of total microalgal fuel potential as fatty acid methyl esters by in situ transesterification. *Anal. Bioanal. Chem.* **2012**, *403*, 167–178. [CrossRef] [PubMed]

37. Kuhn, N.; Guan, L.; Dai, Z.W.; Wu, B.-H.; Lauvergeat, V.; Gomès, E.; Li, S.-H.; Godoy, F.; Arce-Johnson, P.; Delrot, S. Berry ripening: Recently heard through the grapevine. *J. Exp. Bot.* **2013**, *65*, 4543–4559. [CrossRef] [PubMed]

38. Rienth, M.; Torregrosa, L.; Sarah, G.; Ardisson, M.; Brillouet, J.-M.; Romieu, C. Temperature desynchronizes sugar and organic acid metabolism in ripening grapevine fruits and remodels their transcriptome. *BMC Plant Biol.* **2016**, *16*, 164. [CrossRef]

39. Dal Santo, S.; Fasoli, M.; Negri, S.; D'Inca, E.; Vicenzi, N.; Guzzo, F.; Tornielli, G.B.; Pezzotti, M.; Zenoni, S. Plasticity of the Berry Ripening Program in a White Grape Variety. *Front. Plant Sci.* **2016**, *7*, 17. [CrossRef]

40. Trought, M.C.T.; Naylor, A.P.; Frampton, C. Effect of row orientation, trellis type, shoot and bunch position on the variability of Sauvignon Blanc (*Vitis vinifera* L.) juice composition. *Aust. J. Grape Wine Res.* **2017**, *23*, 240–250. [CrossRef]

41. Vondras, A.M.; Commisso, M.; Guzzo, F.; Deluc, L.G. Metabolite Profiling Reveals Developmental Inequalities in Pinot Noir Berry Tissues Late in Ripening. *Front. Plant Sci.* **2017**, *8*, 14. [CrossRef] [PubMed]

42. Klug, L.; Daum, G. Yeast lipid metabolism at a glance. *FEMS Yeast Res.* **2014**, *14*, 369–388. [CrossRef]

43. Bokulich Nicholas, A.; Collins Thomas, S.; Masarweh, C.; Allen, G.; Heymann, H.; Ebeler Susan, E.; Mills David, A. Associations among Wine Grape Microbiome, Metabolome, and Fermentation Behavior Suggest Microbial Contribution to Regional Wine Characteristics. *Mbio* **2016**, *7*, e00616–e00631. [CrossRef]

44. Stefanini, I.; Carlin, S.; Tocci, N.; Albanese, D.; Donati, C.; Franceschi, P.; Paris, M.; Zenato, A.; Tempesta, S.; Bronzato, A.; et al. Core Microbiota and Metabolome of *Vitis vinifera* L. cv. Corvina Grapes and Musts. *Front. Microbiol.* **2017**, *8*, 457. [CrossRef]

45. Bauman, J.A.; Gallander, J.F.; Peng, A.C. Effect of maturation on the lipid content of Concord grapes. *Am. J. Enol. Vitic.* **1977**, *28*, 241–244. [CrossRef]

46. Le Fur, Y.; Hory, C.; Bard, M.-H.; Olsson, A. Evolution of phytosterols in Chardonnay grape berry skins during last stages of ripening. *Vitis* **1994**, *33*, 127–131.

47. Barron, L.; Santa-María, G. A relationship between triglycerides and grape-ripening indices. *Food Chem.* **1990**, *37*, 37–45. [CrossRef]

48. Girardi Piva, G.; Casalta, E.; Legras, J.-L.; Tesnière, C.; Sablayrolles, J.-M.; Ferreira, D.; Ortiz-Julien, A.; Galeote, V.; Mouret, J.-R. Characterization and Role of Sterols in Saccharomyces cerevisiae during White Wine Alcoholic Fermentation. *Fermentation* **2022**, *8*, 90. [CrossRef]

49. Liu, P.-T.; Zhang, B.-Q.; Duan, C.-Q.; Yan, G.-L. Pre-fermentative supplementation of unsaturated fatty acids alters the effect of overexpressing ATF1 and EEB1 on esters biosynthesis in red wine. *LWT* **2020**, *120*, 108925. [CrossRef]

50. Tumanov, S.; Pinu, F.R.; Greenwood, D.R.; Villas-Bôas, S.G. Effect of free fatty acids and lipolysis on Sauvignon Blanc fermentation. *Aust. J. Grape Wine Res.* **2018**, *24*, 398–405. [CrossRef]

51. Rosi, I.; Bertuccioli, M. Influences of lipid addition on fatty acid composition of Saccharomyces cerevisiae and aroma characteristics of experimental wines. *J. Inst. Brew.* **1992**, *98*, 305–314. [CrossRef]

 *fermentation*

Review

Fermented Beverages Revisited: From Terroir to Customized Functional Products

Spiros Paramithiotis [1,*], **Jayanta Kumar Patra** [2], **Yorgos Kotseridis** [3] **and Maria Dimopoulou** [4]

[1] Department of Biological Applications and Technology, University of Ioannina, 45110 Ioannina, Greece
[2] Research Institute of Integrative Life Sciences, Dongguk University-Seoul, Goyangsi 10326, Republic of Korea; jkpatra@dongguk.edu
[3] Department of Food Science & Human Nutrition, Agricultural University of Athens, 11855 Athens, Greece; ykotseridis@aua.gr
[4] Department of Wine, Vine and Beverage Sciences, University of West Attica, 12243 Athens, Greece; mdimopoulou@uniwa.gr
* Correspondence: paramithiotis@uoi.gr

Abstract: Fermented beverages have been a constant companion of humans throughout their history. A wide range of products have been developed with time, depending on the availability of raw materials and ambient conditions. Their differentiation was based on the specific characteristics of each product, resulting from the cultivation of different varieties and the variability of environmental conditions and agricultural practices, collectively described by the term 'terroir' that was developed in winemaking. The health benefits that have been associated with their consumption, which include the control of blood pressure and glycemic control, along with immunomodulatory, hypocholesterolemic, hepatoprotective, and antiproliferative activities, directed their re-discovery that occurred over the last few decades. Thus, the dynamics of the microbial communities of fermented beverages during fermentation and storage have been thoroughly assessed. The functional potential of fermented beverages has been attributed to the chemical composition of the raw materials and the bioconversions that take place during fermentation and storage, due to the metabolic capacity of the driving microbiota. Thus, the proper combination of raw materials with certain microorganisms may allow for the modulation of the organoleptic properties, as well as enrichment with specific functional ingredients, enabling targeted nutritional interventions. This plasticity of fermented beverages is their great advantage that offers limitless capabilities. The present article aims to critically summarize and present the current knowledge on the microbiota and functional potential of fermented beverages and highlight the great potential of these products.

Keywords: wine; kefir; kombucha; functional ingredients

check for updates

Citation: Paramithiotis, S.; Patra, J.K.; Kotseridis, Y.; Dimopoulou, M. Fermented Beverages Revisited: From Terroir to Customized Functional Products. *Fermentation* **2024**, *10*, 57. https://doi.org/10.3390/fermentation10010057

Academic Editors: Giuseppe Italo Francesco Perretti and Francesco Grieco

Received: 25 November 2023
Revised: 20 December 2023
Accepted: 12 January 2024
Published: 14 January 2024

1. Introduction

Fermented beverages are produced since antiquity, with wine being the most characteristic example. They can be classified according to the nature of the raw materials, or the type of fermentation employed. According to the first, two classes of fermented beverages are distinguished: plant-based and dairy. The plant-based ones can be further subdivided into cereal-based products, such as boza, cheka, pozol, kvass, the various types of beer, etc. [1–6]; fruit-based products, such as the various types of wine, cider, gilaburu, etc. [7–11]; and herbal-based products, such as kombucha [12,13]. As far as fermented dairy beverages are concerned, several types have been described, such as kefir, kumis, viili, acidophilus milk, etc. [14], each with a unique history, production procedure and microecosystem composition. If the predominant type of fermentation is taken as a criterion, fermented beverages can be classified into acidic, alcoholic, and mixed fermented products; kefir and kombucha can serve as examples of acidic beverages, wine of alcoholic, and boza of mixed fermented beverages.

The term 'terroir' has been developed and is currently in use in wine technology. It indicates a specific geographical area along with the associated grape cultivars and oenological practices, which altogether define the specific features of the produced wines. Nowadays, the health benefits that accompany food consumption have also been at the epicenter of consumer interest. This trend has led to the intensive study of fermented beverages throughout the world, their functional ingredients as well as the molecular mechanisms that are implicated. Apart from grape wine, two fermented beverages have also been distinguished for their functional potential, namely kefir and kombucha.

Thus, moderate wine consumption has been correlated with reduced risk of cardiovascular events [15,16], neurodegenerative diseases [17,18], and type 2 diabetes [19,20], as well as systolic blood pressure decrease [21–23], improvement of the gastrointestinal tract function [24–26] and the main symptoms of fibromyalgia [27]. These have all been principally attributed to the phenolic compounds that it contains. Kombucha consumption has been correlated with a series of effects, such as antidiabetic, antihypertensive, antimicrobial, antiproliferative, hepatoprotective, hypocholesterolemic and immunomodulatory [28–37]. As in the case of grape wine, these activities have been attributed to kombucha phenolic content. Finally, kefir consumption has been correlated with a series of effects, including antiproliferative and immunomodulatory capacity, effective glycemic control, prevention and treatment of atherosclerosis and liver damage, as well as control of blood pressure [38–44]. These activities have been attributed to the amino acid and peptide content of kefir.

The present study aimed to comprehensively and critically present the microbiota of fermented beverages, the functional properties that have been associated with the consumption of the most prominent ones, along with the underlying molecular mechanisms, as well as the strategies that have been employed towards their customization, in order to summarize the current knowledge and facilitate the identification of research gaps.

2. The Microbiota of Traditional Fermented Beverages

In Table 1, the microbiota of traditional fermented beverages around the world is presented. Three aspects are immediately noticed: (1) the diversity of the raw materials employed, (2) the diversity of the microorganisms implicated in each type of fermentation, and (3) some products have attracted scientific attention and are therefore heavily studied, while the microecosystem of others has been less intensively assessed.

The availability of raw materials combined with suitable environmental conditions has led to the development of a wide range of fermented beverages, some of which are considered as characteristic of certain geographical areas. Indeed, products such as kombucha and bhaati jaanr, which are based on tea and rice, respectively, have Asian origins; products based on maize, such as pozol, chicha, and atole agrio originate from America, whereas milk and cereal-based fermented beverages are abundant in Africa and Europe. Despite the current globalization of production, traditional knowledge may still be derived from the cradle in which each product was originally developed.

In the majority of the cases, the microbial consortium that drives fermentation of the products reported in Table 1 consists of yeasts and bacteria, and more specifically, lactic and acetic acid bacteria. This can be attributed to the composition of the raw materials and the conditions used for fermented product manufacture but is biased by the number of samples analyzed and the identification methodology employed. In general, the higher the number of samples analyzed, the more likely is to enrich the biodiversity already reported in the literature. A wide range of techniques are currently available, the correct application of which may lead to reliable identification at species or subspecies level. The specific nutritional and growth requirements of each microorganism, at strain level, define the nature of the relationships that will be developed within the microecosystem. Despite the large number of studies, at least in the case of kombucha and kefir, the trophic relationships between the microorganisms are still far from being understood. The only exception seems to be wine must fermentation. In that case, the requirements and the capacity of the implicated yeast species have been extensively studied, and this knowledge is already considered as established. On the other hand, research is currently focused on understanding the physiological attributes of the lactic acid bacteria that carry out malolactic fermentation [45].

Table 1. The microbiota of representative traditional fermented beverages around the world.

Product	Main Ingredient	Microbiota	References
Kombucha	Sweetened black or green tea	**LAB**: *O. oeni, Lq. nagelii* **AAB**: *A. aceti, A. musti, A. okinawensis, A. pasteurianus, A. peroxydans, A. senegalensis, A. tropicalis, A. xylinoides, A. xylinum, Ga. europaeus, Ga. hansenii, Ga. intermedius, Ga. xylinus, Gb. oxydans, Kb. europaeus, Kb. hansenii, Kb. intermedius, Kb. rhaeticus, Kb. saccharivorans, Kb. xylinus* **Yeasts**: *Br. anomalus, Br. bruxellensis, Br. lambicus, Ca. albicans, Ca. boidinii, Ca. colleculosa, Ca. guilliermondii, Ca. kefyr, Ca. krusei, Ca. sake, Ca. stellata, De. anomala, De. bruxellensis, H. valbyensis, K. marxianus, Kz. unispora, Lh. fermentati, Pi. fermentans, Pi. membranifaciens, R. mucilaginosa, S. cerevisiae, S. uvarum, Sd. ludwigii, Sz. pombe, T. delbrueckii, Z. bailii, Z. lentus, Z. parabaillii, Zt. fiorentina*	[12,13,46–54]
Kefir	Milk	**LAB**: *E. durans, E. faecalis, La. casei, La. paracasei, Lp. plantarum, Lb. acidophilus, Lb. amylovorus, Lb. delbrueckii, Lb. crispatus, Lb. helveticus, Lb. kefiranofaciens, Lc. lactis, Lt. buchneri, Lt. kefiri, Lt. parabuchneri, Lt. parakefiri, Le. parakefiri, Ln. mesenteroides, Ln. paramesenteroides, Ln. pseudomesenteroides, Lq. uvarum, Lq. satsumensis, Lv. brevis, Str. durans, Str. thermophilus* **AAB**: *A. aceti, A. fabarum, A. lovaniensis, A. okinawensis, A. orientalis, A. rancens, A. syzygii, Gb. frateurii, Gb. japonicus* **Yeasts**: *Br. anomalus, Ca. colliculosa, Ca. inconspicua, Ca. kefyr, Ca. krusei, Ca. lambica, Ca. maris, De. anomala, K. lactis, K. marxianus, Kz. aerobia, Kz. exigua, Kz. kefir, Kz. unispora, Lh. meyersii, M. guilliermondii, Pi. kudriavzevii, Pi. guilliermondii, S. cerevisiae, S. fragilis, S. turicensis, T. delbrueckii*	[55–81]
Wine	Fruits	**Yeasts**: *Au. pullulans, Ca. fermentati, Ca. intermedia, Ca. parapsilopsis, Ca. pulcherrima, Ca. quercitrusa, Ca. zemplinina, H. uvarum, H. guillermondii, H. uvarum, H. valbyensis, I. occidentalis, I. orientalis, I. terricola, Kc. apiculata, Lh. thermotolerans, Pi. fermentans, R. graminis, R. mucilaginosa, S. bayanus, S. cerevisiae, S. italicus, S. pastorianus, S. uvarum, Sd. ludwigii, Sz. pombe, T. delbrueckii, T. globispora, Y. lipolytica, Z. bailii, Z. fermentati*	[9,10,82–86]
Apple cider	Apples	**Yeasts**: *H. osmophila, H. uvarum, H. valbyensis, M. pulcherrima, Pi. guillermondii, S. bayanus, S. cerevisiae*	[8]
Amabere amaruranu	Milk	**LAB**: *Ln. mesenteroides, Lp. plantarum, Str. thermophilus* **Yeasts**: *Ca. albicans, Ca. famata, S. cerevisiae, Tr. mucoides*	[87]
Andean chicha	Cereals	**Yeasts**: *H. guiermondii, H. opuntiae, H. uvarum, Ko. ohmeri, R. slooffiae, Mz. guillermondii, Pi. kluyveri, Pi. kudriavzevii, R. mucilaginosa, Wi. anomalus, S. cerevisiae, Y. lipolytica*	[88]
Atole agrio	Maize	**LAB**: *Ag. composti, E. hirae, La. casei, La. paracasei, La. rhamnosus, Lc. lactis, Lc. piscium, Li. aviarius, Ln. garlicum, Ln. mesenteroides, Ln. pseudomesenteroides, Lo. coryniformis, Lp. fabifermentans, Lp. paraplantarum, Lp. pentosus, Lp. plantarum, Lt. curvatus, Lv. brevis, P. pentosaceus, P. stilesii, W. cibaria, W. confusa, W. hellenica, W. paramesenteroides, Str. equinus* **AAB**: *A. estunensis, A. indonesiensis, A. pasteurianus, A. tropicalis, Gb. frateurii*	[89]
Bacaba chicha	*Oenocarpus bacaba*	**LAB**: *E. durans, E. hirae, Ln. lactis* **Yeasts**: *Pi. caribbica, Pi. guillermondii*	[90]
Bhaati jaanr	Rice	**LAB**: *Lo. bifermentans, P. pentosaceus* **Yeasts**: *Ca. glabrata, Pi. anomala, S. cerevisiae, Sp. fibuligera*	[91]
Bili bili	Sorghum	**Yeasts**: *Ca. melibiosica, Cr. albidius, D. hansenii, De. bruxelensis, K. marxianus, R. mucilaginosa, S. cerevisiae, T. delbrueckii*	[92]
Borde	Cereals	**LAB**: *Lv. brevis, P. pentosaceus, W. confusa, W. viridescens*	[93]

Table 1. *Cont.*

Product	Main Ingredient	Microbiota	References
Boza	Cereals	**LAB**: *Fr. sanfransiscensis, La. casei, La. paracasei, Lb. acidophilus, Lc. lactis, Li. salivarius, Lm. fermentum, Ln. amelibiosum, Ln. mesenteroides, Ln. paramesenteroides, Ln. pseudomesenteroides, Lo. coryniformis, Lp. plantarum, Lt. buchneri, Lt. parabuchneri, Lt. sakei, P. parvulus, W. confusa* **Yeasts**: *Ca. glabrata, Ca. tropicalis, Pi. fermentans, Pi. guillermondii, Pi. norvegensis, S. cerevisiae, S. uvarum*	[94–96]
Burukutu	Sorghum	**LAB**: *Lb. acidopilus, Lc. lactis, Lm. fermentum, Lp. plantarum, Lv. brevis* **Yeasts**: *S. cerevisiae*	[97]
Chicha	Maize	**LAB**: *Lm. fermentum, Lp. plantarum, W. cibaria, Str. alactolyticus, Str. luteciae* **AAB**: *A. okinawensis*	[98]
Ikigage	Sorghum	**Yeasts**: *Ca. humilis, Ca. inconspicua, Ca. magnolia, S. cerevisae, I. orientalis* **LAB**: *Le. buchneri, Lm. fermentum*	[99]
Gilaburu	European cranberry	**LAB**: *La. casei, La. pantheris, Le. buchneri, Le. parabuchneri, Ln. pseudomesenteroides, Lp. plantarum, Lv. brevis, Sc. harbinensis*	[11]
Mahewu	Cereals	**LAB**: *E. hermanniensis, E. lactis, Fu. rossiae, Lc. lactis, Lm. fermentum, Ln. holzapfelii, Ln. pseudomesenteroides, Lp. plantarum, P. pentosaceus, W. cibaria, W. confusa* **Yeasts**: *Ca. glabrata, S. cerevisiae*	[100]
Fermented masau	*Ziziphus mauritiana*	**LAB**: *Cb. divergens, Le. hilgardii, Li. agilis, Lo. bifermentans, Lm. fermentum, Lp. plantarum, W. minor* **Yeasts**: *Ca. glabrata, H. opuntiae, I. orientalis, Pi. fabianii, S. cerevisiae, Sp. fibuligera*	[7]
Pito	Cereals	**Yeasts**: *Ca. tropicalis, Ha. anomala, K. africanus, Kc. apiculata, S. cerevisiae, Sz. pombe, T. delbrueckii*	[1]
Pozol	Maize	**LAB**: *C. alimentarius, E. saccharolyticus, La. casei, Lb. delbrueckii, Lc. lactis, Lm. fermentum, Lp. plantarum, Str. bovis, Str. suis* **Yeasts**: *Ca. guilliermondii, Cs. cladosporioides, D. hansenii, Ge. candidum, K. lactis, Pe. fellutanum, Ph. fimeti, Ph. glomerata, R. minuta, R. mucilaginosa*	[2,101,102]
Pulque	*Agave* spp.	**LAB**: *Fr. sanfranciscensis, Lb. acetotolerans, Lb. acidophilus, Lb. delbrueckii, Lc. lactis, Le. hilgardii, Le. kefiri, Ln. citreum, Ln. gasocomitatum, Ln. kimchi, Ln. mesenteroides, Ln. pseudomesenteroides, Lp. plantarum, P. urinaeequi, Se. paracollinoides, Str. deviesei* **AAB**: *A. aceti, A. malorum, A. orientalis, A. pomorum, Gb. oxydans* **Yeasts**: *Ca. parapsilosis, Ca. valida, Cl. lusitaniae, D. carsonii, H. uvarum, Ge. candidum, K. lactis, K. marxianus, Pi. guilliermondii, Pi. membranifaciens, R. mucilaginosa, S. bayanus, S. cerevisiae, S. pastorianus, T. delbrueckii*	[103–106]

AAB: *Acetic acid bacteria*; LAB: *lactic acid bacteria*. *A.: Acetobacter; Ag.: Agrilactobacillus; Au.: Aureobasidium; Br.: Brettanomyces; C.: Companilactobacillus; Ca.: Candida; Cb.: Carnobacterium; Cl.: Clavispora; Cr.: Cryptococcus; Cs.: Cladosporium; D.: Debaryomyces; De.: Dekkera; E.: Enterococcus; Fr.: Fructilactobacillus; Fu.: Furfurilactobacillus; Ga: Gluconacetobacter; Gb., Gluconobacter; Ge.: Geotrichum; I.: Issatschenkia; Kb.: Komagataeibacter; H.: Hanseniaspora; Ha.: Hansenula; K.: Kluyveromyces; Kc.: Kloeckera; Ko.: Kodamaea; Kz.: Kazachstania; La.: Lacticaseibacillus; Lb.: Lactobacillus; Lc.: Lactococcus; Le.: Lentilactobacillus; Lh., Lachancea; Li.: Ligilactobacillus; Lm.: Limosilactobacillus; Ln.: Leuconostoc; Lo.: Loigolactobacillus; Lp.: Lactiplantibacillus; Lq. Liquorilactobacillus; Lt.: Latilactobacillus; Lv.: Levilactobacillus; M.: Metschnikowia; Mz.: Meyerozyma; O.: Oenococcus; P.: Pediococcus; Pe.: Penicillium; Ph.: Phoma; Pi.: Pichia; R.: Rhodotorula; S.: Saccharomyces; Sc.: Schleiferilactobacillus; Sd.: Saccharomycodes; Se.: Secundilactobacillus; Sp.: Saccharomycopsis; Str.: Streptococcus; Sz.: Schizosaccharomyces; T.: Torulospora; Tr.: Trichosporon; W.: Weissella; Wi.: Wickeramomyces; Y.: Yarrowia; Z.: Zygosaccharomyces; Za.: Zygoascus; Zt.: Zygotorulaspora.*

3. Functional Properties of Fermented Beverages

The consumption of fermented beverages has been associated with a series of functional properties. These have been attributed to the chemical composition of the raw materials employed and the bioconversions that take place during the fermentation process. Thus, the same functional properties may be assigned to the same or different bioactive compounds, the formation of which depends upon the metabolic capacity of the microorganisms that constitute the driving micro-community. In the next paragraphs, the functional properties of wine, kombucha and kefir, the most studied fermented beverages, are summarized.

3.1. Functional Properties of Wine

A series of functional properties have been associated with wine and are mainly attributed to the phenolic compounds that it contains. The type and amount of phenolic compounds depend upon factors such as grape variety, environmental conditions, agricultural practices and winemaking technology [107]. In general, the major phenolic compounds of wine are distinguished into flavonoids and non-flavonoids. Flavonols (quercetin, kaempferol and myricetin), anthocyanins (cyanin, petunin, peonin and malvin), and flavan-3-ols (catechin, epicatechin, gallocatechin, procyanidins and condensed tannins) belong to the first category, while phenolic acids (hydroxybenzoic and hydroxycinnamic acids), volatile phenols (ethyl phenol, vinyl phenol, guaiacol, etc.), and stilbenes (resveratrol and its polymers) belong to the second one [108,109]. The total amount of flavonoids in white and red wines has been reported to range between 25–30 and 700–1000 mg of gallic acid equivalent (GAE/L), respectively. Catechins and soluble tannins are quantitatively the most abundant classes of compounds in white and red wines, respectively. On the other hand, the total amount of non-flavonoids has been reported to range between 160–260 and 230–500 mg GAE/L in white and red wines, respectively, with cinnamates being, in both cases, the most abundant class of compounds, with approximately 150 mg of GAE/L [108]. The bioavailability of the aforementioned compounds is a key issue as it affects their biological function. In general, wine phenolic compounds are only partially bioavailable, not only because of their chemical structure but also due to the biotransformations that take place during digestion [110]; it has been reported to range between 2–25% [111–113].

Excessive alcohol consumption has been correlated with an increased incidence of disease [114]. On the contrary, moderate alcohol consumption, particularly wine, seems to have a protective role on human health [115]. The concept of moderate or low-risk wine consumption has been exhaustively debated [115,116]. From a quantitative point of view, the consumption of up to 25–40 g alcohol per day for males and 13–25 g for females is generally accepted as moderate [117]. A range of health benefits have been associated with moderate wine consumption, such as lowering the risk of cardiovascular disease and neurodegenerative disease development, protection against type 2 diabetes, and generally life span prolongation. The capacity of wine to confer these health benefits has been assessed through in vitro, in vivo and clinical studies. The next paragraphs focus on the underlying mechanisms that validate the latter.

An association between moderate wine consumption and reduced risk of cardiovascular events, even among persons with established heart diseases, has been reported [15,118]. This has been attributed to the modulation of circulating cholesterol and anti-platelet activity of alcohol and to the antioxidant, anti-inflammatory and anti-platelet activities of the phenolic compounds [119,120]. The antioxidant activity is expressed through free radical scavenging and through the upregulation of Nrf2, which in turn induces antioxidant gene expression [121]. Regarding their anti-inflammatory activities, a series of mechanisms have been proposed, such as switching off the NF-κB pathway [122], blocking oxysterol-related NOX1 induction [123], suppression of NLRP3 inflammasome activation [124], suppression of the JAK/STAT inflammatory pathway and modulation of Nrf2 activity [125], as well as decrease in IL-1β, IL-6 and IL-8 secretion [126,127]. Finally, the capacity of red wine to inhibit thrombin, ADP- and PAF-induced platelet aggregation has been reported [128–133] and attributed to ethanol and polyphenols, particularly quercetin, tyrosol and trans-resveratrol [132,133]. In addition, the inhibition of PAF biosynthesis by tyrosol and resveratrol has also been reported in U-937 cells under inflammatory conditions [134].

An association between moderate wine consumption and reduction in the risk of neurodegenerative diseases has also been developed [17]. The mode through which wine consumption may affect the onset of Alzheimer's and Parkinson's diseases has been extensively studied. In the first case, protection may take place through the antioxidant and anti-inflammatory activities of wine, as well as through more specific functions such as the modulation of secretase enzymes, the enhancement of amyloid clearance, the inhibition of amyloid aggregation and the prevention of tau protein hyperphosphorylation [135–137]. Indeed, the activation of α-secretase activity by 6% Cabernet Sauvignon, myricetin, quercetin, resveratrol, and caffeic acid [138–143], along with the inhibition of BACE1 activity by resveratrol and some of its oligomers, epicatechin, myricetin, quercetin, kaempherol and caffeic acid [142–154] and inhibition of γ-secretase by resveratrol, oxy-resveratrol, and piceatannol [155,156], have been reported. In addition, amyloid accumulation and aggregation seem to be prevented by a variety of mechanisms. Amyloid clearance or the induction of degradation mechanisms, activities that have been reported for resveratrol [157–159] and quercetin [160], have been reported to prevent amyloid accumulation. On the other hand, resveratrol, quercetin, and grape seed pro-anthocyanidin consisting of catechin, epicatechin, and epicatechin gallate, have presented an anti-aggregation capacity [161–165]. Finally, the inhibition of tau protein hyperphosphorylation has been reported for resveratrol, quercetin and caffeic acid [166–169]. Regarding Parkinson's disease, protection may take place through the antioxidant activity and neuroprotective effects of resveratrol, which seem to be related to its SIRT-activating potential [170–174], and quercetin, which seems to be related to the induction of the PKD1/CREB/BDNF axis [175,176]. In addition, other constituents such as caffeic acid, gallic acid and catechins have also been reported to contribute to the aforementioned activities [177–179].

The association between wine consumption and a reduced risk of type 2 diabetes has been repeatedly reported [19,20]. This association was further improved by Ma et al. [180], which highlighted that this protective action takes place when moderate alcohol drinking, especially wine, takes place with meals. The mode of action includes the inhibition of α-amylase and α-glucosidase activities, the inhibition of sodium-dependent glucose transporter 1 (SGLT1) and the activation of 5-adenosine monophosphate-activated protein kinase (AMPK) [181–184]. Regarding diabetic patients, wine consumption has been reported to attenuate insulin resistance, with no effect on vascular reactivity and nitric oxide production [185], to reduce the diastolic blood pressure and total cholesterol but not glucose parameters and other cardiovascular risk factors [186], and to reduce the risks of cardiovascular events and all-cause mortality [187]. In addition, the initiation of red wine consumption has been associated with increased high-density lipoprotein cholesterol (HDL-C) and apolipoprotein(a)1 level and decreased the ratio of total cholesterol to HDL-C [188]. In addition, wine consumption improved glycemic control, in terms of fasting plasma glucose, homeostatic model assessment of insulin resistance and hemoglobin A1c, but only in patients carrying the alcohol dehydrogenase alleles [ADH1B*1], i.e., the slow ethanol metabolizers.

Several other health benefits have been correlated with moderate wine consumption, such as the decrease in systolic blood pressure [21–23], the improvement of gastrointestinal tract function [24–26] and the improvement of the main symptoms of fibromyalgia [27]. However, further research is still necessary in order to identify the responsible molecular mechanisms.

3.2. Functional Properties of Kombucha

Kombucha is commercially available as a non-alcoholic beverage; therefore, the ethanol content should not exceed 0.5% alcohol by volume (ABV). However, in many cases, this limit is not respected [189–191] and the production of kombucha with ethanol content as high as 5.83 mg/mL has been reported [49]. A series of functional properties, such as antioxidant, antiproliferative, immunomodulatory, antihypertensive, antidiabetic, hypocholesterolemic, hepatoprotective and antimicrobial, have been attributed to kombucha. In all cases, the functional properties have been attributed, at least partially, to specific compounds that are either present in the raw materials or are formed during fermentation. Therefore, every factor that affects the production of the raw materials or the fermentation procedure is expected to affect the functional properties of the final product, to a greater or lesser degree [32,192–196]. Below, a short description of studies assessing these properties is offered.

The antioxidant capacity of kombucha has been principally attributed to the presence of phenolic compounds. The phenolic concentration and diversity in black-tea kombucha have been reported to be greater than those of green-tea kombucha [35]. This has been attributed to the processing steps that are necessary for black tea production. During these steps, the concentration of theaflavins and thearubigins increases, and they become the main polyphenols in black tea [197]. During fermentation, these compounds are subjected to enzymatic or chemical biotransformation, resulting in the formation of a wealth of lower-molecular-weight phenolic compounds. Indeed, Cardoso et al. [35] identified 126 phenolic compounds in the samples of green and black-tea kombucha that they analyzed, of which, 50 compounds were common, 75 were unique to black-tea kombucha, and only one, namely verbascoside, was unique to green-tea kombucha. Interestingly, the occurrence of five phenolic compounds in green-tea kombucha and 30 phenolic compounds in black-tea kombucha was solely assigned to the fermentation process.

The antioxidant activity of kombucha is usually assessed in vitro through free-radical scavenging assays such as DPPH, FRAP, ABTS, MCA, Curpac, etc. [35,198,199]. In vivo studies are generally lacking; only a few are currently available, the main findings of which are described in the following lines. Dipti et al. [200] used a lead acetate solution to induce oxidative stress on male albino (Sprague Dawley) rats and studied the antioxidant effect of black-tea kombucha. The results of kombucha oral administration included the reduction of DNA damage and lipid peroxidation, as well as the increase of glutathione level and GPx activity. Yang et al. [201] used mice from the Institute of Cancer Research and fed them with a hypocholesterolemic diet (HCD) combined with 66 mL Kg^{-1} DW of traditional kombucha tea (TKT) or modified kombucha tea (MKT) (sweetened black tea fermented with *Gluconoacetobacter* sp. strain A4) or 60 mg Kg^{-1} DW of D-saccharic acid-1,4-lactone (DSL) for 12 weeks. The antioxidant activity, measured as total antioxidant capacity (TAOC), superoxide dismutase (SOD) and malonaldehyde (MDA) was assessed in the serum after the end of the 12 weeks of the treatment. The hypocholesterolemic diet resulted in a statistically significant decrease in TAOC and SOD and an increase in MDA values. These values were restored when the HCD was supplemented with TKT, MKT, or DSL. Vazquez-Cabral et al. [202] measured the antioxidant activities of kombucha and a kombucha analog (KAO) made of *Quercus resinosa* leaves against the oxidative damage caused by H_2O_2 in activated THP-1 human monocyte cells and reported the capacity of KAO to decrease oxidative stress. Finally, Gaggia et al. [49] studied the capacity of kombucha made from *Aspalathus linearis* leaves, fermented for 7 and 14 days, to decrease oxidative stress in L929 mouse fibroblasts caused by H_2O_2. When the treatment with the kombucha preceded the H_2O_2 application, the cell viability was partially restored by both products. When the treatment with the kombucha followed the application of H_2O_2, only the kombucha fermented for 14 days was able to restore the viability of the cells.

The antioxidant capacity of kombucha has also been reported to result in hypocholesterolemic and antidiabetic effects. Indeed, kombucha administration has been reported to decrease total and LDL cholesterol in rabbits and mice fed a high-cholesterol diet, as well as attenuate histological effects, such as lesions in the intima [30,34,201–203]. As far as the antidiabetic effect is concerned, this is attributed not only to the antioxidant capacity of kombucha, which addresses the oxidative stress caused by diabetes, but also to the reduction in blood glucose and the increase in plasma insulin, which have been reported as the effects of kombucha administration in experimental rats with induced diabetic consequences [31,33,204,205].

The antiproliferative capacity of kombucha has been demonstrated in vitro using human cancer cell lines. More precisely, the cytotoxicity against lung carcinoma (A549), osteosarcoma (U2OS), renal carcinoma (786-O), ileocecal colorectal adenocarcinoma (HCT8), colorectal adenocarcinoma (CACO-2), rhabdomyosarcoma (RD), cervix carcinoma (Hep2c), prostate cancer (PC-3), colon cancer (HCT-116), breast cancer (MCF-7) and a murine fibroblast (L2OB) has been exhibited [32,35,206,207]. Jayabalan et al. [206] proposed that Dimethyl 2-(2-hydroxy-2-methoxypropylidene) malonate and vitexin may contribute to these cytotoxic effects. Cardoso et al. [35] attributed the higher antiproliferative capacity of green-tea kombucha, compared to that of black-tea kombucha, to the presence of higher concentrations of catechins and verbascoside, the antitumor activity of which has already been reported [208,209]. However, not all cell lines were affected by black-tea kombucha [210]. In addition, Srihari et al. [211] reported that the survival of the prostate cancer cell line (PC-3) decreased after treatment with lyophilized kombucha extract, most likely due to the downregulation of the angiogenesis-associated genes HIF-1α, VEGF, IL-8, COX-2, MMP-2, and MMP-9. Despite these promising results, clinical studies are still lacking.

Strong indications of the immunomodulatory capacity of kombucha have been reported. More precisely, black-tea kombucha administration in male Swiss albino mice with indomethacin-induced stomach ulceration resulted in effective healing, which was attributed to the antioxidant activity and the reduction of gastric acid secretion [28]. The delay in the onset and severity of experimental autoimmune encephalomyelitis induced in female C57BL/6 mice through black-tea kombucha administration was reported by Marzban et al. [212]. In addition, the suppression of TNF-α and IL-6 levels in lipopolysaccharide-stimulated macrophages, as well as the in vitro inhibition of 5-LOX enzyme activity by black-tea kombucha extract, has also been reported [32,201].

As far as the antihypertensive activity is concerned, this is indicated by the detection of ACE inhibitory capacity. Certain flavonoids [213] with certain structural features [214] have an excellent antihypertensive capacity. The ACE inhibitory activity of green- and black-tea kombucha, as well as a series of analogues, has been reported [37,215].

The hepatoprotective activity of kombucha has been repeatedly exhibited in animal models. Indeed, the protective effect of black-tea kombucha against tertiary butyl hydroperoxide-induced cytotoxicity in murine hepatocytes of male albino Swiss mice, by reducing ROS generation, as well as through the inhibition of glutathione depletion and the attenuation of malonaldehyde levels, was reported by Bhattacharya et al. [216]. Abshenas et al. [29] induced hepatotoxicity in male Balb/c mice through acetaminophen treatment. Kombucha consumption for 7 days before acetaminophen treatment reduced its toxicity through the reduction of the serum aspartate aminotransferase, alanine aminotransferase, lactate dehydrogenase, and alkaline phosphatase levels. In addition, the decrease of histopathological changes, such as hepatocellular glycogen storage degeneration and necrosis, mononuclear cell infiltration in the portal area, dilation of central veins and capillarization, was also reported. Acetaminophen was also used by Wang et al. [160] to induce hepatotoxicity in male ICR mice. The administration of traditional black-tea kombucha as well as kombucha fermented only by *Gluconoacetobacter* sp. strain A4 effectively inhibited the increase of alanine aminotransferase, alkaline phosphatase, triglyceride and malondialdehyde, which were induced by acetaminophen treatment. These positive effects were largely attributed to the D-saccharic acid-1,4-lactone produced by the bacterial strain. Kabiri et al. [217] induced hepatotoxicity in male Wistar rats with thioacetamide and studied the effect of black-tea kombucha. Administration of kombucha for 3 weeks before TAA or after TAA treatment had all biochemical parameters assessed (alanine aminotransferase, aspartate aminotransferase, alkaline phosphatase, lactate dehydrogenase, total, LDL and HDL cholesterol, triglycerides, and bilirubin) at comparable levels to the control group, accompanied by normal histology. Hyun et al. [218] induced hepatic steatosis in male C57BLKS and C57BLKS db/db mice through a methionine/choline-deficient diet. Black-tea kombucha administration was reported to reduce the liver weight/body weight ratio to control-group levels through the reduction of fatty acids uptake and triglyceride synthesis, which were also indicated through the application of reverse-transcription quantitative PCR.

The antimicrobial activity of kombucha, as well as a series of analogues, has been extensively assessed. The methods of choice are diffusion and microdilution. The inhibition of the growth of a series of microorganisms, including molds such as *Aspergillus flavus* and *As. niger*; yeasts such as *Candida albicans*, *Ca. glabrata*, *Ca. krusei*, and *Ca. tropicalis*; Gram-positive bacteria such as *Alicyclobacillus acidoterrestris*, *Bacillus cereus*, *Listeria monocytogenes*, *Micrococcus luteus*, *Staphylococcus aureus*, and *St. epidermidis*; and Gram-negative bacteria such as *Aeromonas hydrophila*, *Agrobacterium tumefaciens*, *Campylobacter jejuni*, *Enterobacter cloacae*, *Escherichia coli*, *Esch. coli* O157:H7, *Haemophilus influenzae*, *Klebsiella pneumoniae*, *Proteus mirabilis*, *Pr. vulgaris*, *Pseudomonas aeruginosa*, *Salmonella* Enteritidis, *Sa.* Typhi, *Sa.* Typhimurium, *Shigella dysenteriae*, *Sh. sonnei*, *Vibrio cholerae*, and *Yersinia enterocolitica* has been reported [35,36,207,219–224].

The ingredients to which this antimicrobial activity has been primarily attributed are organic acids, mainly acetic acid, as well as a series of compounds with antimicrobial capacity, such as catechins, unsaturated lactones, hydroxylactones and verbascoside [225–227]. Qualitative and quantitative differences between different types of kombucha have also been attributed to differences in their chemical composition [35,220,224].

In summary, kombucha seems to possess a wealth of functional properties, which have been assessed either in vitro or using animal models. Further study is still necessary, including clinical trials [228].

3.3. Functional Properties of Kefir

Kefir is another thoroughly studied non-alcoholic beverage. As in the case of kombucha, the limit of 0.5% ABV is not always respected [229], and the production of kefir with as much as 10% ethanol has been reported [230]. Many functional properties have been described for kefir and attributed to its amino acid and peptide content. The type of milk, the proteolytic capacity of the micro-consortium that drives fermentation, and the storage time [231] greatly affect these properties. The most important kefir activities that play an essential role in its functional properties are the antioxidant and anti-inflammatory ones. The first is attributed to a series of enzymatic and non-enzymatic systems. Catalase, superoxide dismutase, and glutathione peroxidase account for the enzymatic ones, while vitamin E and β-carotene—along with peptides resulting from the proteolytic breakdown of casein and amino acids, especially methionine, lysine and tryptophan, which seem to possess higher antioxidant activity than threonine, serine, alanine, valine, isoleucine and phenylalanine—account for the non-enzymatic ones [232–235]. The anti-inflammatory activity of kefir is evidenced through the reduction of the levels of proinflammatory mediators, such as TNF-a and IL-6β, and the increase of the levels of anti-inflammatory cytokines such as IL-10 [236–242]. The bacteria themselves [238], their extracellular vesicles (as in the case of *Lactobacillus kefiranofaciens* subsp. *kefirgranum* PRCC-1301 [243]), their micro integral membrane protein (such as the one of *Lactiplantibacillus plantarum* [244]) as well as kefir peptides [44,245] seem to contribute to this activity.

The association between kefir consumption and effective glycemic control through reduction of IL-1β and increase of IL-10 expression, as well as a reduction of fasting glucose and insulin levels and reduction of insulin resistance, has been reported [39,40,239,246,247]. Although the exact mode of action is still to be elucidated, it includes the release of bioactive peptides from caseins, most likely through the proteolytic activity of the kefir microbiota [247]. These bioactive peptides may also include ACE-inhibitory ones [38,248], which contribute to the effective control of blood pressure. Indeed, a series of studies have highlighted the antihypertensive capacity of kefir and indicated that ACE-inhibitory activity is among the most important mechanisms [249–251]. Kefir has also been reported to have an immunomodulatory capacity [43] and act towards the prevention and treatment of atherosclerosis [44] and liver damage [41], mostly due to its antioxidant and anti-inflammatory activities. In addition, it may also act as a psychobiotic due to the occurrence of LAB capable of producing gamma-aminobutyric acid, a major inhibitory neurotransmitter of the mammalian central nervous system [252], combined with anti-inflammatory activities [253,254].

The anti-carcinogenic capacity of kefir has been assessed in a variety of human cancer cell lines, such as the gastric AGS and SGC7901, mammary MCF-7, myeloid leukemia HL60/AR, colorectal Caco-2 and HT-29, etc. [255–259]. In the majority of cases, inhibition of proliferation and apoptosis induction were reported, mediated by molecular mechanisms that included downregulation of TGF-α and Bcl-2, decreased polarization of mitochondrial membrane potential, as well as the upregulation of TGF-β1, Bax, caspase 3, caspase 8, caspase 9, etc. [42,256–258,260–262].

4. Customization of Fermented Beverages

In the previous paragraphs, the importance of raw materials in terms of their chemical composition, as well as the importance of the microbiota involved in fermentation in terms of metabolic capacity, have been highlighted. The functional properties of fermented beverages are achieved through the combination of these two parameters.

Product customization is as old as fermentation itself. The use of starter cultures is a form of customization, which, before the dawn of microbiology, took place through back-slopping. Nowadays, customization is facilitated by a wealth of information regarding the molecular mechanisms of disease and the mode of action of bioactive compounds. The customization of fermented beverages aims at achieving reproducibility, and therefore, the standardization of the product, and enhancing its functional properties. The first is principally achieved through the use of starter cultures. Indeed, the use of defined monocultures or micro-consortia allows the acceleration of the fermentation procedure, as well as the predictability of the outcome [263]. On the other hand, the enhancement of functional properties has been achieved mostly through the use of alternative and/or supplementary raw materials. In Table 2, representative studies on the customization of fermented beverages are exhibited.

Table 2. Representative studies on customization of fermented beverages.

Product	Customization Strategy-Outcome	References
Fruit-based fermented beverage	Improvement in the antioxidant activity of kiwifruit pulp through fermentation with a *Lp. plantarum* strain. The increase in DPPH and ABTS scavenging activities were correlated with the increase in total phenolic and flavonoid content.	[264]
Fruit-based fermented beverage	Pomegranate juice was fermented with *Lp. paraplantarum* CRL2051 and *Lp. plantarum* CRL2030 and administered to C57BL/6 mice fed a high-fat diet. The fermented juice offered protection against weight gain, liver damage, and dyslipidemia.	[265]
Fermented mango juice	Different mango cultivars were subjected to lactic acid fermentation with two LAB strains, namely *Lp. plantarum* 75 and *Ln. pseudomesenteroides* 56. The latter strain improved the retention of carotenoids, while the former enhanced the phenolic content and the antioxidant activity of all mango cultivars.	[266]
Whey-based fermented beverage	Commercially available probiotic LAB cultures, capable of producing conjugated linoleic acid (CLA), were used to ferment whey that was enriched with walnut oil lipolyzed by endogenous lipases as a source of free linoleic acid. After the optimization of the fermentation conditions, the whey-based beverage, apart from the CLA content, which could reach 36 mg/g fat, also presented a remarkable antioxidant capacity, most likely due to the presence of phenolic compounds and tocopherol in the walnut oil.	[267]
Kombucha analog	A kombucha analog with the use of coffee (*Coffea arabica*) by-product infusion, instead of *Camellia sinensis* infusion, was developed. The antioxidant activity (as estimated through the reduction in intracellular ROS and uric acid concentration in HK-2 model cells) and the anti-inflammatory activity (as estimated through a reduction in NO formation in LPS-induced macrophages of the kombucha analog and black-tea kombucha) were comparable.	[268]
Kombucha analog	Hops (*Humulus lupulus* L.), madimak (*Polygonum cognatum*), and hawthorn (*Crataegus monogyna*) were used to supplement black-tea kombucha, using the same SCOBY. After fermentation, the antioxidant activity of traditional kombucha was higher than that of the ones supplemented with herbs. All kombuchas exhibited comparative antiproliferative capacity against two cancer cell lines, namely HCT116 and Mahlavu.	[269]

Table 2. *Cont.*

Product	Customization Strategy-Outcome	References
Kombucha analog	*Hibiscus sabdarifa* L. leaves and stems were used to develop kombucha analogs. The products exhibited similar antioxidant capacity and no cytotoxicity against noncancer cells.	[270]
Kombucha analog	Infusions of blackcurrant (*Ribes nigrum*), black chokeberry (*Aronia melanocarpa*), and blueberry (*Vaccinium myrtillus*) were fermented using the same SCOBY, resulting in products with a significant content of polyphenolic compounds. The kombucha analogs exhibited significant antioxidant activity, assessed both in vitro and with the use of human keratinocytes (HaCaT) and fibroblasts (BJ).	[271]
Enhanced kombucha	Incorporation of *Echium amoenum* in kombucha resulted in a significant increase in the total phenol, anthocyanin, and flavonoid content, as well as the antioxidant activity. The kombucha prepared solely via *E. amoenum* infusion exhibited enhanced cytotoxicity against the human prostate cancer cell line (PC3) compared to the products containing both black tea and *E. amoenum* infusions.	[272]
Fermented soy beverage	Commercially available probiotic strains *La. rhamnosus* GG and *B. longum* BB536, along with isolates with probiotic potential, namely *B. breve* INIA P734, *B. longum* INIA P132, *La. paracasei* INIA P272 and *La. rhamnosus* INIA P344, were used to ferment a commercially available soy beverage. The product obtained with *La. rhamnosus* GG and *La. rhamnosus* INIA P344 contained high levels of bioactive isoflavone aglycones. The viability of the strains, along with the bioactive compounds, was maintained during refrigerated storage for 28 d.	[273]

B.: Bifidobacterium; La.: Lacticaseibacillus; Ln.: Leuconostoc; Lp.: Lactiplantibacillus.

The employment of alternative raw materials, as partial or complete replacements of traditional ones, has been extensively exercised in the case of kefir and kombucha beverages. In fact, this strategy has been so extensively employed that it has led to a whole new class of products, namely kefir analogs and kombucha analogs. In both cases, the aim was to meet the consumer needs of nutritionally dense and organoleptically appealing products. Especially the latter seems quite a challenge, taking into consideration the varying taste preferences [274]. Therefore, a variety of raw materials have been employed, resulting in variable nutritional and sensorial outcomes. In the case of kefir analogs, the utilization of fruits, vegetables, and sugar solutions is most commonly reported [275–282], while for the production of kombucha analogs, the employment of herbal infusions and fruits is most frequently encountered [283–297]. Apart from using kefir or kombucha cultures to ferment their analogs, the development of fruit- and vegetable-based beverages through the fermentation of substrates not traditionally considered for that purpose, using cultures that could effectively carry out fermentation and enhance their functional potential, has been extensively assessed. Indeed, a series of fruit or vegetable juices and pulps have been subjected to fermentation, principally lactic acid fermentation [298–308]. Apart from the organoleptically interesting products, in the majority of cases, the results exhibited an increase in the concentration of bioactive compounds, such as the total phenolic content, vitamin C, shikimic acid, etc., along with an increase in the antioxidant capacity, which, in some cases, was further verified through in vitro experimentation.

The enhancement of functional properties has also been achieved through the direct addition of compounds or their precursor molecules. An example of the first strategy is the study by Frolova et al. [309], in which pre-dissolved inulin and a vitamin premix consisting of thiamine, riboflavin, pyridoxine, folic acid and niacin were added to the already fermented black-tea kombucha at concentrations corresponding to 100% of the recommended daily intake (RDI) of inulin and 29–44% of the RDI of the vitamins. In addition, an infusion of frozen strawberries and lime leaves was created and added after the primary fermentation, and a secondary one was allowed, at 23 °C, for 24 h. The final product exhibited a 82% DPPH inhibitory activity and it was highly accepted by the sensory evaluation panel. Another example of the direct addition of functional compounds is the study by Shahbazi et al. [292]. In that study, medicinal plants, namely cinnamon, cardamom, and shirazi thyme, were added to the green-tea concoction and allowed to ferment using the same SCOBY. The cinnamon-flavored kombucha exhibited higher antioxidant and antimicrobial activity, as well as better sensorial scores, than the green-tea kombucha that served as the control and the cardamom- and shirazi thyme-flavored ones. Based on these results, cinnamon was used to partially or completely replace green tea. Increasing the cinnamon concentration resulted in an increased total phenolic content and radical scavenging activity and had a variable effect on the minimum inhibitory concentration against the Gram-positive and Gram-negative pathogenic bacteria that were examined; the authors noted that the Gram-negative ones seemed to be more susceptible. Similarly, Ozturk et al. [269] combined black tea leaves with hops (*Humulus lupulus* L.), madimak (*Polygonum cognatum*), or hawthorn (*Crataegus monogyna*) dry leaves and created a concoction that was left to ferment into kombucha. The herbs employed had no additive effect on the antioxidant capacity of the black-tea kombucha that served as the control and all products had a comparable antiproliferative activity against the human colorectal carcinoma cell line HCT116 and the human hepatocellular carcinoma cell line Mahlavu. Both studies, along with many similar ones, can be considered as a non-targeted attempt to improve the functionality of the final product. The term 'non-targeted' is used to highlight that the aim of the studies was to enhance the total phenolic content, and therefore, the antioxidant capacity, and not the concentration of a specific compound. In the case of kombucha, an example of such a compound would be epigallocatechin gallate (ECGC), the most studied bioactive compound of green tea. There is a strong indication that ECGC exhibits significant antiproliferative and antihypertensive activity through a variety of mechanisms [310,311]. Therefore, targeting the increase in this compound would create a product with specific capacities. However, research is still necessary in order to verify these actions and elucidate the underlying molecular mechanisms. This targeted approach was employed by Moslemi et al. [267], aiming to enhance the conjugated linoleic acid (CLA) concentration of a whey-based beverage. CLA is a group of linoleic acid isomers, the consumption of which has been correlated with a series of health benefits [312,313]. In that study, walnut oil that was already lipolyzed by endogenous lipases was added to a whey-based formulation, homogenized, and allowed to ferment with commercially available starter and probiotic cultures. The lipolysis of the walnut oil was necessary in order to liberate the esterified linoleic acid and thus enable the microorganisms to use it as a precursor for conjugated linoleic acid synthesis. Although the maximum produced amount of 36 mg/g of fat does not meet the recommended daily intake, this study proved that supplementation with substrates used by microorganisms for the production of bioactive compounds is an effective strategy and definitely worth further assessment.

Finally, the valorization of market surplus food, especially bread, into fermented beverages with functional potential has also been considered. Indeed, Massa et al. [314] reported the development of a non-alcoholic beverage using *Saccharomyces bayanus* 995, a SCOBY, or water kefir grains. The authors proposed a saccharification pre-treatment with *Aspergillus oryzae* and the supplementation of the thermally treated infusion with 1% *w/v* multiflora honey. The final product was sensorially evaluated, and the beverage prepared with *S. bayanus* was more preferred. On the other hand, Nguyen et al. [315] inoculated a sterilized slurry made after the homogenization of finely cut bread and water with *La. rhamnosus* GG and/or *S. cerevisiae* CNCM I-3856. Before sterilization, the addition of commercially available zero-calorie sweetener mix and a stabilizer took place. After fermentation at 37 °C for 72 h, the beverage fermented with a consortium of both strains contained the highest amount of amino acids, such as leucine, valine, glycine and GABA, throughout storage at 5 and 30 °C for 6 w. More recently, Siguenza-Andres et al. [316] applied desalting and treatment with a-amylase and glucoamylase to dried and milled surplus bread before inoculation with *La. rhamnosus* GG or a microconsortium consisting of *Bifidobacterium* sp., *Lb. delbrueckii* subsp. *bulgaricus*, and *Streptococcus thermophilus*. Fermentation took place at 38 °C for 24 h. The authors reported that the enzyme treatment allowed faster acidification to occur, whereas desalting restricted the maximum rates of growth, pH reduction and acidification.

5. Conclusions

Fermented beverages have a long tradition and a very promising future due to their great capabilities, spanning from their capacity to fit into the mentality of subsequent generations, including the current 'on the go' generation, to their customization potential. Especially regarding the latter, the wide range of raw materials that can be used, combined with the metabolic potential of food-grade microorganisms, can give birth to customized products that meet the extensive range of organoleptic preferences and enable targeted nutritional interventions. Although a lot of research is still necessary, this exciting future seems to be within reach.

Author Contributions: Conceptualization, S.P. and M.D.; investigation, S.P., M.D., J.K.P. and Y.K.; resources, S.P., M.D., J.K.P. and Y.K.; data curation, S.P., M.D., J.K.P. and Y.K.; writing—original draft preparation, S.P., M.D., J.K.P. and Y.K.; writing—review and editing, S.P., M.D., J.K.P. and Y.K. All authors have read and agreed to the published version of the manuscript.

Funding: This research received no external funding.

Institutional Review Board Statement: Not applicable.

Informed Consent Statement: Not applicable.

Data Availability Statement: The data presented in this study are available in the manuscript.

Conflicts of Interest: The authors declare no conflicts of interest.

References

1. Sefa-Dedeh, S.; Sanni, A.I.; Tetteh, G.; Sakyi-Dawson, E. Yeasts in the traditional brewing of pito in Ghana. *World J. Microbiol. Biotechnol.* **1999**, *15*, 593–597. [CrossRef]
2. Wacher-Rodarte, C.; Canas, A.; Barzana, E.; Lappe, P.; Ulloa, M.; Owens, J.D. Microbiology of Indian and Mestizo pozol fermentations. *Food Microbiol.* **2000**, *17*, 251–256. [CrossRef]
3. Van der Aa Kuhle, A.; Jespersen, L.; Glover, R.L.; Diawara, B.; Jakobsen, M. Identification and characterization of *Saccharomyces cerevisiae* strains isolated from West African sorghum beer. *Yeast* **2001**, *18*, 1069–1079. [CrossRef]
4. Osimani, A.; Garofalo, C.; Aquilanti, L.; Milanovic, V.; Clementi, F. Unpasteurised commercial boza as a source of microbial diversity. *Int. J. Food Microbiol.* **2015**, *194*, 62–70. [CrossRef] [PubMed]
5. Binitu Worku, B.; Gemede, H.F.; Woldegiorgis, A.Z. Nutritional and alcoholic contents of cheka: A traditional fermented beverage in Southwestern Ethiopia. *Food Sci. Nutr.* **2018**, *6*, 2466–2472. [CrossRef]
6. Wang, P.; Wu, J.; Wang, T.; Zhang, Y.; Yao, X.; Li, J.; Wang, X.; Lu, X. Fermentation process optimization, chemical analysis, and storage stability evaluation of a probiotic barley malt kvass. *Bioprocess Biosyst. Eng.* **2022**, *45*, 1175–1188. [CrossRef]
7. Nyanga, L.K.; Nout, M.J.R.; Gadaga, T.H.; Theelen, B.; Boekhout, T.; Zwietering, M.H. Yeasts and lactic acid bacteria microbiota from masau (*Ziziphus mauritiana*) fruits and their fermented fruit pulp in Zimbabwe. *Int. J. Food Microbiol.* **2007**, *120*, 159–166. [CrossRef] [PubMed]
8. Valles, B.S.; Bedrinana, R.P.; Tascon, N.F.; Simon, A.Q.; Madrera, R.R. Yeast species associated with the spontaneous fermentation of cider. *Food Microbiol.* **2007**, *24*, 25–31. [CrossRef]
9. Duarte, W.F.; Dias, D.R.; de Melo Pereira, G.V.; Gervásio, I.M.; Schwan, R.F. Indigenous and inoculated yeast fermentation of gabiroba (*Campomanesia pubescens*) pulp for fruit wine production. *J. Ind. Microbiol. Biotechnol.* **2009**, *36*, 557–569. [CrossRef]
10. Stringini, M.; Comitini, F.; Taccari, M.; Ciani, M. Yeast diversity during tapping and fermentation of palm wine from Cameroon. *Food Microbiol.* **2009**, *26*, 415–420. [CrossRef]
11. Sagdic, O.; Ozturk, I.; Yapar, N.; Yetim, H. Diversity and probiotic potentials of lactic acid bacteria isolated from Gilaburu, a traditional Turkish fermented European cranberrybush (*Viburnum opulus* L.) fruit drink. *Food Res. Int.* **2014**, *64*, 537–545. [CrossRef] [PubMed]
12. Marsh, A.J.; O'Sullivan, O.; Hill, C.; Ross, R.P.; Cotter, P.D. Sequence-based analysis of the bacterial and fungal compositions of multiple kombucha (tea fungus) samples. *Food Microbiol.* **2013**, *38*, 171–178. [CrossRef] [PubMed]
13. Ramadani, A.S.; Abulreesh, H.H. Isolation and identification of yeast flora in local kombucha sample: Al Nabtah. *J. App. Sci.* **2010**, *2*, 42–51.
14. Shiby, V.K.; Mishra, H.N. Fermented milks and milk products as functional foods—A review. *Crit. Rev. Food Sci. Nutr.* **2013**, *53*, 482–496. [CrossRef] [PubMed]

15. Levantesi, G.; Marfisi, R.; Mozaffarian, D.; Franzosi, M.G.; Maggioni, A.; Nicolosi, G.L.; Schweiger, C.; Silletta, M.; Tavazzi, L.; Tognoni, G.; et al. Wine consumption and risk of cardiovascular events after myocardial infarction: Results from the GISSI-Prevenzione trial. *Int. J. Cardiol.* **2013**, *163*, 282–287. [CrossRef] [PubMed]

16. Chiva-Blanch, G.; Badimon, L. Benefits and risks of moderate alcohol consumption on cardiovascular disease: Current findings and controversies. *Nutrients* **2020**, *12*, 108. [CrossRef]

17. Luceron-Lucas-Torres, M.; Cavero-Redondo, I.; Martinez-Vizcaino, V.; Saz-Lara, A.; Pascual-Morena, C.; Alvarez-Bueno, C. Association between wine consumption and cognitive decline in older people: A systematic review and meta-analysis of longitudinal studies. *Front. Nutr.* **2022**, *9*, 863059. [CrossRef]

18. Restani, P.; Fradera, U.; Ruf, J.-C.; Stockley, C.; Teissedre, P.L.; Biella, S.; Colombo, F.; Di Lorenzo, C. Grapes and their derivatives in modulation of cognitive decline: A critical review of epidemiological and randomized-controlled trials in humans. *Crit. Rev. Food Sci. Nutr.* **2021**, *61*, 566–576. [CrossRef]

19. Vazquez-Ruiz, Z.; Martinez-Gonzalez, M.A.; Vitelli-Storelli, F.; Bes-Rastrollo, M.; Basterra-Gortari, F.J.; Toledo, E. Effect of dietary phenolic compounds on incidence of type 2 diabetes in the "Seguimiento Universidad de Navarra" (SUN) cohort. *Antioxidants* **2023**, *12*, 507. [CrossRef]

20. Baliunas, D.O.; Taylor, B.J.; Irving, H.; Roerecke, M.; Patra, J.; Mohapatra, S.; Rehm, J. Alcohol as a risk factor for type 2 diabetes: A systematic review and meta-analysis. *Diabetes Care* **2009**, *32*, 2123–2132. [CrossRef]

21. Karatzi, K.N.; Papamichael, C.M.; Karatzis, E.N.; Papaioannou, T.G.; Aznaouridis, K.A.; Katsichti, P.P.; Stamatelopoulos, K.S.; Zampelas, A.; Lekakis, J.P.; Mavrikakis, M.E. Red wine acutely induces favorable effects on wave reflections and central pressures in coronary artery disease patients. *Am. J. Hypertens.* **2005**, *18*, 1161–1167. [CrossRef]

22. Chiva-Blanch, G.; Urpi-Sarda, M.; Ros, E.; Arranz, S.; Valderas-Martinez, P.; Casas, R.; Sacanella, E.; Llorach, R.; Lamuela-Raventos, R.M.; Andres-Lacueva, C.; et al. Dealcoholized red wine decreases systolic and diastolic blood pressure and increases plasma nitric oxide. *Circ. Res.* **2012**, *111*, 1065–1068. [CrossRef]

23. Fantin, F.; Bulpitt, C.J.; Zamboni, M.; Cheek, E.; Rajkumar, C. Arterial compliance may be reduced by ingestion of red wine. *J. Hum. Hypertens.* **2015**, *30*, 68–72. [CrossRef] [PubMed]

24. Queipo-Ortuno, M.I.; Boto-Ordonez, M.; Murri, M.; Gomez-Zumaquero, J.M.; Clemente-Postigo, M.; Estruch, R.; Cardona Diaz, F.; Andres-Lacueva, C.; Tinahones, F.J. Influence of red wine polyphenols and ethanol on the gut microbiota ecology and biochemical biomarkers. *Am. J. Clin. Nutr.* **2012**, *95*, 1323–1334. [CrossRef] [PubMed]

25. Kasicka-Jonderko, A.; Jonderko, K.; Gajek, E.; Piekielniak, A.; Zawislan, R. Sluggish gallbladder emptying and gastrointestinal transit after intake of common alcoholic beverages. *J. Physiol. Pharmacol.* **2014**, *65*, 55–66.

26. Moreno-Indias, I.; Sanchez-Alcoholado, L.; Perez-Martinez, P.; Andres-Lacueva, C.; Cardona, F.; Tinahones, F.; Queipo-Ortuno, M.I. Red wine polyphenols modulate fecal microbiota and reduce markers of the metabolic syndrome in obese patients. *Food Funct.* **2016**, *7*, 1775–1787. [CrossRef]

27. Gonzalez-Lopez-Arza, M.V.; Trivino-Palomo, J.V.; Montanero-Fernandez, J.; Garrido-Ardila, E.M.; Gonzalez-Sanchez, B.; Jimenez-Palomares, M.; Rodriguez-Mansilla, J. Benefits of the light consumption of red wine in pain, tender points, and anxiety in women with fibromyalgia: A pilot study. *Nutrients* **2023**, *15*, 3469. [CrossRef] [PubMed]

28. Banerjee, D.; Hassarajani, S.A.; Maity, B.; Narayan, G.; Bandyopadhyay, S.K.; Chattopadhyay, S. Comparative healing property of kombucha tea and black tea against indomethacin-induced gastric ulceration in mice: Possible mechanism of action. *Food Funct.* **2010**, *1*, 284–293. [CrossRef]

29. Abshenas, J.; Derakhshanfar, A.; Ferdosi, M.H.; Hasanzadeh, S. Protective effect of kombucha tea against acetaminophen-induced hepatotoxicity in mice: A biochemical and histopathological study. *Comp. Clin. Pathol.* **2012**, *21*, 1243–1248. [CrossRef]

30. Bellassoued, K.; Ghrab, F.; Makni-Ayadi, F.; van Pelt, J.; Elfeki, A.; Ammar, E. Protective effect of kombucha on rats fed a hypercholesterolemic diet is mediated by its antioxidant activity. *Pharm. Biol.* **2015**, *53*, 1699–1709. [CrossRef]

31. Hosseini, S.; Gorjian, M.; Rasouli, L.; Shirali, S. A comparison between the effect of green tea and kombucha prepared from green tea on the weight of diabetic rats. *Biosci. Biotechnol. Res. Asia* **2015**, *12*, 141–146. [CrossRef]

32. Villarreal-Soto, S.A.; Beaufort, S.; Bouajila, J.; Souchard, J.P.; Renard, T.; Rollan, S.; Taillandier, P. Impact of fermentation conditions on the production of bioactive compounds with anticancer, anti-inflammatory and antioxidant properties in kombucha tea extracts. *Process Biochem.* **2019**, *83*, 44–54. [CrossRef]

33. Zubaidah, E.; Afagni, C.A.; Kalsum, U.; Srianta, I.; Blanc, P.J. Comparison of in vivo antidiabetes activity of snake fruit kombucha, black tea kombucha and metformin. *Biocatal. Agric. Biotechnol.* **2019**, *17*, 465–469. [CrossRef]

34. Alaei, Z.; Doudi, M.; Setorki, M. The protective role of kombucha extract on the normal intestinal microflora, high-cholesterol diet caused hypercholesterolemia, and histological structures changes in New Zealand white rabbits. *Avicenna J. Phytomed.* **2020**, *10*, 604–614. [PubMed]

35. Cardoso, R.R.; Neto, R.O.; D'Almeida, C.T.S.; Nascimento, T.P.; Pressete, C.G.; Azevedo, L.; Martino, H.S.D.; Cameron, L.C.; Ferreira, M.S.L.; Barros, F.A.R. Kombuchas from green and black teas have different phenolic profile, which impacts their antioxidant capacities, antibacterial and antiproliferative activities. *Food Res. Int.* **2020**, *128*, 108782. [CrossRef]

36. Mizuta, A.G.; de Menezes, J.L.; Dutra, T.V.; Ferreira, T.V.; Castro, J.C.; da Silva, C.A.J.; Pilau, E.J.; Junior, M.M.; Filho, A.A.A. Evaluation of antimicrobial activity of green tea kombucha at two fermentation time points against *Alicyclobacillus* spp. *LWT-Food Sci. Technol.* **2020**, *130*, 109641. [CrossRef]

37. Vitas, J.; Vukmanovic, S.; Cakarevic, J.; Popovic, L.; Malbasa, R. Kombucha fermentation of six medicinal herbs: Chemical profile and biological activity. *Chem. Ind. Chem. Eng. Q.* **2020**, *26*, 157–170. [CrossRef]

38. Ebner, J.; Asçi Arslan, A.; Fedorova, M.; Hoffmann, R.; Küçükçetin, A.; Pischetsrieder, M. Peptide profiling of bovine kefir reveals 236 unique peptides released from caseins during its production by starter culture or kefir grains. *J. Proteom.* **2015**, *117*, 41–57. [CrossRef]

39. Golunuk, S.B.; Oztasan, N.; Sozen, H.; Koca, H.B. Effects of traditional fermented beverages on some blood parameters in aerobic exercises. *Biomed. Res.* **2017**, *28*, 9475–9480.

40. Bellikci-Koyu, E.; Sarer-Yurekli, B.P.; Akyon, Y.; Aydin-Kose, F.; Karagozlu, C.; Ozgen, A.G.; Brinkmann, A.; Nitsche, A.; Ergunay, K.; Yilmaz, E.; et al. Effects of regular kefir consumption on gut microbiota in patients with metabolic syndrome: A parallel-group, randomized, controlled study. *Nutrients* **2019**, *11*, 2089. [CrossRef]

41. Golli-Bennour, E.E.; Timoumi, R.; Annaibi, E.; Mokni, M.; Omezzine, A.; Bacha, H.; Abid-Essefi, S. Protective effects of kefir against deltamethrin-induced hepatotoxicity in rats. *Environ. Sci. Pollut. Res. Int.* **2019**, *26*, 18856–18865. [CrossRef]

42. Riaz Rajoka, M.S.; Mehwish, H.M.; Fang, H.; Padhiar, A.A.; Zeng, X.; Khurshid, M.; He, Z.; Zhao, L. Characterization and anti-tumor activity of exopolysaccharide produced by *Lactobacillus kefiri* isolated from Chinese kefir grains. *J. Funct. Foods* **2019**, *63*, 103588. [CrossRef]

43. Praznikar, Z.J.; Kenig, S.; Vardjan, T.; Cernelic-Bizjak, M.C.; Petelin, A. Effects of kefir or milk supplementation on zonulin in overweight subjects. *J. Dairy Sci.* **2020**, *103*, 3961–3970. [CrossRef] [PubMed]

44. Tung, M.C.; Lan, Y.W.; Li, H.H.; Chen, H.L.; Chen, S.Y.; Chen, Y.H.; Lin, C.C.; Tu, M.Y.; Chen, C.M. Kefir peptides alleviate high-fat diet-induced atherosclerosis by attenuating macrophage accumulation and oxidative stress in ApoE knockout mice. *Sci. Rep.* **2020**, *10*, 8802. [CrossRef]

45. Paramithiotis, S.; Stasinou, V.; Tzamourani, A.; Kotseridis, Y.; Dimopoulou, M. Malolactic fermentation-theoretical advances and practical considerations. *Fermentation* **2022**, *8*, 521. [CrossRef]

46. Belloso-Morales, G.; Hernandez-Sanchez, H. Manufacture of a beverage from cheese whey using a tea fungus fermentation. *Rev. Latinoam. Microbiol.* **2003**, *45*, 5.

47. Amarasinghe, H.; Weerakkody, N.; Waisundara, V.Y. Evaluation of physicochemical properties and antioxidant activities of kombucha "tea fungus" during extended periods of fermentation. *Food Sci. Nutr.* **2018**, *6*, 659–665. [CrossRef] [PubMed]

48. Barbosa, C.D.; Uetanabaro, A.P.T.; Santos, W.C.R.; Caetano, R.G.; Albano, H.; Kato, R.; Cosenza, G.P.; Azeredo, A.; Goes-Neto, A.; Rosa, C.A.; et al. Microbial–physicochemical integrated analysis of kombucha fermentation. *LWT-Food Sci. Tehnol.* **2021**, *148*, 111788. [CrossRef]

49. Gaggia, F.; Baffoni, L.; Galiano, M.; Nielsen, D.S.; Jakobsen, R.R.; Castro-Mejía, J.L.; Bosi, S.; Truzzi, F.; Musumeci, F.; Dinelli, G.; et al. Kombucha beverage from green, black and rooibos teas: A comparative study looking at microbiology, chemistry and antioxidant activity. *Nutrients* **2019**, *11*, 1. [CrossRef]

50. Teoh, A.L.; Heard, G.; Cox, J. Yeast ecology of kombucha fermentation. *Int. J. Food Microbiol.* **2004**, *95*, 119–126. [CrossRef]

51. Coton, M.; Pawtowski, A.; Taminiau, B.; Burgaud, G.; Deniel, F.; Coulloumme-Labarthe, L.; Fall, A.; Daube, G.; Coton, E. Unraveling microbial ecology of industrial-scale kombucha fermentations by metabarcoding and culture-based methods. *FEMS Microbiol. Ecol.* **2017**, *93*, fix048. [CrossRef] [PubMed]

52. Wang, S.; Zhang, L.; Qi, L.; Liang, H.; Lin, X.; Li, S.; Yu, C.; Ji, C. Effect of synthetic microbial community on nutraceutical and sensory qualities of kombucha. *Int. J. Food Sci. Technol.* **2020**, *55*, 3327–3333. [CrossRef]

53. Mayser, P.; Fromme, S.; Leitzmann, G.; Grunder, K. The yeast spectrum of the 'tea fungus kombucha'. *Mycoses* **1995**, *38*, 289–295. [CrossRef]

54. Huang, X.; Xin, Y.; Lu, T. A systematic, complexity-reduction approach to dissect the kombucha tea microbiome. *eLife* **2022**, *11*, e76401. [CrossRef]

55. Angulo, L.; Lopez, E.; Lema, C. Microflora present in kefir grains of the Galician region (North-West of Spain). *J. Dairy Res.* **1993**, *60*, 263–267. [CrossRef] [PubMed]

56. Wyder, M.-T.; Meile, L.; Teuber, M. Description of *Saccharomyces turicensis* sp. nov., a new species from kefyr. *Syst. Appl. Microbiol.* **1999**, *22*, 420–425. [CrossRef]

57. Wyder, M.T.; Spillmann, H.; Puhan, Z. Investigation of the yeast flora in dairy products: A case study of kefyr. *Food Technol. Biotechnol.* **1997**, *35*, 299–304.

58. Assadi, M.; Pourahmad, R.; Moazami, N. Use of isolated kefir starter cultures in kefir production. *World J. Microbiol. Biotechnol.* **2000**, *16*, 541–543. [CrossRef]

59. Garrote, G.L.; Abraham, A.G.; De Antoni, G.L. Chemical and microbiological characterisation of kefir grains. *J. Dairy Res.* **2001**, *68*, 639–652. [CrossRef]

60. Simova, E.; Beshkova, D.; Angelov, A.; Hristozova, T.; Frengova, G.; Spasov, Z. Lactic acid bacteria and yeasts in kefir grains and kefir made from them. *J. Ind. Microbiol. Biotechnol.* **2002**, *28*, 1–6. [CrossRef]

61. Dobson, A.; O'Sullivan, O.; Cotter, P.D.; Ross, R.; Hill, C. High-throughput sequence-based analysis of the bacterial composition of kefir and an associated kefir grain. *FEMS Microbiol. Lett.* **2011**, *320*, 56–62. [CrossRef] [PubMed]

62. Diosma, G.; Romanin, D.E.; Rey-Burusco, M.F.; Londero, A.; Garrote, G.L. Yeasts from kefir grains: Isolation, identification, and probiotic characterization. *World J. Microbiol. Biotechnol.* **2013**, *30*, 43–53. [CrossRef] [PubMed]

63. Chen, Z.; Shi, J.; Yang, X.; Nan, B.; Liu, Y.; Wang, Z. Chemical and physical characteristics and antioxidant activities of the exopolysaccharide produced by Tibetan kefir grains during milk fermentation. *Int. Dairy J.* **2015**, *43*, 15–21. [CrossRef]
64. Londero, A.; Hamet, M.F.; De Antoni, G.L.; Garrote, G.L.; Abraham, A.G. Kefir grains as a starter for whey fermentation at different temperatures: Chemical and microbiological characterisation. *J. Dairy Res.* **2012**, *79*, 262–271. [CrossRef] [PubMed]
65. Hamet, M.F.; Londero, A.; Medrano, M.; Vercammen, E.; Van Hoorde, K.; Garrote, G.L.; Huys, G.; Vandamme, P.; Abraham, A.G. Application of culture-dependent and culture-independent methods for the identification of *Lactobacillus kefiranofaciens* in microbial consortia present in kefir grains. *Food Microbiol.* **2013**, *36*, 327–334. [CrossRef]
66. Korsak, N.; Taminiau, B.; Leclercq, M.; Nezer, C.; Crevecoeur, S.; Ferauche, C.; Detry, E.; Delcenserie, V.; Daube, G. Short communication: Evaluation of the microbiota of kefir samples using metagenetic analysis targeting the 16S and 26S ribosomal DNA fragments. *J. Dairy Sci.* **2015**, *98*, 3684–3689. [CrossRef] [PubMed]
67. Miguel, M.G.C.P.; Cardoso, P.G.; Lago, L.A.; Schwan, R.F. Diversity of bacteria present in milk kefir grains using culture-dependent and culture-independent methods. *Food Res. Int.* **2010**, *43*, 1523–1528. [CrossRef]
68. Leite, A.M.O.; Mayoa, B.; Rachid, C.T.; Peixoto, R.S.; Silva, J.T.; Paschoalin, V.M.F.; Delgado, S. Assessment of the microbial diversity of Brazilian kefir grains by PCR-DGGE and pyrosequencing analysis. *Food Microbiol.* **2012**, *31*, 215–221. [CrossRef]
69. Zanirati, D.F.; Abatemarco, M.; Cicco Sandesb, S.H.; Nicolia, J.R.; Nunes, A.C.; Neumanna, E. Selection of lactic acid bacteria from Brazilian kefir grains for potential use as starter or probiotic cultures. *Anarobe* **2015**, *32*, 70–76. [CrossRef]
70. Magalhães, K.T.; Pereira, G.V.D.M.; Campos, C.R.; Dragone, G.; Schwan, R.F. Brazilian kefir: Structure, microbial communities and chemical composition. *Braz. J. Microbiol.* **2011**, *42*, 693–702. [CrossRef]
71. Jianzhong, Z.; Xiaoli, L.; Hanhu, J.; Mingsheng, D. Analysis of the microflora in Tibetan kefir grains using denaturing gradient gel electrophoresis. *Food Microbiol.* **2009**, *26*, 770–775.
72. Gao, J.; Gu, F.; Abdella, N.H.; Ruan, H.; He, G. Investigation on culturable microflora in Tibetan kefir grains from different areas of China. *J. Food Sci.* **2012**, *77*, 425–433. [CrossRef] [PubMed]
73. Gao, J.; Gu, F.; He, J.; Xiao, J.; Chen, Q.; Ruan, H.; He, G. Metagenome analysis of bacterial diversity in Tibetan kefir grains. *Eur. Food Res. Technol.* **2013**, *236*, 549–556. [CrossRef]
74. Garofalo, C.; Osimani, A.; Milanovic, V.; Aquilanti, L.; De Filippis, F.; Stellato, G.; Di Mauro, S.; Turchetti, B.; Buzzini, P.; Ercolini, D.; et al. Bacteria and yeast microbiota in milk kefir grains from different Italian regions. *Food Microbiol.* **2015**, *49*, 123–133. [CrossRef] [PubMed]
75. Yang, X.J.; Fan, M.T.; Shi, J.L.; Dang, B. Isolation and identification of preponderant flora in Tibetan kefir. *China Brew.* **2007**, *171*, 52–55.
76. Witthuhn, R.C.; Schoeman, T.; Britz, T.J. Characterisation of the microbial population at different stages of kefir production and kefir grain mass cultivation. *Int. Dairy J.* **2005**, *15*, 383–389. [CrossRef]
77. Guzel-Seydim, Z.; Wyffels, J.T.; Seydim, A.C.; Greene, A.K. Turkish kefir and kefir grains: Microbial enumeration and electron microscopic observation. *Int. J. Dairy Technol.* **2005**, *58*, 25–29. [CrossRef]
78. Kesmen, Z.; Kacmaz, N. Determination of lactic microflora of kefir grains and kefir beverage by using culture-dependent and culture-independent methods. *J. Food Sci.* **2011**, *76*, 276–283. [CrossRef]
79. Kok-Tas, T.; Ekinci, F.Y.; Guzele Seydim, Z.B. Identification of microbial flora in kefir grains produced in Turkey using PCR. *Int. J. Dairy Technol.* **2012**, *65*, 126–131.
80. Nalbantoglu, U.; Cakar, A.; Dogan, H.; Abaci, N.; Ustek, D.; Sayood, K.; Can, H. Metagenomic analysis of the microbial community in kefir grains. *Food Microbiol.* **2014**, *41*, 42–51. [CrossRef]
81. Yüksekdag, Z.N.; Beyatli, Y.; Aslim, B. Determination of some characteristic coccoid forms of lactic acid bacteria isolated from Turkish kefirs with natural probiotic. *Food Sci. Technol.* **2004**, *37*, 663–667. [CrossRef]
82. Satora, P.; Tuszynski, T. Biodiversity of yeasts during plum Wegierka Zwykla spontaneous fermentation. *Food Technol. Biotechnol.* **2005**, *43*, 277–282.
83. Chanprasartsuk, O.O.; Prakitchaiwattana, C.; Sanguandeekul, R.; Fleet, G.H. Autochthonous yeasts associated with mature pineapple fruits, freshly crushed juice and their ferments; and the chemical changes during natural fermentation. *Bioresour. Technol.* **2010**, *101*, 7500–7509. [CrossRef] [PubMed]
84. Maragatham, C.; Panneerselvam, A. Isolation, identification and characterization of wine-yeast from rotten papaya fruits for wine production. *Adv. Appl. Sci. Res.* **2011**, *2*, 93–98.
85. Padilla, B.; Garcia- Fernandez, D.; Gonzalez, B.; Izidoro, I.; Esteve-Zarzoso, B.; Beltran, G.; Mas, A. Yeast biodiversity from DOQ priorat uninoculated fermentations. *Front. Microbiol.* **2016**, *7*, 930. [CrossRef] [PubMed]
86. Bougreau, M.; Ascencio, K.; Bugarel, M.; Nightingale, K.; Loneragan, G. Yeast species isolated from Texas High Plains vineyards and dynamics during spontaneous fermentations of Tempranillo grapes. *PLoS ONE* **2019**, *14*, e0216246. [CrossRef]
87. Nyambane, B.; Thari, W.M.; Wangoh, J.; Njage, P.M.K. Lactic acid bacteria and yeasts involved in the fermentation of Amabere Amaruranu, a Kenyan fermented milk. *Food Sci. Nutr.* **2014**, *2*, 692–699. [CrossRef] [PubMed]
88. Grijalva-Vallejos, N.; Aranda, A.; Matallana, E. Evaluation of yeasts from Ecuadorian Chicha by their performance as starters for alcoholic fermentations in the food industry. *Int. J. Food Microbiol.* **2020**, *317*, 108462. [CrossRef]
89. Pérez-Cataluña, A.; Elizaquível, P.; Carrasco, P.; Espinosa, J.; Reyes, D.; Wacher, C.; Aznar, R. Diversity and dynamics of lactic acid bacteria in Atole Agrio, a traditional maize-based fermented beverage from South-Eastern Mexico, analysed by high throughput sequencing and culturing. *Antonie Van Leeuwenhoek* **2017**, *111*, 385–399. [CrossRef]

90. Puerari, C.; Magalhães-Guedes, K.T.; Schwan, R.F. Bacaba beverage produced by Umutina Brazilian Amerindians: Microbiological and chemical characterization. *Braz. J. Microbiol.* **2015**, *46*, 1207–1216. [CrossRef]

91. Tamang, J.; Thapa, S. Fermentation dynamics during production of Bhaati Jaanr, a traditional fermented rice beverage of the Eastern Himalayas. *Food Biotechnol.* **2006**, *20*, 251–261. [CrossRef]

92. Maoura, N.; Mbaiguinam, M.; Nguyen, H.V.; Gaillardin, C.; Pourquie, J. Identification and typing of the yeast strains isolated from Bili Bili, a traditional sorghum beer of Chad. *Afr. J. Biotechnol.* **2005**, *4*, 646–656.

93. Abegaz, K. Isolation, characterization and identification of lactic acid bacteria involved in traditional fermentation of Borde, an Ethiopian cereal beverage. *Afr. J. Biotechnol.* **2007**, *6*, 1469–1478.

94. Hancioglu, O.; Karapinar, M. Microflora of Boza, a traditional fermented Turkish beverage. *Int. J. Food Microbiol.* **1997**, *35*, 271–274. [CrossRef] [PubMed]

95. Moncheva, P.; Chipeva, V.; Kujumdzieva, A.; Ivanova, I.; Dousset, X.; Gocheva, B. The composition of the microflora of Boza, an original Bulgarian beverage. *Biotechnol. Biotechnol. Equip.* **2003**, *17*, 164–168. [CrossRef]

96. Gotcheva, V.; Pandiella, S.S.; Angelov, A.; Roshkova, Z.; Webb, C. Monitoring the fermentation of the traditional Bulgarian beverage boza. *Int. J. Food Sci. Technol.* **2001**, *36*, 129–134. [CrossRef]

97. Atter, A.; Obiri-Danso, K.; Amoa-Awua, W.K. Microbiological and chemical processes associated with the production of Burukutu a traditional beer in Ghana. *Int. Food Res. J.* **2014**, *21*, 1769–1776.

98. Bassi, D.; Orru, L.; Vasquez, J.C.; Cocconcelli, P.S.; Fontana, C. Peruvian chicha: A focus on the microbial populations of this ancient maize-based fermented beverage. *Microorganisms* **2020**, *8*, 93. [CrossRef]

99. Lyumugabe, L.; Kamaliza, G.; Bajyana, E.; Thonart, P. Microbiological and physico-chemical characteristics of Rwandese traditional beer "Ikigage". *Afr. J. Biotechnol.* **2010**, *9*, 4241–4246.

100. Pswarayi, F.; Ganzle, M.G. Composition and origin of the fermentation microbiota of Mahewu, a Zimbabwean fermented cereal beverage. *Appl. Environ. Microbiol.* **2019**, *85*, e03130-18. [CrossRef]

101. ben Omar, N.; Ampe, F. Microbial community dynamics during production of the Mexican fermented maize dough pozol. *Appl. Environ. Microbiol.* **2000**, *66*, 3664–3673. [CrossRef] [PubMed]

102. Escalante, A.; Wacher, C.; Farres, A. Lactic acid bacterial diversity in the traditional Mexican fermented dough pozol as determined by 16S rDNA sequence analysis. *Int. J. Food Microbiol.* **2001**, *64*, 21–31. [CrossRef] [PubMed]

103. Estrada-Godina, A.R.; Cruz-Guerrero, A.E.; Lappe, P.; Ulloa, M.; Garcia-Garibay, M.; Gomez-Ruiz, L. Isolation and identification of killer yeasts from *Agave* sap (aguamiel) and pulque. *World J. Microbiol. Biotechnol.* **2001**, *17*, 557–560. [CrossRef]

104. Lappe-Oliveras, P.; Moreno-Terrazas, R.; Arrizon-Gavino, J.; Herrera-Suarez, T.; Garcia-Mendoza, A.; Gschaedler-Mathis, A. Yeasts associated with the production of Mexican alcoholic nondistilled and distilled Agave beverages. *FEMS Yeast Res.* **2008**, *8*, 1037–1052. [CrossRef] [PubMed]

105. Escalante, A.; Giles-Gomez, M.; Hernandez, G.; Cordova-Aguilar, M.; Lopez Munguia, A.; Gosset, G.; Bolivar, F. Analysis of bacterial community during the fermentation of pulque, a traditional Mexican alcoholic beverage, using a polyphasic approach. *Int. J. Food Microbiol.* **2008**, *124*, 126–134. [CrossRef] [PubMed]

106. Escalante, A.; Rodriguez, M.E.; Martinez, A.; Lopez-Munguia, A.; Bolivar, F.; Gosset, G. Characterization of bacterial diversity in Pulque, a traditional Mexican alcoholic fermented beverage, as determined by 16S rDNA analysis. *FEMS Microbiol. Lett.* **2004**, *235*, 273–279. [CrossRef]

107. Sikuten, I.; Stambuk, P.; Andabaka, Z.; Tomaz, I.; Markovic, Z.; Stupic, D.; Maletic, E.; Kontic, J.K.; Preiner, D. Grapevine as a rich source of polyphenolic compounds. *Molecules* **2020**, *25*, 5604. [CrossRef]

108. Jackson, R.S. *Wine Science, Principles and Applications*, 2nd ed.; Academic Press: San Diego, CA, USA, 2000; pp. 270–332.

109. Rentzsch, M.; Wilkens, A.; Winterhalter, P. Non-flavonoid phenolic compounds. In *Wine Chemistry and Biochemistry*; Moreno-Arribas, M.V., Polo, M.C., Eds.; Springer: New York, NY, USA, 2009; pp. 509–527.

110. Fernandes, I.; Perez-Gregorio, R.; Soares, S.; Mateus, N.; Freitas, V. Wine flavonoids in health and disease prevention. *Molecules* **2017**, *22*, 292. [CrossRef]

111. Manach, C.; Williamson, G.; Morand, C.; Scalbert, A.; Remesy, C. Bioavailability and bioefficacy of polyphenols in humans. I. Review of 97 bioavailability studies. *Am. J. Clin. Nutr.* **2005**, *81*, 230–242. [CrossRef]

112. Tian, X.J.; Yang, X.W.; Yang, X.; Wang, K. Studies of intestinal permeability of 36 flavonoids using Caco-2 cell monolayer model. *Int. J. Pharm.* **2009**, *367*, 58–64. [CrossRef]

113. Xu, C.; Kong, L.; Tian, Y. Investigation of the phenolic component bioavailability using the in vitro digestion/Caco-2 cell model, as well as the antioxidant activity in Chinese red wine. *Foods* **2022**, *11*, 3108. [CrossRef] [PubMed]

114. Poli, A.; Marangoni, F.; Avogaro, A.; Barba, G.; Bellentani, S.; Bucci, M.; Cambieri, R.; Catapano, A.L.; Costanzo, S.; Cricelli, C.; et al. Moderate alcohol use and health: A consensus document. *Nutr. Metab. Cardiovasc. Dis.* **2013**, *23*, 487–504. [CrossRef] [PubMed]

115. Hrelia, S.; Di Renzo, L.; Bavaresco, L.; Bernardi, E.; Malaguti, M.; Giacosa, A. Moderate wine consumption and health: A narrative review. *Nutrients* **2023**, *15*, 175. [CrossRef] [PubMed]

116. Giacosa, A.; Barale, R.; Bavaresco, L.; Faliva, M.A.; Gerbi, V.; La Vecchia, C.; Negri, E.; Opizzi, A.; Perna, S.; Pezzotti, M.; et al. Mediterranean way of drinking and longevity. *Crit. Rev. Food Sci. Nutr.* **2016**, *56*, 635–640. [CrossRef]

117. Poli, A. Is drinking wine in moderation good for health or not? *Eur. Heart J. Suppl.* **2022**, *24*, I119–I122. [CrossRef]

118. Jani, B.D.; McQueenie, R.; Nicholl, B.I.; Field, R.; Hanlon, P.; Gallacher, K.I.; Mair, F.S.; Lewsey, J. Association between patterns of alcohol consumption (beverage type, frequency and consumption with food) and risk of adverse health outcomes: A prospective cohort study. *BMC Med.* **2021**, *19*, 8. [CrossRef]
119. Fragopoulou, E.; Petsini, F.; Choleva, M.; Detopoulou, M.; Arvaniti, O.S.; Kallinikou, E.; Sakantani, E.; Tsolou, A.; Nomikos, T.; Samaras, Y. Evaluation of anti-inflammatory, anti-platelet and anti-oxidant activity of wine extracts prepared from ten different grape varieties. *Molecules* **2020**, *25*, 5054. [CrossRef]
120. Sperkowska, B.; Murawska, J.; Przybylska, A.; Gackowski, M.; Kruszewski, S.; Durmowicz, M.; Rutkowska, D. Cardiovascular effects of chocolate and wine-narrative review. *Nutrients* **2021**, *13*, 4269. [CrossRef]
121. Xu, J.; Yang, Y.; Liu, Y.; Cao, S. Natural Nrf2 activators from juices, wines, coffee, and cocoa. *Beverages* **2020**, *6*, 68.
122. Magrone, T.; Jirillo, E. Polyphenols from red wine are potent modulators of innate and adaptive immune responsiveness. *Proc. Nutr. Soc.* **2010**, *69*, 279–285. [CrossRef]
123. Biasi, F.; Guina, T.; Maina, M.; Cabboi, B.; Deiana, M.; Tuberoso, C.I.; Calfapietra, S.; Chiarpotto, E.; Sottero, B.; Gamba, P.; et al. Phenolic compounds present in Sardinian wine extracts protect against the production of inflammatory cytokines induced by oxysterols in Caco-2 human enterocyte-like cells. *Biochem. Pharmacol.* **2013**, *86*, 138–145. [CrossRef]
124. Chang, Y.-P.; Ka, S.-M.; Hsu, W.-H.; Chen, A.; Chao, L.K.; Lin, C.-C.; Hsieh, C.-C.; Chen, M.-C.; Chiu, H.-W.; Ho, C.-L.; et al. Resveratrol inhibits NLRP3 inflammasome activation by preserving mitochondrial integrity and augmenting autophagy. *J. Cell Physiol.* **2015**, *230*, 1567–1579. [CrossRef] [PubMed]
125. Nunes, C.; Teixeira, N.; Serra, D.; Freitas, V.; Almeida, L.; Laranjinha, J. Red wine polyphenol extract efficiently protects intestinal epithelial cells from inflammation via opposite modulation of JAK/STAT and Nrf2 pathways. *Toxicol. Res.* **2016**, *5*, 53–65. [CrossRef] [PubMed]
126. Chalons, P.; Amor, S.; Courtaut, F.; Cantos-Villar, E.; Richard, T.; Auger, C.; Chabert, P.; Schni-Kerth, V.; Aires, V.; Delmas, D. Study of potential anti-inflammatory effects of red wine extract and resveratrol through a modulation of interleukin-1-beta in macrophages. *Nutrients* **2018**, *10*, 1856. [CrossRef]
127. Cornebise, C.; Perus, M.; Hermetet, F.; Valls-Fonayet, J.; Richard, T.; Aires, V.; Delmas, D. Red wine extract prevents oxidative stress and inflammation in ARPE-19 retinal cells. *Cells* **2023**, *12*, 1408. [CrossRef] [PubMed]
128. Struck, M.; Watkins, T.; Tomeo, A.; Halley, J.; Bierenbaum, M. Effect of red and white wine on serum lipids, platelet aggregation, oxidation products and antioxidants: A preliminary report. *Nutr. Res.* **1994**, *14*, 1811–1819. [CrossRef]
129. Xia, J.; Allenbrand, B.; Sun, G.Y. Dietary supplementation of grape polyphenols and chronic ethanol administration on LDL oxidation and Platelet function in rats. *Life Sci.* **1998**, *63*, 383–390. [CrossRef] [PubMed]
130. Muñoz-Bernal, O.A.; de la Rosa, L.A.; Rodrigo-García, J.; Martínez-Ruiz, N.R.; Sáyago-Ayerdi, S.; Rodriguez, L.; Fuentes, E.; Palomo, I.; Alvarez-Parrilla, E. Phytochemical characterization and antiplatelet activity of Mexican red wines and their by-products. *S. Afr. J. Enol. Vitic.* **2021**, *42*, 77–90. [CrossRef]
131. Fragopoulou, E.; Antonopoulou, S.; Nomikos, T.; Demopoulos, C.A. Structure elucidation of phenolic compounds from red/white wine with antiatherogenic properties. *Biochim. Biophys. Acta* **2003**, *1632*, 90–99. [CrossRef]
132. Fragopoulou, E.; Nomikos, T.; Antonopoulou, S.; Mitsopoulou, C.A.; Demopoulos, C.A. Separation of biologically active lipids from red wine. *J. Agric. Food Chem.* **2000**, *48*, 1234–1238. [CrossRef]
133. Xanthopoulou, M.N.; Asimakopoulos, D.; Antonopoulou, S.; Demopoulos, C.A.; Fragopoulou, E. Effect of Robola and Cabernet Sauvignon extracts on platelet activating factor enzymes activity on U937 cells. *Food Chem.* **2014**, *165*, 50–59. [CrossRef] [PubMed]
134. Vlachogianni, I.C.; Fragopoulou, E.; Stamatakis, G.M.; Kostakis, I.K.; Antonopoulou, S. Platelet activating factor (PAF) biosynthesis is inhibited by phenolic compounds in U-937 cells under inflammatory conditions. *Prostaglandins Other Lipid Mediat.* **2015**, *121*, 176–183. [CrossRef] [PubMed]
135. Silva, P.; Vauzour, D. Wine polyphenols and neurodegenerative diseases: An update on the molecular mechanisms underpinning their protective effects. *Beverages* **2018**, *4*, 96. [CrossRef]
136. Reale, M.; Costantini, E.; Jagarlapoodi, S.; Khan, H.; Belwal, T.; Cichelli, A. Relationship of wine consumption with Alzheimer's disease. *Nutrients* **2020**, *12*, 206. [CrossRef]
137. Chinraj, V.; Raman, S. Neuroprotection by resveratrol: A review on brain delivery strategies for Alzheimer's and Parkinson's disease. *J. Appl. Pharm. Sci.* **2022**, *12*, 1–17. [CrossRef]
138. Porquet, D.; Casadesus, G.; Bayod, S.; Vicente, A.; Canudas, A.M.; Vilaplana, J.; Pelegri, C.; Sanfeliu, C.; Camins, A.; Pallas, M.; et al. Dietary resveratrol prevents Alzheimer's markers and increases life span in SAMP8. *Age* **2013**, *35*, 1851–1865. [CrossRef] [PubMed]
139. Corpas, R.; Grinan-Ferre, C.; Rodriguez-Farre, E.; Pallas, M.; Sanfeliu, C. Resveratrol induces brain resilience against Alzheimer neurodegeneration through proteostasis enhancement. *Mol. Neurobiol.* **2019**, *56*, 1502–1516. [CrossRef]
140. Wang, J.; Ho, L.; Zhao, Z.; Seror, I.; Humala, N.; Dickstein, D.L.; Thiyagarajan, M.; Percival, S.S.; Talcott, S.T.; Pasinetti, G.M. Moderate consumption of Cabernet Sauvignon attenuates a beta neuropathology in a mouse model of Alzheimer's disease. *FASEB J. Off. Pub. Fed. Am. Soc. Exp. Biol.* **2006**, *20*, 2313–2320.
141. Martin-Aragon, S.; Jimenez-Aliaga, K.L.; Benedi, J.; Bermejo-Bescos, P. Neurohormetic responses of quercetin and rutin in a cell line over-expressing the amyloid precursor protein (APPswe cells). *Phytomed. Inter. J. Phytother. Phytopharmacol.* **2016**, *23*, 1285–1294. [CrossRef]

142. Shimmyo, Y.; Kihara, T.; Akaike, A.; Niidome, T.; Sugimoto, H. Multifunction of myricetin on a beta: Neuroprotection via a conformational change of a beta and reduction of a beta via the interference of secretases. *J. Neurosci. Res.* **2008**, *86*, 368–377. [CrossRef]

143. Huang, Y.; Jin, M.; Pi, R.; Zhang, J.; Chen, M.; Ouyang, Y.; Liu, A.; Chao, X.; Liu, P.; Liu, J.; et al. Protective effects of caffeic acid and caffeic acid phenethyl ester against acrolein-induced neurotoxicity in HT22 mouse hippocampal cells. *Neurosci. Lett.* **2013**, *535*, 146–151. [CrossRef] [PubMed]

144. Choi, Y.H.; Yoo, M.Y.; Choi, C.W.; Cha, M.R.; Yon, G.H.; Kwon, D.Y.; Kim, Y.S.; Park, W.K.; Ryu, S.Y. A new specific BACE-1 inhibitor from the stembark extract of *Vitis vinifera*. *Planta Med.* **2009**, *75*, 537–540. [CrossRef] [PubMed]

145. Seino, S.; Kimoto, T.; Yoshida, H.; Tanji, K.; Matsumiya, T.; Hayakari, R.; Seya, K.; Kawaguchi, S.; Tsuruga, K.; Tanaka, H.; et al. Gnetin c, a resveratrol dimer, reduces amyloid-beta 1-42 (Aβ42) production and ameliorates Aβ42-lowered cell viability in cultured SH-SY5Y human neuroblastoma cells. *Biomed. Res.* **2018**, *39*, 105–115. [CrossRef] [PubMed]

146. Hu, J.; Lin, T.; Gao, Y.; Xu, J.; Jiang, C.; Wang, G.; Bu, G.; Xu, H.; Chen, H.; Zhang, Y.W. The resveratrol trimer miyabenol c inhibits beta-secretase activity and beta-amyloid generation. *PLoS ONE* **2015**, *10*, e0115973.

147. Cox, C.J.; Choudhry, F.; Peacey, E.; Perkinton, M.S.; Richardson, J.C.; Howlett, D.R.; Lichtenthaler, S.F.; Francis, P.T.; Williams, R.J. Dietary (-)-epicatechin as a potent inhibitor of βγ-secretase amyloid precursor protein processing. *Neurobiol. Aging* **2015**, *36*, 178–187. [CrossRef]

148. Shimmyo, Y.; Kihara, T.; Akaike, A.; Niidome, T.; Sugimoto, H. Flavonols and flavones as BACE-1 inhibitors: Structure-activity relationship in cell-free, cell-based and in silico studies reveal novel pharmacophore features. *Biochim. Biophys. Acta* **2008**, *1780*, 819–825. [CrossRef] [PubMed]

149. De Simone, A.; Mancini, F.; Real Fernandez, F.; Rovero, P.; Bertucci, C.; Andrisano, V. Surface plasmon resonance, fluorescence, and circular dichroism studies for the characterization of the binding of BACE-1 inhibitors. *Anal. Bioanal.Chem.* **2013**, *405*, 827–835. [CrossRef]

150. Chakraborty, S.; Kumar, S.; Basu, S. Conformational transition in the substrate binding domain of beta-secretase exploited by NMA and its implication in inhibitor recognition: BACE1-myricetin a case study. *Neurochem. Inter.* **2011**, *58*, 914–923. [CrossRef]

151. Jimenez-Aliaga, K.; Bermejo-Bescos, P.; Benedi, J.; Martin-Aragon, S. Quercetin and rutin exhibit antiamyloidogenic and fibril-disaggregating effects in vitro and potent antioxidant activity in APPswe cells. *Life Sci.* **2011**, *89*, 939–945. [CrossRef]

152. Lu, J.; Wu, D.M.; Zheng, Y.L.; Hu, B.; Zhang, Z.F.; Shan, Q.; Zheng, Z.H.; Liu, C.M.; Wang, Y.J. Quercetin activates amp-activated protein kinase by reducing PP2C expression protecting old mouse brain against high cholesterol-induced neurotoxicity. *J. Pathol.* **2010**, *222*, 199–212. [CrossRef]

153. Cheng, J.; Rui, Y.; Qin, L.; Xu, J.; Han, S.; Yuan, L.; Yin, X.; Wan, Z. Vitamin D combined with resveratrol prevents cognitive decline in SAMP8 mice. *Curr. Alzheimer Res.* **2017**, *14*, 820–833. [CrossRef] [PubMed]

154. Eom, T.K.; Ryu, B.; Lee, J.K.; Byun, H.G.; Park, S.J.; Kim, S.K. Beta-secretase inhibitory activity of phenolic acid conjugated chitooligosaccharides. *J. Enzym. Inhib. Med. Chem.* **2013**, *28*, 214–217. [CrossRef] [PubMed]

155. Choi, B.; Kim, S.; Jang, B.-G.; Kim, M.-J. Piceatannol, a natural analogue of resveratrol, effectively reduces beta-amyloid levels via activation of alpha-secretase and matrix metalloproteinase-9. *J. Funct. Foods* **2016**, *23*, 124–134. [CrossRef]

156. Ohta, K.; Mizuno, A.; Ueda, M.; Li, S.; Suzuki, Y.; Hida, Y.; Hayakawa-Yano, Y.; Itoh, M.; Ohta, E.; Kobori, M.; et al. Autophagy impairment stimulates PS1 expression and gamma-secretase activity. *Autophagy* **2010**, *6*, 345–352. [CrossRef]

157. Marambaud, P.; Zhao, H.; Davies, P. Resveratrol promotes clearance of Alzheimer's disease amyloid-beta peptides. *J. Biol. Chem.* **2005**, *280*, 37377–37382. [CrossRef]

158. Harada, N.; Zhao, J.; Kurihara, H.; Nakagata, N.; Okajima, K. Resveratrol improves cognitive function in mice by increasing production of insulin-like growth factor-I in the hippocampus. *J. Nutr. Biochem.* **2011**, *22*, 1150–1159. [CrossRef]

159. Deng, H.; Mi, M.T. Resveratrol attenuates Aβ25-35 caused neurotoxicity by inducing autophagy through the TyrRS-PARP1-SIRT1 signaling pathway. *Neurochem. Res.* **2016**, *41*, 2367–2379. [CrossRef]

160. Wang, D.M.; Li, S.Q.; Wu, W.L.; Zhu, X.Y.; Wang, Y.; Yuan, H.Y. Effects of long-term treatment with quercetin on cognition and mitochondrial function in a mouse model of Alzheimer's disease. *Neurochem. Res.* **2014**, *39*, 1533–1543. [CrossRef]

161. Ribeiro, C.A.; Saraiva, M.J.; Cardoso, I. Stability of the transthyretin molecule as a key factor in the interaction with a-beta peptide–relevance in Alzheimer's disease. *PLoS ONE* **2012**, *7*, e45368. [CrossRef]

162. Espargaro, A.; Ginex, T.; Vadell, M.D.; Busquets, M.A.; Estelrich, J.; Munoz-Torrero, D.; Luque, F.J.; Sabate, R. Combined in vitro cell-based/in silico screening of naturally occurring flavonoids and phenolic compounds as potential anti-Alzheimer drugs. *J. Nat. Prod.* **2017**, *80*, 278–289. [CrossRef]

163. Sabogal-Guaqueta, A.M.; Munoz-Manco, J.I.; Ramirez-Pineda, J.R.; Lamprea-Rodriguez, M.; Osorio, E.; Cardona-Gomez, G.P. The flavonoid quercetin ameliorates Alzheimer's disease pathology and protects cognitive and emotional function in aged triple transgenic Alzheimer's disease model mice. *Neuropharmacology* **2015**, *93*, 134–145. [CrossRef] [PubMed]

164. Ho, L.; Ferruzzi, M.G.; Janle, E.M.; Wang, J.; Gong, B.; Chen, T.Y.; Lobo, J.; Cooper, B.; Wu, Q.L.; Talcott, S.T.; et al. Identification of brain-targeted bioactive dietary quercetin-3-o-glucuronide as a novel intervention for Alzheimer's disease. *FASEB J. Off. Pub. Fed. Am. Soc. Exp. Biol.* **2013**, *27*, 769–781. [CrossRef] [PubMed]

165. Lian, Q.; Nie, Y.; Zhang, X.; Tan, B.; Cao, H.; Chen, W.; Gao, W.; Chen, J.; Liang, Z.; Lai, H.; et al. Effects of grape seed proanthocyanidin on Alzheimer's disease in vitro and in vivo. *Exp. Ther. Med.* **2016**, *12*, 1681–1692. [CrossRef] [PubMed]

166. Jhang, K.A.; Park, J.S.; Kim, H.S.; Chong, Y.H. Resveratrol ameliorates tau hyperphosphorylation at Ser396 site and oxidative damage in rat hippocampal slices exposed to vanadate: Implication of ERK1/2 and GSK-3β signaling cascades. *J. Agric. Food. Chem.* **2017**, *65*, 9626–9634. [CrossRef] [PubMed]

167. Shen, X.Y.; Luo, T.; Li, S.; Ting, O.Y.; He, F.; Xu, J.; Wang, H.Q. Quercetin inhibits okadaic acid-induced tau protein hyperphosphorylation through the Ca^{2+}-calpain-p25-CDK5 pathway in HT22 cells. *Int. J. Mol. Med.* **2018**, *41*, 1138–1146. [CrossRef] [PubMed]

168. Ali, T.; Kim, M.J.; Rehman, S.U.; Ahmad, A.; Kim, M.O. Anthocyanin-loaded PEG-gold nanoparticles enhanced the neuroprotection of anthocyanins in an Aβ$_{1-42}$ mouse model of Alzheimer's disease. *Mol. Neurobiol.* **2017**, *54*, 6490–6506. [CrossRef]

169. Sul, D.; Kim, H.S.; Lee, D.; Joo, S.S.; Hwang, K.W.; Park, S.Y. Protective effect of caffeic acid against beta-amyloid-induced neurotoxicity by the inhibition of calcium influx and tau phosphorylation. *Life Sci.* **2009**, *84*, 257–262. [CrossRef] [PubMed]

170. Okawara, M.; Katsuki, H.; Kurimoto, E.; Shibata, H.; Kume, T.; Akaike, A. Resveratrol protects dopaminergic neurons in midbrain slice culture from multiple insults. *Biochem. Pharmacol.* **2007**, *73*, 550–560. [CrossRef]

171. Wang, H.; Dong, X.; Liu, Z.; Zhu, S.; Liu, H.; Fan, W.; Hu, Y.; Hu, T.; Yu, Y.; Li, Y.; et al. Resveratrol suppresses rotenone-induced neurotoxicity through activation of SIRT1/AKT1 signaling pathway. *Anat. Rec.* **2018**, *301*, 1115–1125. [CrossRef]

172. Ferretta, A.; Gaballo, A.; Tanzarella, P.; Piccoli, C.; Capitanio, N.; Nico, B.; Annese, T.; Di Paola, M.; Dell'aquila, C.; De Mari, M.; et al. Effect of resveratrol on mitochondrial function: Implications in Parkin-associated familiar Parkinson's disease. *Biochim. Biophys. Acta* **2014**, *1842*, 902–915. [CrossRef]

173. Wu, Y.; Li, X.; Zhu, J.X.; Xie, W.; Le, W.; Fan, Z.; Jankovic, J.; Pan, T. Resveratrol-activated AMPK/SIRT1/autophagy in cellular models of Parkinson's disease. *Neurosignals* **2011**, *19*, 163–174. [CrossRef] [PubMed]

174. Guo, Y.J.; Dong, S.Y.; Cui, X.X.; Feng, Y.; Liu, T.; Yin, M.; Kuo, S.H.; Tan, E.K.; Zhao, W.J.; Wu, Y.C. Resveratrol alleviates MPTP-induced motor impairments and pathological changes by autophagic degradation of α-synuclein via SIRT1-deacetylated LC3. *Mol. Nutr. Food Res.* **2016**, *60*, 2161–2175. [CrossRef] [PubMed]

175. Ay, M.; Luo, J.; Langley, M.; Jin, H.; Anantharam, V.; Kanthasamy, A.; Kanthasamy, A.G. Molecular mechanisms underlying protective effects of quercetin against mitochondrial dysfunction and progressive dopaminergic neurodegeneration in cell culture and MitoPark transgenic mouse models of Parkinson's disease. *J. Neurochem.* **2017**, *141*, 766–782. [CrossRef] [PubMed]

176. Zhu, M.; Han, S.; Fink, A.L. Oxidized quercetin inhibits alpha-synuclein fibrillization. *Biochim. Biophys. Acta* **2013**, *1830*, 2872–2881. [CrossRef] [PubMed]

177. Ardah, M.T.; Paleologou, K.E.; Lv, G.; Abul Khair, S.B.; Kazim, A.S.; Minhas, S.T.; Al-Tel, T.H.; Al-Hayani, A.A.; Haque, M.E.; Eliezer, D.; et al. Structure activity relationship of phenolic acid inhibitors of alpha-synuclein fibril formation and toxicity. *Front. Aging Neurosci.* **2014**, *6*, 197. [CrossRef] [PubMed]

178. Fazili, N.A.; Naeem, A. Anti-fibrillation potency of caffeic acid against an antidepressant induced fibrillogenesis of human alpha-synuclein: Implications for Parkinson's disease. *Biochimie* **2015**, *108*, 178–185. [CrossRef] [PubMed]

179. Jimenez-Del-Rio, M.; Guzman-Martinez, C.; Velez-Pardo, C. The effects of polyphenols on survival and locomotor activity in *Drosophila melanogaster* exposed to iron and paraquat. *Neurochem. Res.* **2010**, *35*, 227–238. [CrossRef]

180. Ma, H.; Wang, X.; Li, X.; Heianza, Y.; Qi, L. Moderate alcohol drinking with meals is related to lower incidence of type 2 diabetes. *Am. J. Clin. Nutr.* **2022**, *116*, 1507–1514. [CrossRef]

181. Xia, X.; Sun, B.; Li, W.; Zhang, X.; Zhao, Y. Anti-diabetic activity phenolic constituents from red wine against α-glucosidase and α-amylase. *J. Food Process. Preserv.* **2017**, *41*, e12942. [CrossRef]

182. Welsch, C.A.; Lachance, P.A.; Wasserman, B.P. Dietary phenolic compounds: Inhibition of Na$^+$-dependent d-glucose uptake in rat intestinal brush border membrane vesicles. *J. Nutr.* **1989**, *119*, 1698–1704. [CrossRef]

183. Cermak, R.; Landgraf, S.; Wolffram, S. Quercetin glucosides inhibit glucose uptake into brush-border-membrane vesicles of porcine jejunum. *Br. J. Nutr.* **2004**, *91*, 849–855. [CrossRef] [PubMed]

184. Gasparrini, M.; Giampieri, F.; Suarez, J.M.A.; Mazzoni, L.; Hernandez, T.Y.F.; Quiles, J.L.; Bullon, P.; Battino, M. AMPK as a new attractive therapeutic target for disease prevention: The role of dietary compounds AMPK and disease prevention. *Curr. Drug Tar.* **2016**, *17*, 865–889. [CrossRef] [PubMed]

185. Napoli, R.; Cozzolino, D.; Guardasole, V.; Angelini, V.; Zarra, E.; Matarazzo, M.; Cittadini, A.; Saccà, L.; Torella, R. Red wine consumption improves insulin resistance but not endothelial function in type 2 diabetic patients. *Metabolism* **2005**, *54*, 306–313. [CrossRef]

186. Ye, J.; Chen, X.; Bao, L. Effects of wine on blood pressure, glucose parameters, and lipid profile in type 2 diabetes mellitus. *Medicine* **2019**, *90*, e15771. [CrossRef]

187. Blomster, J.I.; Zoungas, S.; Chalmers, J.; Li, Q.; Chow, C.K.; Woodward, M.; Mancia, G.; Poulter, N.; Williams, B.; Harrap, S.; et al. The relationship between alcohol consumption and vascular complications and mortality in individuals with type 2 diabetes. *Diabetes Care* **2014**, *37*, 1353–1359. [CrossRef]

188. Gepner, Y.; Golan, R.; Harman-Boehm, I.; Henkin, Y.; Schwarzfuchs, D.; Shelef, I.; Durst, R.; Kovsan, J.; Bolotin, A.; Leitersdorf, E.; et al. Effects of initiating moderate alcohol intake on cardiometabolic risk in adults with type 2 diabetes: A 2-year randomized, controlled trial. *Ann. Intern Med.* **2015**, *163*, 569–579. [CrossRef]

189. Jang, S.S.; McIntyre, L.; Chan, M.; Brown, P.N.; Finley, J.; Chen, S.X. Ethanol concentration of Kombucha teas in British Columbia, Canada. *J. Food Prot.* **2021**, *84*, 1878–1883. [CrossRef]

190. Ivory, R.; Delaney, E.; Mangan, D.; McCleary, B.V. Determination of ethanol concentration in Kombucha beverages: Single-laboratory validation of an enzymatic method, first action method 2019.08. *J. AOAC Int.* **2021**, *104*, 422–430. [CrossRef]
191. Rossini, D.; Bogsan, C. Is it possible to brew non-alcoholic Kombucha? Brazilian scenario after restrictive legislation. *Fermentation* **2023**, *9*, 810. [CrossRef]
192. Malbasa, R.V.; Loncar, E.S.; Vitas, J.S.; Canadanovic-Brunet, J.M. Influence of starter cultures on the antioxidant activity of kombucha beverage. *Food Chem.* **2011**, *127*, 1727–1731. [CrossRef]
193. Fu, C.; Yan, F.; Cao, Z.; Xie, F.; Lin, J. Antioxidant activities of kombucha prepared from three different substrates and changes in content of probiotics during storage. *Food Sci. Technol.* **2014**, *34*, 123–126. [CrossRef]
194. Jayabalan, R.; Subathradevi, P.; Marimuthu, S.; Sathishkumar, M.; Swaminathan, K. Changes in free-radical scavenging ability of kombucha tea during fermentation. *Food Chem.* **2008**, *109*, 227–234. [CrossRef] [PubMed]
195. Chu, S.-C.; Chen, C. Effects of origins and fermentation time on the antioxidant activities of kombucha. *Food Chem.* **2006**, *98*, 502–507. [CrossRef]
196. Bortolomedi, B.M.; Paglarini, C.S.; Brod, F.C.A. Bioactive compounds in kombucha: A review of substrate effect and fermentation conditions. *Food Chem.* **2022**, *385*, 132719. [CrossRef]
197. Tanaka, T.; Kouno, I. Oxidation of tea catechins: Chemical structures and reaction mechanism. *Food Sci. Technol. Res.* **2003**, *9*, 128–133. [CrossRef]
198. Xia, X.; Dai, Y.; Wu, H.; Liu, X.; Wang, Y.; Yin, L.; Wang, Z.; Li, X.; Zhou, J. Kombucha fermentation enhances the health-promoting properties of soymilk beverage. *J. Funct. Foods* **2019**, *62*, 103549. [CrossRef]
199. Gulhan, M.F. A new substrate and nitrogen source for traditional kombucha beverage: *Stevia rebaudiana* leaves. *Appl. Biochem. Biotechnol.* **2023**, *195*, 4096–4115. [CrossRef] [PubMed]
200. Dipti, P.; Yogesh, B.; Kain, A.K.; Pauline, T.; Anju, B.; Sairam, M.; Singh, B.; Mongia, S.S.; Kumar, G.I.; Selvamurthy, W. Lead induced oxidative stress: Beneficial effects of kombucha tea. *Biomed. Environ. Sci.* **2003**, *16*, 276–282.
201. Yang, Z.-W.; Ji, B.-P.; Zhou, F.; Li, B.; Luo, Y.; Yang, L.; Li, T. Hypocholesterolaemic and antioxidant effects of kombucha tea in high-cholesterol fed mice. *J. Sci. Food Agric.* **2009**, *89*, 150–156. [CrossRef]
202. Vazquez-Cabral, B.D.; Larrosa-Perez, M.; Gallegos-Infante, J.A.; Moreno-Jimenez, M.R.; Gonzalez-Laredo, R.F.; Rutiaga-Quinones, J.G.; Gamboa-Gomez, C.I.; Rocha-Guzman, N.E. Oak kombucha protects against oxidative stress and inflammatory processes. *Chem. Biol. Interact.* **2017**, *272*, 1–9. [CrossRef]
203. Aloulou, A.; Hamden, K.; Elloumi, D.; Ali, M.B.; Hargafi, K.; Jaouadi, B.; Ayadi, F.; Elfeki, A.; Ammar, E. Hypoglycemic and antilipidemic properties of kombucha tea in alloxan-induced diabetic rats. *BMC Complement. Med. Ther.* **2012**, *12*, 63–72. [CrossRef] [PubMed]
204. Bhattacharya, S.; Gacchui, R.; Sil, P.C. Effect of kombucha, a fermented black tea in attenuating oxidative stress mediated tissue damage in alloxan induced diabetic rats. *Food Chem. Toxicol.* **2013**, *60*, 328–340. [CrossRef] [PubMed]
205. Srihari, T.; Karthikesan, K.; Ashokkumar, N.; Satyanarayana, U. Antihyperglycaemic efficacy of kombucha in streptozotocin-induced rats. *J. Funct. Foods* **2013**, *5*, 1794–1802. [CrossRef]
206. Jayabalan, R.; Chen, P.N.; Hsieh, Y.S.; Prabhakaran, K.; Pitchai, P.; Marimuthu, S.; Yun, S.E. Effect of solvent fractions of kombucha tea on viability and invasiveness of cancer cells-characterization of dimethyl 2-(2-hydroxy-2-methoxypropylidine)malonate and vitexin. *Indian J. Biotechnol.* **2011**, *10*, 75–82.
207. Vitas, J.S.; Cvetanovic, A.D.; Maskovic, P.Z.; Svarc-Gajic, J.V.; Malbasa, R.V. Chemical composition and biological activity of novel types of kombucha beverages with yarrow. *J. Funct. Foods* **2018**, *44*, 95–102. [CrossRef]
208. Yang, C.S.; Wang, H. Cancer preventive activities of tea catechins. *Molecules* **2016**, *21*, 12. [CrossRef] [PubMed]
209. Zhang, Y.; Yuan, Y.; Wu, H.; Xie, Z.; Wu, Y.; Song, X.; Wang, J.; Shu, W.; Xu, J.; Liu, B.; et al. Effect of verbascoside on apoptosis and metastasis in human oral squamous cell carcinoma. *Int. J. Cancer* **2018**, *143*, 980–991. [CrossRef]
210. Cetojevic-Simin, D.D.; Bogdanovic, G.M.; Cvetkovic, D.D.; Velicanski, A.S. Antiproliferative and antimicrobial activity of traditional kombucha and *Satureja montana* L. kombucha. *J. BUON* **2008**, *13*, 395–401.
211. Srihari, T.; Arunkumar, R.; Arunakaran, J.; Satyanarayana, U. Downregulation of signalling molecules involved in angiogenesis of prostate cancer cell line (PC-3) by kombucha (lyophilized). *Biomed. Prev. Nutr.* **2013**, *3*, 53–58. [CrossRef]
212. Marzban, F.; Azizi, G.; Afraei, S.; Sedaghat, R.; Seyedzadeh, M.H.; Razavi, A.; Mirshafiey, A. Kombucha tea ameliorates experimental autoimmune encephalomyelitis in mouse model of multiple sclerosis. *Food Agric. Immunol.* **2015**, *26*, 782–793. [CrossRef]
213. Nileeka Balasuriya, B.; Vasantha Rupasinghe, H. Plant flavonoids as angiotensin converting enzyme inhibitors in regulation of hypertension. *Funct. Food Health Dis.* **2011**, *5*, 172–188. [CrossRef]
214. Guerrero, L.; Castillo, J.; Quiñones, M.; Garcia-Vallvé, S.; Arola, L.; Pujadas, G.; Muguerza, B. Inhibition of angiotensin-converting enzyme activity by flavonoids: Structure-activity relationship studies. *PLoS ONE* **2012**, *7*, e49493. [CrossRef] [PubMed]
215. Gamboa-Gomez, C.I.; Gonzalez-Laredo, R.F.; Gallegos-Infante, J.A.; Perez, M.M.L.; Moreno-Jimenez, M.R.; Flores-Rueda, A.G.; Rocha-Guzmam, N.E. Antioxidant and angiotensin-converting enzyme inhibitory activity of *Eucalyptus camaldulensis* and *Litsea glaucescens* infusions fermented with kombucha consortium. *Food Technol. Biotechnol.* **2016**, *54*, 367–373. [CrossRef] [PubMed]
216. Bhattacharya, S.; Gachhui, R.; Sil, P.C. Hepatoprotective properties of kombucha tea against TBHP-induced oxidative stress via suppression of mitochondria dependent apoptosis. *Pathophysiology* **2011**, *18*, 221–234. [CrossRef] [PubMed]

217. Kabiri, N.; Ahangar-Darabi, M.; Rafieian-Kopaei, M.; Setorki, M.; Doudi, M. Protective effect of kombucha tea on liver damage induced by thioacetamide in rats. *J. Biol. Sci.* **2014**, *14*, 343–348. [CrossRef]
218. Hyun, J.; Lee, Y.; Wang, S.; Kim, J.; Kim, J.; Cha, J.; Seo, Y.S.; Jung, Y. Kombucha tea prevents obese mice from developing hepatic steatosis and liver damage. *Food Sci. Biotechnol.* **2016**, *25*, 861–866. [CrossRef]
219. Al-Mohammadi, A.R.; Ismaiel, A.A.; Ibrahim, R.A.; Moustafa, A.H.; Abou Zeid, A.; Enan, G. Chemical constitution and antimicrobial activity of kombucha fermented beverage. *Molecules* **2021**, *26*, 5026. [CrossRef]
220. Greenwalt, C.J.; Ledford, R.A.; Steinkraus, K.H. Determination and characterization of the antimicrobial activity of the fermented tea kombucha. *LWT-Food Sci. Technol.* **1998**, *31*, 291–296. [CrossRef]
221. Sreeramulu, G.; Zhu, Y.; Knol, W. Characterization of antimicrobial activity in kombucha fermentation. *Acta Biotechnol.* **2001**, *21*, 49–56. [CrossRef]
222. Sreeramulu, G.; Zhu, Y.; Knol, W. Kombucha fermentation and its antimicrobial activity. *J. Agric. Food Chem.* **2000**, *48*, 2589–2594. [CrossRef]
223. Battikh, H.; Bakhrouf, A.; Ammar, E. Antimicrobial effect of kombucha analogues. *LWT-Food Sci. Technol.* **2012**, *47*, 71–77. [CrossRef]
224. Kaewkod, T.; Bovonsombut, S.; Tragoolpua, Y. Efficacy of Kombucha obtained from green, oolong, and black teas on inhibition of pathogenic bacteria, antioxidation, and toxicity on colorectal cancer cell line. *Microorganisms* **2019**, *7*, 700. [CrossRef] [PubMed]
225. Bhattacharya, D.; Bhattacharya, S.; Patra, M.M.; Chakravorty, S.; Sarkar, S.; Chakraborty, W.; Koley, H.; Gachhui, R. Antibacterial activity of polyphenolic fraction of kombucha against enteric bacterial pathogens. *Curr. Microbiol.* **2016**, *3*, 885–896. [CrossRef] [PubMed]
226. Winska, K.; Grabarczyk, M.; Maczka, W.; Zarowska, B.; Maciejewska, G.; Aniol, M. Antimicrobial activity of new bicyclic lactones with three or four methyl groups obtained both synthetically and biosynthetically. *J. Saudi Chem. Soc.* **2016**, *22*, 363–371. [CrossRef]
227. Milkyas Endale, F.D.; Melaku, Y. Antibacterial and antioxidant phenylpropanoid derivative from the leaves of plantago lanceolate. *Nat. Prod. Chem. Res.* **2018**, *6*, 3. [CrossRef]
228. Morales, D. Biological activities of kombucha beverages: The need of clinical evidence. *Trends Food Sci. Technol.* **2020**, *105*, 323–333. [CrossRef]
229. Muzaifa, M.; Abubakar, Y.; Nilda, C.; Sapitri, M. Alcohol content and chemical characteristics of fermented beverages in Aceh Province-Indonesia. *IOP Conf. Ser. Earth Environ. Sci.* **2023**, *1241*, 012095. [CrossRef]
230. Fiorda, F.A.; Pereira, G.V.M.; Thomaz-Soccol, V.; Rakshit, S.K.; Pagnoncelli, M.G.B.; Vandenberghe, L.P.S.; Soccol, C.R. Microbiological, biochemical, and functional aspects of sugary kefir fermentation—A review. *Food Microbiol.* **2017**, *66*, 86–95. [CrossRef]
231. Yilmaz-Ersan, L.; Ozcan, T.; Akpinar-Bayizit, A.; Sahin, S. The antioxidative capacity of kefir produced from goat milk. *Int. J. Chem. Eng. Appl.* **2016**, *7*, 22–26. [CrossRef]
232. Suetsuna, K.; Ukeda, H.; Ochi, H. Isolation and characterization of free radical scavenging activities peptides derived from casein. *J. Nutr. Biochem.* **2000**, *11*, 128–131. [CrossRef]
233. Arslan, S. A review: Chemical, microbiological and nutritional characteristics of kefir. *CyTA-J. Food* **2015**, *13*, 340–345. [CrossRef]
234. Xu, N.; Chen, G.; Liu, H. Antioxidative categorization of twenty amino acids based on experimental evaluation. *Molecules* **2017**, *22*, 2066. [CrossRef] [PubMed]
235. Yilmaz-Ersan, L.; Ozcan, T.; Akpinar-Bayizit, A.; Sahin, S. Comparison of antioxidant capacity of cow and ewe milk kefirs. *J. Dairy Sci.* **2018**, *101*, 3788–3798. [CrossRef] [PubMed]
236. Hadisaputro, S.; Djokomoeljanto, R.R.; Judiono; Soesatyo, M.H.N.E. The effects of oral plain kefir supplementation on proinflammatory cytokine properties of the hyperglycemia Wistar rats induced by streptozotocin. *Acta Med. Indones.* **2012**, *44*, 100–104.
237. Chen, Y.P.; Hsiao, P.J.; Hong, W.S.; Dai, T.Y.; Chen, M.J. *Lactobacillus kefiranofaciens* M1 isolated from milk kefir grains ameliorates experimental colitis in vitro and in vivo. *J. Dairy Sci.* **2012**, *95*, 63–74. [CrossRef]
238. Carasi, P.; Racedo, S.M.; Jacquot, C.; Romanin, D.E.; Serradell, M.A.; Urdaci, M.C. Impact of kefir derived *Lactobacillus kefiri* on the mucosal immune response and gut microbiota. *J. Immunol. Res.* **2015**, *2015*, 361604. [CrossRef]
239. Rosa, D.D.; Grzeskowiak, L.M.; Ferreira, C.L.L.F.; Fonseca, A.C.M.; Reis, B.E.M.; Dias, M.M.; Siqueira, N.P.; Silva, L.L.; Neves, C.A.; Oliveira, L.L.; et al. Kefir reduces insulin resistance and inflammatory cytokine expression in an animal model of metabolic syndrome. *Food Funct.* **2016**, *7*, 3390–3401. [CrossRef]
240. Zamberi, N.R.; Abu, N.; Mohamed, N.E.; Nordin, N.; Keong, Y.S.; Beh, B.K.; Zakaria, Z.A.B.; Rahman, N.M.A.N.A.; Alitheen, N.B. The antimetastatic and antiangiogenesis effects of kefir water on murine breast cancer cells. *Integr. Cancer Ther.* **2016**, *15*, 53–66. [CrossRef]
241. Seo, M.K.; Park, E.J.; Ko, S.Y.; Choi, E.W.; Kim, S. Therapeutic effects of kefir grain *Lactobacillus*-derived extracellular vesicles in mice with 2,4,6-trinitrobenzene sulfonic acid-induced inflammatory bowel disease. *J. Dairy Sci.* **2018**, *101*, 8662–8671. [CrossRef]
242. Vieira, L.V.; Sousa, L.M.; Maia, T.A.C.; Gusmao, J.N.F.M.; Goes, P.; Pereira, K.M.A.; Miyajima, F.; Gondim, D.V. Milk Kefir therapy reduces inflammation and alveolar bone loss on periodontitis in rats. *Biomed. Pharmacother.* **2021**, *139*, 111677. [CrossRef]
243. Kang, E.A.; Choi, H.I.; Hong, S.W.; Kang, S.; Jegal, H.Y.; Choi, E.W.; Park, B.S.; Kim, J.S. Extracellular vesicles derived from kefir grain *Lactobacillus* ameliorate intestinal inflammation via regulation of proinflammatory pathway and tight junction integrity. *Biomedicines* **2020**, *8*, 522. [CrossRef] [PubMed]

244. Yin, M.; Yan, X.; Weng, W.; Yang, Y.; Gao, R.; Liu, M.; Pan, C.; Zhu, Q.; Li, H.; Wei, Q.; et al. Micro integral membrane protein (MIMP), a newly discovered anti-inflammatory protein of *Lactobacillus plantarum*, enhances the gut barrier and modulates microbiota and inflammatory cytokines. *Cell. Physiol. Biochem.* **2018**, *45*, 474–490. [CrossRef]

245. Chen, H.L.; Hung, K.F.; Yen, C.C.; Laio, C.H.; Wang, J.L.; Lan, Y.W.; Chong, K.Y.; Fan, H.C.; Chen, C.M. Kefir peptides alleviate particulate matter <4 m (PM4.0)-induced pulmonary inflammation by inhibiting the NF-B pathway using luciferase transgenic mice. *Sci. Rep.* **2019**, *9*, 11529. [PubMed]

246. Judiono, J.; Hadisaputro, S.; Indranila, K.; Cahyono, B.; Suzery, M.; Widiastuti, Y. Effects of clear kefir on biomolecular aspects of glycemic status of type 2 diabetes mellitus (T2DM) patients in Bandung, West Java [study on human blood glucose, c peptide and insulin]. *Funct. Foods Health Dis.* **2014**, *4*, 340–348. [CrossRef]

247. Ostadrahimi, A.; Taghizadeh, A.; Mobasseri, M.; Farrin, N.; Payahoo, L.; Gheshlaghi, Z.B.; Vahedjabbari, M. Effect of probiotic fermented milk (kefir) on glycemic control and lipid profile in type 2 diabetic patients: A randomized double-blind placebo controlled clinical trial. *Iran. J. Public Health* **2015**, *44*, 228. [PubMed]

248. Quirós, A.; Hernández-Ledesma, B.; Ramos, M.; Amigo, L.; Recio, I. Angiotensin-converting enzyme inhibitory activity of peptides derived from caprine kefir. *J. Dairy Sci.* **2005**, *88*, 3480–3487. [CrossRef]

249. Aihara, K.; Nakamura, Y.; Kajimoto, O.; Hirata, H.; Takahashi, R. Effect of powdered fermented milk with *Lactobacillus helveticus* on subjects with high-normal blood pressure or mild hypertension. *J. Am. Coll. Nutr.* **2005**, *24*, 257–265. [CrossRef] [PubMed]

250. Friques, A.G.F.; Arpini, C.M.; Kalil, I.C.; Gava, A.L.; Leal, M.A.; Porto, M.L.; Nogueira, B.V.; Dias, A.T.; Andrade, T.U.; Pereira, T.M.C.; et al. Chronic administration of the probiotic kefir improves the endothelial function in spontaneously hypertensive rats. *J. Transl. Med.* **2015**, *13*, 390. [CrossRef]

251. Brasil, G.A.; Silva-Cutini, M.d.A.; Moraes, F.d.S.A.; Pereira, T.d.M.C.; Vasquez, E.C.; Lenz, D.; Bissoli, N.S.; Endringer, D.C.; de Lima, E.M.; Biancardi, V.C.; et al. The benefits of soluble non-bacterial fraction of kefir on blood pressure and cardiac hypertrophy in hypertensive rats are mediated by an increase in baroreflex sensitivity and decrease in angiotensin-converting enzyme activity. *Nutrition* **2018**, *51–52*, 66–72. [CrossRef]

252. Ochoa-de la Paz, L.; Gulias-Canizo, R.; Ruiz-Leyja, E.A.; Sanchez-Castillo, H.; Parodi, J. The role of GABA neurotransmitter in the human central nervous system, physiology, and pathophysiology. *Rev. Mex. Neurocienc.* **2021**, *22*, 67–76. [CrossRef]

253. Noori, N.; Bangash, M.Y.; Motaghinejad, M.; Hosseini, P.; Noudoost, B. Kefir protective effects against nicotine cessation-induced anxiety and cognition impairments in rats. *Adv. Biomed. Res.* **2014**, *3*, 251. [CrossRef] [PubMed]

254. Van de Wouw, M.; Walsh, A.M.; Crispie, F.; Leuven, L.; Lyte, J.M.; Boehme, M.; Clarke, G.; Dinan, T.G.; Cotter, P.D.; Cryan, J.F. Distinct actions of the fermented beverage kefir on host behaviour, immunity and microbiome gut-brain modules in the mouse. *Microbiome* **2020**, *8*, 67. [CrossRef] [PubMed]

255. Chen, C.; Hing, M.C.; Kubow, S. Kefir extracts suppress in vitro proliferation of estrogen-dependent human breast cancer cells but not normal mammary epithelial cells. *J. Med. Food* **2007**, *10*, 416–422. [CrossRef] [PubMed]

256. Gao, J.; Gu, F.; Ruan, H.; Chen, Q.; He, J.; He, G. Induction of apoptosis of gastric cancer cells SGC7901 in vitro by a cell-free fraction of Tibetan kefir. *Int. Dairy J.* **2013**, *30*, 14–18. [CrossRef]

257. Ghoneum, M.; Felo, N. Selective induction of apoptosis in human gastric cancer cells by *Lactobacillus kefiri* (PFT), a novel kefir product. *Oncol. Rep.* **2015**, *34*, 1659–1666. [CrossRef] [PubMed]

258. Ghoneum, M.; Gimzewski, J. Apoptotic effect of a novel kefir product, PFT, on multidrug-resistant myeloid leukemia cells via a hole-piercing mechanism. *Int. J. Oncol.* **2014**, *44*, 830–837. [CrossRef]

259. Khoury, N.; El-Hayek, S.; Tarras, O.; El-Sabban, M.; El-Sibai, M.; Rizk, S. Kefir exhibits anti-proliferative and pro-apoptotic effects on colon adenocarcinoma cells with no significant effects on cell migration and invasion. *Int. J. Oncol.* **2014**, *45*, 2117–2127. [CrossRef] [PubMed]

260. Rizk, S.; Maalouf, K.; Baydoun, E. The antiproliferative effect of kefir cell-free fraction on HuT-102 malignant T lymphocytes. *Clin. Lymphoma Myeloma Leuk.* **2009**, *9*, S198–S203. [CrossRef]

261. Rizk, S.; Maalouf, K.; Nasser, H.; El-Hayek, S. The pro-apoptotic effect of kefir in malignant T-lymphocytes involves a p53 dependent pathway. *Clin. Lymphoma Myeloma Leuk.* **2013**, *13*, S367. [CrossRef]

262. Maalouf, K.; Baydoun, E.; Rizk, S. Kefir induces cell-cycle arrest and apoptosis in HTLV-l-negative malignant T-lymphocytes. *Cancer Manag. Res.* **2011**, *3*, 39–47.

263. Laranjo, M. Starter cultures and their role in fermented foods. In *Microbial Fermentations in Nature and as Designed Processes*; Hurst, C.J., Ed.; John Wiley & Sons: Hoboken, NJ, USA, 2023; pp. 281–292.

264. Zhou, Y.; Wang, R.; Zhang, Y.; Yang, Y.; Sun, X.; Zhang, Q.; Yang, N. Biotransformation of phenolics and metabolites and the change in antioxidant activity in kiwifruit induced by *Lactobacillus plantarum* fermentation. *J. Sci. Food Agric.* **2020**, *100*, 3283–3290. [CrossRef] [PubMed]

265. Isas, A.S.; Escobar, F.; Alvarez-Villamil, E.; Molina, V.; Mateos, R.; Lizarraga, E.; Mozzi, F.; Van Nieuwenhove, C. Fermentation of pomegranate juice by lactic acid bacteria and its biological effect on mice fed a high-fat diet. *Food Biosci.* **2023**, *53*, 102516. [CrossRef]

266. Cele, N.P.; Akinola, S.A.; Manhivi, V.E.; Shoko, T.; Remize, F.; Sivakumar, D. Influence of lactic acid bacterium strains on changes in quality, functional compounds and volatile compounds of mango juice from different cultivars during fermentation. *Foods* **2022**, *11*, 682. [CrossRef] [PubMed]

267. Moslemi, M.; Moayedi, A.; Khomeiri, M.; Maghsoudlou, Y. Development of a whey-based beverage with enhanced levels of conjugated linoleic acid (CLA) as facilitated by endogenous walnut lipase. *Food Chem. X* **2023**, *17*, 100547. [CrossRef] [PubMed]

268. Sales, A.L.; Iriondo-DeHond, A.; DePaula, J.; Ribeiro, M.; Ferreira, I.M.P.L.V.O.; Miguel, M.A.L.; del Castillo, M.D.; Farah, A. Intracellular antioxidant and anti-inflammatory effects and bioactive profiles of coffee cascara and black tea kombucha beverages. *Foods* **2023**, *12*, 1905. [CrossRef]

269. Ozturk, B.E.T.; Eroglu, B.; Delik, E.; Cicek, M.; Cicek, E. Comprehensive evaluation of three important herbs for kombucha fermentation. *Food Technol. Biotechnol.* **2023**, *61*, 127–137. [CrossRef] [PubMed]

270. Mendonca, G.R.; Pinto, R.A.; Praxedes, E.A.; Abreu, V.K.G.; Dutra, R.P.; Pereira, A.F.; Lemos, T.O.; Reis, A.S.; Pereira, A.L.F. Kombucha based on unconventional parts of the Hibiscus sabdariffa L.: Microbiological, physico-chemical, antioxidant activity, cytotoxicity and sensorial characteristics. *Int. J. Gastron. Food Sci.* **2023**, *34*, 100804. [CrossRef]

271. Ziemlewska, A.; Zagorska-Dziok, M.; Niziol-Lukaszewska, Z.; Kielar, P.; Molon, M.; Szczepanek, D.; Sowa, I.; Wojciak, M. In vitro evaluation of antioxidant and protective potential of kombucha-fermented black berry extracts against H_2O_2-Induced oxidative stress in human skin cells and yeast model. *Int. J. Mol. Sci.* **2023**, *24*, 4388. [CrossRef]

272. Amjadi, S.; Armanpour, V.; Ghorbani, M.; Tabibiazar, M.; Soofi, M.; Roufegarinejad, L. Determination of phenolic composition, antioxidant activity, and cytotoxicity characteristics of kombucha beverage containing *Echium amoenum*. *J. Food Meas. Charact.* **2023**, *17*, 3162–3172. [CrossRef]

273. de la Bastida, A.R.; Peiroten, A.; Langa, S.; Rodríguez-Mínguez, E.; Curiel, J.A.; Arques, J.L.; Landete, J.M. Fermented soy beverages as vehicle of probiotic lactobacilli strains and source of bioactive isoflavones: A potential double functional effect. *Heliyon* **2023**, *9*, e14991. [CrossRef]

274. Chong, A.Q.; Lau, S.W.; Chin, N.L.; Talib, R.A.; Basha, R.K. Fermented beverage benefits: A comprehensive review and comparison of kombucha and kefir microbiome. *Microorganisms* **2023**, *11*, 1344. [CrossRef]

275. Corona, O.; Randazzo, W.; Miceli, A.; Guarcello, R.; Francesca, N.; Erten, H.; Moschetti, G.; Settanni, L. Characterization of kefir-like beverages produced from vegetable juices. *LWT-Food. Sci. Technol.* **2016**, *66*, 572–581. [CrossRef]

276. Koh, W.Y.; Utra, U.; Rosma, A.; Effarizah, M.E.; Rosli, W.I.W.; Park, Y.-H. Development of a novel fermented pumpkin-based beverage inoculated with water kefir grains: A response surface methodology approach. *Food Sci. Biotechnol.* **2018**, *27*, 525–535. [CrossRef] [PubMed]

277. Laureys, D.; Aerts, M.; Vandamme, P.; De Vuyst, L. Oxygen and diverse nutrients influence the water kefir fermentation process. *Food Microbiol.* **2018**, *73*, 351–361. [CrossRef]

278. Laureys, D.; Aerts, M.; Vandamme, P.; De Vuyst, L. The buffer capacity and calcium concentration of water influence the microbial species diversity, grain growth, and metabolite production during water kefir fermentation. *Front. Microbiol.* **2019**, *10*, 2876. [CrossRef] [PubMed]

279. Syrokou, M.K.; Papadelli, M.; Ntaikou, I.; Paramithiotis, S.; Drosinos, E.H. Sugary kefir: Microbial identification and technological properties. *Beverages* **2019**, *5*, 61. [CrossRef]

280. Verce, M.; De Vuyst, L.; Weckx, S. Shotgun metagenomics of a water kefir fermentation ecosystem reveals a novel *Oenococcus* species. *Front. Microbiol.* **2019**, *10*, 479. [CrossRef] [PubMed]

281. Zannini, E.; Lynch, K.M.; Nyhan, L.; Sahin, A.W.; Riordan, P.O.; Luk, D.; Arendt, E.K. Influence of substrate on the fermentation characteristics and culture-dependent microbial composition of water kefir. *Fermentation* **2022**, *9*, 28. [CrossRef]

282. Tavares, P.P.L.G.; Mamona, C.T.P.; Nascimento, R.Q.; dos Anjos, E.A.; de Souza, C.O.; Almeida, R.C.d.C.; Mamede, M.E.d.O.; Magalhães-Guedes, K.T. Non-conventional sucrose-based substrates: Development of non-dairy kefir beverages with probiotic potential. *Fermentation* **2023**, *9*, 384. [CrossRef]

283. Velicanski, A.; Cvetkovic, D.; Markov, S.; Tumbas, V.; Savatovic, S. Antimicrobial and antioxidant activity of lemon balm kombucha. *Acta Period. Technol.* **2007**, *38*, 165–172. [CrossRef]

284. Sirirat, D.; Jelena, P. Bacterial inhibition and antioxidant activity of kefir produced from Thai jasmine rice milk. *Biotechnology* **2010**, *9*, 332–337. [CrossRef]

285. Yavari, N.; Assadi, M.M.; Larijani, K.; Moghadam, M.B. Response surface methodology for optimization of glucuronic acid production using kombucha layer on sour cherry juice. *Aust. J. Basic Appl. Sci.* **2010**, *4*, 3250–3256.

286. Liamkaew, D.R.; Chattrawanit, J.; Danvirutai, D.P. Kombucha production by combinations of black tea and apple juice. *Prog. Appl. Sci. Technol.* **2016**, *6*, 139–146.

287. Pure, A.E.; Pure, M.E. Antioxidant and antibacterial activity of kombucha beverages prepared using banana peel, common nettles and black tea infusions. *Appl. Food Biotechnol.* **2016**, *3*, 125–130.

288. Watawana, M.I.; Jayawardena, N.; Gunawardhana, C.B.; Waisundara, V.Y. Enhancement of the antioxidant and starch hydrolase inhibitory activities of king coconut water (*Cocos nucifera* var. Aurantiaca) by fermentation with kombucha 'tea fungus'. *Int. J. Food Sci. Technol.* **2016**, *51*, 490–498. [CrossRef]

289. Ayed, L.; Ben Abid, S.; Hamdi, M. Development of a beverage from red grape juice fermented with the kombucha consortium. *Ann. Microbiol.* **2017**, *67*, 111–121. [CrossRef]

290. Ayed, L.; Hamdi, M. Manufacture of a beverage from cactus pear juice using "tea fungus" fermentation. *Ann. Microbiol.* **2015**, *65*, 2293–2299. [CrossRef]

291. Zubaidah, E.; Yurista, S.; Rahmadani, N.R. Characteristic of physical, chemical, and microbiological kombucha from various varieties of apples. *IOP Conf. Ser. Earth Environ. Sci.* **2017**, *131*, 012040. [CrossRef]

292. Shahbazi, H.; Gahruie, H.H.; Golmakani, M.-T.; Eskandari, M.H.; Movahedi, M. Effect of medicinal plant type and concentration on physicochemical, antioxidant, antimicrobial, and sensorial properties of kombucha. *Food Sci. Nutr.* **2018**, *6*, 2568–2577. [CrossRef]

293. Vitas, J.; Malbasa, R.; Grahovac, J.; Loncar, E. The antioxidant activity of kombucha fermented milk products with stinging nettle and winter savory. *Chem. Ind. Chem. Eng. Q.* **2013**, *19*, 129–139. [CrossRef]

294. Abuduaibifu, A.; Tamer, C.E. Evaluation of physicochemical and bioaccessibility properties of goji berry kombucha. *J. Food Process Preserv.* **2019**, *43*, e14077. [CrossRef]

295. Sharifudin, S.A.; Ho, W.Y.; Yeap, S.K.; Abdullah, R.; Koh, S.P. Fermentation and characterisation of potential kombucha cultures on papaya-based substrates. *LWT-Food Sci. Technol.* **2021**, *151*, 112060. [CrossRef]

296. Majid, A.A.; Suroto, D.A.; Utami, T.; Rahayu, E.S. Probiotic potential of kombucha drink from butterfly pea (*Clitoria ternatea* L.) flower with the addition of *Lactiplantibacillus plantarum* subsp. *plantarum* Dad-13. *Biocatal. Agric. Biotechnol.* **2023**, *51*, 102776. [CrossRef]

297. Kilic, G.; Sengun, I.Y. Bioactive properties of Kombucha beverages produced with Anatolian hawthorn (*Crataegus orientalis*) and nettle (*Urtica dioica*) leaves. *Food Biosci.* **2023**, *53*, 102631. [CrossRef]

298. Ankolekar, C.; Pinto, M.; Greene, D.; Shetty, K. In vitro bioassay based screening of antihyperglycemia and antihypertensive activities of *Lactobacillus acidophilus* fermented pear juice. *Innov. Food Sci. Emerg. Technol.* **2012**, *13*, 221–230. [CrossRef]

299. Bontsidis, C.; Mallouchos, A.; Terpou, A.; Nikolaou, A.; Batra, G.; Mantzourani, I.; Alexopoulos, A.; Plessas, S. Microbiological and chemical properties of chokeberry juice fermented by novel lactic acid bacteria with potential probiotic properties during fermentation at 4 °C for 4 weeks. *Foods* **2021**, *10*, 768. [CrossRef]

300. Chen, R.; Chen, W.; Chen, H.; Zhang, G.; Chen, W. Comparative evaluation of the antioxidant capacities, organic acids, and volatiles of papaya juices fermented by *Lactobacillus acidophilus* and *Lactobacillus plantarum*. *J. Food Qual.* **2018**, *2018*, 9490435. [CrossRef]

301. Chen, X.; Yuan, M.; Wang, Y.; Zhou, Y.; Sun, X. Influence of fermentation with different lactic acid bacteria and in vitro digestion on the change of phenolic compounds in fermented kiwifruit pulps. *Int. J. Food Sci. Technol.* **2022**, *57*, 2670–2679. [CrossRef]

302. Filannino, P.; Bai, Y.; Di Cagno, R.; Gobbetti, M.; Ganzle, M.G. Metabolism of phenolic compounds by *Lactobacillus* spp. *during fermentation of cherry juice and broccoli puree*. *Food Microbiol.* **2015**, *46*, 272–279.

303. Hashemi, S.M.B.; Jafarpour, D. Fermentation of bergamot juice with Lactobacillus plantarum strains in pure and mixed fermentations: Chemical composition, antioxidant activity and sensorial properties. *LWT-Food Sci. Technol.* **2020**, *131*, 109803. [CrossRef]

304. Isas, A.S.; Celis, M.S.M.; Correa, J.R.P.; Fuentes, E.; Rodríguez, L.; Palomo, I.; Mozzi, F.; Van Nieuwenhove, C. Functional fermented cherimoya (*Annona cherimola* Mill.) juice using autochthonous lactic acid bacteria. *Food Res. Int.* **2020**, *138*, 109729. [CrossRef]

305. Liu, Y.; Chen, H.; Chen, W.; Zhong, Q.; Zhang, G.; Chen, W. Beneficial effects of tomato juice fermented by *Lactobacillus plantarum* and *Lactobacillus casei*: Antioxidation, antimicrobial effect, and volatile profiles. *Molecules* **2018**, *23*, 2366. [CrossRef]

306. Pontonio, E.; Montemurro, M.; Pinto, D.; Marzani, B.; Trani, A.; Ferrara, G.; Mazzeo, A.; Gobbetti, M.; Rizzello, C.G. Lactic acid fermentation of pomegranate juice as a tool to improve antioxidant activity. *Front. Microbiol.* **2019**, *10*, 1550. [CrossRef] [PubMed]

307. Quan, Q.; Liu, W.; Guo, J.; Ye, M.; Zhang, J. Effect of six lactic acid bacteria strains on physicochemical characteristics, antioxidant activities and sensory properties of fermented orange juices. *Foods* **2022**, *11*, 1920. [CrossRef]

308. Yang, W.; Liu, J.; Zhang, Q.; Liu, H.; Lv, Z.; Zhang, C.; Jiao, Z. Changes in nutritional composition, volatile organic compounds and antioxidant activity of peach pulp fermented by *Lactobacillus*. *Food Biosci.* **2022**, *49*, 101894. [CrossRef]

309. Frolova, Y.; Vorobyeva, V.; Vorobyeva, I.; Sarkisyan, V.; Malinkin, A.; Isakov, V.; Kochetkova, A. Development of fermented kombucha tea beverage enriched with inulin and B vitamins. *Fermentation* **2023**, *9*, 552. [CrossRef]

310. Tauber, A.L.; Schweiker, S.S.; Levonis, S.M. From tea to treatment; epigallocatechin gallate and its potential involvement in minimizing the metabolic changes in cancer. *Nutr. Res.* **2020**, *74*, 23–36. [CrossRef]

311. Parn, K.W.; Ling, W.C.; Chin, J.H.; Lee, S.-K. Safety and efficacy of dietary epigallocatechin gallate supplementation in attenuating hypertension via its modulatory activities on the intrarenal renin–angiotensin system in spontaneously hypertensive rats. *Nutrients* **2022**, *14*, 4605. [CrossRef] [PubMed]

312. Benjamin, S.; Spener, F. Conjugated linoleic acids as functional food: An insight into their health benefits. *Nutr. Metab.* **2009**, *6*, 36. [CrossRef]

313. Koba, K.; Yanagita, T. Health benefits of conjugated linoleic acid (CLA). *Obes. Res. Clin. Pract.* **2014**, *8*, e525–e532. [CrossRef]

314. Massa, A.; Axpe, E.; Atxa, E.; Hernandez, I. Sustainable, carbonated, non-alcoholic beverages using leftover bread. *Int. J. Gastron. Food Sci.* **2022**, *30*, 100607. [CrossRef]

315. Nguyen, T.-L.; Toh, M.; Lu, Y.; Ku, S.; Liu, S.-Q. Biovalorization of market surplus bread for development of probiotic-fermented potential functional beverages. *Foods* **2022**, *11*, 250. [CrossRef]

316. Siguenza-Andres, T.; Gomez, M.; Rodríguez-Nogales, J.M.; Caro, I. Development of a fermented plant-based beverage from discarded bread flour. *LWT-Food Sci. Technol.* **2023**, *182*, 114795. [CrossRef]

MDPI

St. Alban-Anlage 66

4052 Basel

Switzerland

www.mdpi.com

Fermentation Editorial Office

E-mail: fermentation@mdpi.com

www.mdpi.com/journal/fermentation